Positivity in Algebraic Geometry I

AF804744

04M

Robert Lazarsfeld

Positivity in Algebraic Geometry I

Classical Setting: Line Bundles and Linear Series

Springer

Prof. Robert Lazarsfeld

Department of Mathematics
University of Michigan
Ann Arbor, MI 48109, USA
e-mail: rlaz@umich.edu

Also available as a hardcover edition as Volume 48 in our series "Ergebnisse der Mathematik": Robert Lazarsfeld: Positivity in Algebraic Geometry I. Classical Setting: Line Bundles and Linear Series. Springer-Verlag 2004, ISBN 3-540-22533-1

Library of Congress Control Number: 2004109578

Mathematics Subject Classification (2000):
Primary: 14-02
Secondary: 14C20, 14F05, 14F17, 32L10, 32J99, 14J17

ISSN 0071-1136
ISBN 3-540-22528-5 Springer Berlin Heidelberg New York

This work is subject to copyright. All rights are reserved, whether the whole or part of the material is concerned, specifically the rights of translation, reprinting, reuse of illustrations, recitation, broadcasting, reproduction on microfilms or in any other ways, and storage in data banks. Duplication of this publication or parts thereof is permitted only under the provisions of the German Copyright Law of September 9, 1965, in its current version, and permission for use must always be obtained from Springer. Violations are liable for prosecution under the German Copyright Law.

Springer is a part of Springer Science+Business Media
springeronline.com
© Springer-Verlag Berlin Heidelberg 2004
Printed in Germany

Typesetting: Computer to film by the authors' data
Printed on acid-free paper 41/3142XT - 5 4 3 2 1 0

To Lee Yen, Sarah, and John

Preface

The object of this book is to give a contemporary account of a body of work in complex algebraic geometry loosely centered around the theme of positivity.

Our focus lies on a number of questions that grew up with the field during the period 1950–1975. The sheaf-theoretic methods that revolutionized algebraic geometry in the fifties — notably the seminal work of Kodaira, Serre, and Grothendieck — brought into relief the special importance of ample divisors. By the mid sixties a very satisfying theory of positivity for line bundles was largely complete, and first steps were taken to extend the picture to bundles of higher rank. In a related direction, work of Zariski and others led to a greatly deepened understanding of the behavior of linear series on algebraic varieties. At the border with topology the classical theorems of Lefschetz were understood from new points of view, and extended in surprising ways. Hartshorne's book [276] and the survey articles in the Arcata proceedings [281] give a good picture of the state of affairs as of the mid seventies.

The years since then have seen continued interest and activity in these matters. Work initiated during the earlier period has matured and found new applications. More importantly, the flowering of higher dimensional geometry has led to fresh perspectives and — especially in connection with vanishing theorems — vast improvements in technology. It seems fair to say that the current understanding of phenomena surrounding positivity goes fundamentally beyond what it was thirty years ago. However, many of these new ideas have remained scattered in the literature, and others up to now have not been worked out in a systematic fashion. The time seemed ripe to pull together some of these developments, and the present volumes represent an attempt to do so.

The book is divided into three parts. The first, which occupies Volume I, focuses on line bundles and linear series. In the second volume, Part Two takes up positivity for vector bundles of higher ranks. Part Three deals with ideas and methods coming from higher-dimensional geometry, in the form of multiplier ideals. A brief introduction appears at the beginning of each of the parts.

I have attempted to aim the presentation at non-specialists. Not conceiving of this work as a textbook, I haven't started from a clearly defined set of prerequisites. But the subject is relatively non-technical in nature, and familiarity with the canonical texts [280] and [248] (combined with occasional faith and effort) is more than sufficient for the bulk of the material. In places — for example, Chapter 4 on vanishing theorems — our exposition is if anything more elementary than the standard presentations.

I expect that many readers will want to access this material in short segments rather than sequentially, and I have tried to make the presentation as friendly as possible for browsing. At least a third of the book is devoted to concrete examples, applications, and pointers to further developments. The more substantial of these are often collected together into separate sections. Others appear as examples or remarks (typically distinguished by the presence and absence respectively of indications of proof). Sources and attributions are generally indicated in the body of the text: these references are supplemented by brief sections of notes at the end of each chapter.

We work throughout with algebraic varieties defined over the complex numbers. Since substantial parts of the book involve applications of vanishing theorems, hypotheses of characteristic zero are often essential. However I have attempted to flag those foundational discussions that extend with only minor changes to varieties defined over algebraically closed fields of arbitrary characteristic. By the same token we often make assumptions of projectivity when in reality properness would do. Again I try to provide hints or references for the more general facts.

Although we use the Hodge decomposition and the hard Lefschetz theorem on several occasions, we say almost nothing about the Hodge-theoretic consequences of positivity. Happily these are treated in several sources, most recently in the beautiful book [600], [599] of Voisin. Similarly, the reader will find relatively little here about the complex analytic side of the story. For this we refer to Demailly's notes [126] or his anticipated book [119].

Concerning matters of organization, each chapter is divided into several sections, many of which are further partitioned into subsections. A reference to Section 3.1 points to the first section of Chapter 3, while Section 3.1.B refers to the second subsection therein. Statements are numbered consecutively within each section: so for example Theorem 3.1.17 indicates a result from Section 3.1 (which, as it happens, appears in 3.1.B). As an aid to the reader, each of the two volumes reproduces the table of contents of the other. The index, glossary, and list of references cover both volumes. Large parts of Volume I can be read without access to Volume II, but Volume II makes frequent reference to Volume I.

Acknowledgements. I am grateful to the National Science Foundation, the Guggenheim Foundation and the University of Michigan for support during the preparation of these volumes.[1]

I have benefited from comments and suggestions from many students and colleagues, including: T. Bauer, G. Bini, F. Bogomolov, A. Bravo, H. Brenner, A. Chen, D. Cutkosky, M. De Cataldo, T. De Fernex, J.-P. Demailly, I. Dolgachev, L. Ein, G. Farkas, R. Friedman, T. Garrity, A. Gibney, C. Hacon, R. Hartshorne, S. Helmke, M. Hering, M. Hochster, J. Howald, C. Huneke, P. Jahnke, M. Jonsson, Y. Kawamata, D. Keeler, M. Kim, F. Knudsen, A. Küronya, J. M. Landsberg, H. Lee, M. Mustaţă, T. Nevins, M. Nori, M. Paun, T. Peternell, H. Pinkham, M. Popa, I. Radloff, M. Ramachandran, M. Reid, J. Ross, M. Roth, H. Schenck, J. Sidman, R. Smith, A. Sommese, T. Stafford, I. Swanson, T. Szemberg, B. Teissier, Z. Teitler, D. Varolin, E. Viehweg, P. Vojta, J. Winkelmann, A. Wolfe, and Q. Zhang.

This project has profited from collaborations with a number of co-authors, including Jean-Pierre Demailly, Mark Green, and Karen E. Smith. Joint work many years ago with Bill Fulton has helped to shape the content and presentation of Chapter 3 and Part Two. I likewise owe a large mathematical debt to Lawrence Ein: I either learned from him or worked out together with him a significant amount of the material in Chapter 5 and Part Three, and our collaboration has had an influence in many other places as well.

I am grateful to several individuals for making particularly valuable contributions to the preparation of these volumes. János Kollár convinced me to start the book in the first place, and Bill Fulton insisted (on several occasions) that I finish it. Besides their encouragement, they contributed detailed suggestions from careful readings of drafts of several chapters. I also received copious comments on different parts of a preliminary draft from Thomas Eckl, Jun-Muk Hwang, Steve Kleiman, and Karen Smith. Olivier Debarre and Dano Kim read through the draft in its entirety, and provided a vast number of corrections and improvements.

Like many first-time authors, I couldn't have imagined when I began writing how long and consuming this undertaking would become. I'd like to take this opportunity to express my profound appreciation to the friends and family members that offered support, encouragement, and patience along the way.

Ann Arbor
May 2004

Robert Lazarsfeld

[1] Specifically, this work was partially supported by NSF Grants DMS 97-13149 and DMS 01-39713.

Contents

Contents of Volume II

Notation and Conventions

For the most part we follow generally accepted notation, as in [280]. We do however adopt a few specific conventions:

- We work throughout over the complex numbers $\mathbf{C}$.
- A *scheme* is a separated algebraic scheme of finite type over $\mathbf{C}$. A *variety* is a reduced and irreducible scheme. We deal exclusively with closed points of schemes.
- If X is a variety, a *modification* of X is a projective birational mapping $\mu : X' \longrightarrow X$ from an irreducible variety X' onto X.
- Given an irreducible variety X, we say that a property holds at a *general* point of X if it holds for all points in the complement of a proper algebraic subset. A property holds at a *very general* point if it is satisfied off the union of countably many proper subvarieties.
- Let $f : X \longrightarrow Y$ be a morphism of varieties or schemes, and $\mathfrak{a} \subseteq \mathcal{O}_Y$ an ideal sheaf on Y. We sometimes denote by $f^{-1}\mathfrak{a} \subseteq \mathcal{O}_X$ the ideal sheaf $\mathfrak{a}\cdot\mathcal{O}_X$ on X determined by $\mathfrak{a}$. While strictly speaking this conflicts with the notation for the sheaf-theoretic pullback of $\mathfrak{a}$, we trust that no confusion will result.
- Given a vector space or vector bundle E, S^kE denotes the kth symmetric power of E, and $\text{Sym}(E) = \oplus\, S^kE$ is the full symmetric algebra on E. The dual of E is written E^*.
- If E is a vector bundle on a scheme X, $\mathbf{P}(E)$ denotes the projective bundle of one-dimensional *quotients* of E. On a few occasions we will want to work with the bundle of one-dimensional *subspaces* of E: this is denoted $\mathbf{P}_{\text{sub}}(E)$. Thus
$$\mathbf{P}_{\text{sub}}(E) \ = \ \mathbf{P}(E^*).$$
A brief review of basic facts concerning projective bundles appears in Appendix A.

- Let X_1, X_2 be varieties or schemes. Without specific comment we write

$$\mathrm{pr}_1 : X_1 \times X_2 \longrightarrow X_1 \quad , \quad \mathrm{pr}_2 : X_1 \times X_2 \longrightarrow X_2$$

  for the two projections of $X_1 \times X_2$ onto its factors. When other notation for the projections seems preferable, we introduce it explicitly.

- Given a real-valued function $f : \mathbf{N} \longrightarrow \mathbf{R}$ defined on the natural numbers, we say that $f(m) = O(m^k)$ if

$$\limsup_{m \to \infty} \frac{|f(m)|}{m^k} < \infty.$$

Part One

Ample Line Bundles and Linear Series

Introduction to Part One

Linear series have long stood at the center of algebraic geometry. Systems of divisors were employed classically to study and define invariants of projective varieties, and it was recognized that varieties share many properties with their hyperplane sections. The classical picture was greatly clarified by the revolutionary new ideas that entered the field starting in the 1950s. To begin with, Serre's great paper [530], along with the work of Kodaira (e.g. [353]), brought into focus the importance of amplitude for line bundles. By the mid 1960s a very beautiful theory was in place, showing that one could recognize positivity geometrically, cohomologically, or numerically. During the same years, Zariski and others began to investigate the more complicated behavior of linear series defined by line bundles that may not be ample. This led to particularly profound insights in the case of surfaces [623]. In yet another direction, the classical theorems of Lefschetz comparing the topology of a variety with that of a hyperplane section were understood from new points of view, and developed in surprising ways in [258] and [30].

The present Part One is devoted to this body of work and its developments. Our aim is to give a systematic presentation, from a contemporary viewpoint, of the circle of ideas surrounding linear series and ample divisors on a projective variety.

We start in Chapter 1 with the basic theory of positivity for line bundles. In keeping with the current outlook, **Q**- and **R**-divisors and the notion of nefness play a central role, and the concrete geometry of nef and ample cones is given some emphasis. The chapter concludes with a section on Castelnuovo–Mumford regularity, a topic that we consider to merit inclusion in the canon of positivity.

Chapter 2 deals with linear series, our focus being the asymptotic geometry of linear systems determined by divisors that may not be ample. We study in particular the behavior of big divisors, whose role in birational geometry is similar to that of ample divisors in the biregular theory. The chapter also contains several concrete examples of the sort of interesting and challenging behavior that such linear series can display.

In Chapter 3 we turn to the theorems of Lefschetz and Bertini and their subsequent developments by Barth, Fulton–Hansen, and others. Here the surprising geometric properties of projective subvarieties of small codimension come into relief. The Lefschetz hyperplane theorem is applied in Chapter 4 to prove the classical vanishing theorems of Kodaira and Nakano. Chapter 4 also contains the vanishing theorem for big and nef divisors discovered by Kawamata and Viehweg, as well as one of the generic vanishing theorems from [242].

Finally, Chapter 5 takes up the theory of local positivity. This is a topic that has emerged only recently, starting with ideas of Demailly for quantifying how much of the positivity of a line bundle can be localized at a given point of a variety. Although some of the results are not yet definitive, the picture is surprisingly rich and structured.

Writing about linear series seems to lead unavoidably to conflict between the additive notation of divisors and the multiplicative language of line bundles. Our policy is to avoid explicitly mixing the two. However, on many occasions we adopt the compromise of speaking about divisors while using notation suggestive of bundles. We discuss this convention — as well as some of the secondary issues it raises — at the end of Section 1.1.A.

1

Ample and Nef Line Bundles

This chapter contains the basic theory of positivity for line bundles and divisors on a projective algebraic variety.

After some preliminaries in Section 1.1 on divisors and linear series, we present in Section 1.2 the classical theory of ample line bundles. The basic conclusion is that positivity can be recognized geometrically, cohomologically, or numerically. Section 1.3 develops the formalism of **Q**- and **R**-divisors, which is applied in Section 1.4 to study limits of ample bundles. These so-called nef divisors are central to the modern view of the subject, and Section 1.4 contains the core of the general theory. Most of the remaining material is more concrete in flavor. Section 1.5 is devoted to examples of ample cones and to further information about their structure, while Section 1.6 focuses on inequalities of Hodge type. After a brief review of the definitions and basic facts surrounding amplitude for a mapping, we conclude in Section 1.8 with an introduction to Castelnuovo–Mumford regularity.

We recall that according to our conventions we deal unless otherwise stated with complex algebraic varieties and schemes, and with closed points on them. However, as we go along we will point out that much of this material remains valid for varieties defined over algebraically closed fields of arbitrary characteristic.

1.1 Preliminaries: Divisors, Line Bundles, and Linear Series

In this section we collect some facts and notation that will be used frequently in the sequel. We start in Section 1.1.A by recalling some constructions involving divisors and line bundles, and turn in the second subsection to linear series. Section 1.1.C deals with intersection numbers and numerical equivalence, and we conclude in 1.1.D by discussing asymptotic formulations of the Riemann–Roch theorem. As a practical matter we assume that much of this material is familiar to the reader.[1] However, we felt it would be useful to include a brief summary in order to fix ideas.

1.1.A Divisors and Line Bundles

We start with a quick review of the definitions and facts concerning Cartier divisors, following [280, p. 140ff], [445, Chapters 9 and 10], and [344]. We take up first the very familiar case of reduced and irreducible varieties, and then pass to more general schemes.

Consider then an irreducible complex variety X, and denote by $\mathcal{M}_X = \mathbf{C}(X)$ the (constant) sheaf of rational functions on X. It contains the structure sheaf $\mathcal{O}_X$ as a subsheaf, and so there is an inclusion $\mathcal{O}_X^* \subseteq \mathcal{M}_X^*$ of sheaves of multiplicative abelian groups.

Definition 1.1.1. (Cartier divisors). A *Cartier divisor* on X is a global section of the quotient sheaf $\mathcal{M}_X^*/\mathcal{O}_X^*$. We denote by $\operatorname{Div}(X)$ the group of all such, so that

$$\operatorname{Div}(X) \;=\; \Gamma\big(X, \mathcal{M}_X^*/\mathcal{O}_X^*\big). \quad \square$$

Concretely, then, a divisor $D \in \operatorname{Div}(X)$ is represented by data $\big\{(U_i, f_i)\big\}$ consisting of an open covering $\{U_i\}$ of X together with elements $f_i \in \Gamma\big(U_i, \mathcal{M}_X^*\big)$, having the property that on $U_{ij} = U_i \cap U_j$ one can write

$$f_i \;=\; g_{ij} f_j \quad \text{for some} \quad g_{ij} \;\in\; \Gamma\big(U_{ij}, \mathcal{O}_X^*\big). \tag{1.1}$$

The function f_i is called a *local equation* for D at any point $x \in U_i$. Two such collections determine the same Cartier divisor if there is a common refinement $\{V_k\}$ of the open coverings on which they are defined so that they are given by data $\big\{(V_k, f_k)\big\}$ and $\big\{(V_k, f_k')\big\}$ with

$$f_k \;=\; h_k f_k' \quad \text{on} \quad V_k \quad \text{for some} \quad h_k \;\in\; \Gamma\big(V_k, \mathcal{O}_X^*\big).$$

The group operation on $\operatorname{Div}(X)$ is always written additively: if $D, D' \in \operatorname{Div}(X)$ are represented respectively by data $\big\{(U_i, f_i)\big\}$ and $\big\{(U_i, f_i')\big\}$, then $D + D'$ is given by $\big\{(U_i, f_i f_i')\big\}$. The *support* of a divisor $D = \big\{(U_i, f_i)\big\}$ is the set of points $x \in X$ at which a local equation of D at x is not a unit in $\mathcal{O}_x X$. D is *effective* if $f_i \in \Gamma\big(U_i, \mathcal{O}_X\big)$ is regular on U_i: this is written $D \succcurlyeq 0$. The notation $D \succcurlyeq D'$ indicates that $D - D'$ is effective.

Suppose now that X is a possibly non-reduced algebraic scheme. Then the same definition works except that one has to be more careful about what one means by $\mathcal{M}_X$, which now becomes the sheaf of total quotient rings of $\mathcal{O}_X$.[2] As

[1] For the novice, we give some suggestions and pointers to the literature.

[2] Grothendieck speaks of the sheaf of "meromorphic functions" on X (cf. [257, 20.1]). However since we are working with complex varieties this seems potentially confusing, and we prefer to follow the terminology of [280].

explained in [445, Chapter 9] there is a unique sheaf $\mathcal{M}_X$ on X characterized by the property that if $U = \operatorname{Spec}(A)$ is an affine open subset of X, then

$$\Gamma(U, \mathcal{M}_X) \;=\; \Gamma(U, \mathcal{O}_X)_{\text{tot}} \;=\; A_{\text{tot}}$$

is the total ring of fractions of A, i.e. the localization of A at the set of non zero-divisors.[3] Similarly, on the stalk level there is an isomorphism $\mathcal{M}_{X,x} = (\mathcal{O}_{X,x})_{\text{tot}}$. As before one has an inclusion $\mathcal{O}_X^* \subseteq \mathcal{M}_X^*$ of multiplicative groups of units, and Definition 1.1.1 — as well as the discussion following it — remains valid without change.

Convention 1.1.2. (Divisors). In Parts One and Two of this work we adopt the convention that when we speak of a "divisor" we always mean a Cartier divisor. (In Part Three it will be preferable to think instead of Weil divisors.) □

One should view Cartier divisors as "cohomological" objects, but one can also define "homological" analogues:

Definition 1.1.3. (Cycles and Weil divisors). Let X be a variety or scheme of pure dimension n. A *k-cycle* on X is a $\mathbf{Z}$-linear combination of irreducible subvarieties of dimension k. The group of all such is written $Z_k(X)$. A *Weil divisor* on X is an $(n-1)$-cycle, i.e. a formal sum of codimension one subvarieties with integer coefficients. We often use $\operatorname{WDiv}(X)$ in place of $Z_{n-1}(X)$ to denote the group of Weil divisors. □

Remark 1.1.4. (Cycle map for Cartier divisors). There is a *cycle map*

$$\operatorname{Div}(X) \longrightarrow \operatorname{WDiv}(X) \quad , \quad D \mapsto [D] \;=\; \sum \operatorname{ord}_V(D) \cdot [V]$$

where $\operatorname{ord}_V(D)$ is the order of D along a codimension-one subvariety. In general this homomorphism is neither injective nor surjective, although it is one-to-one when X is a normal variety and an isomorphism when X is non-singular. (See [208, Chapter 2.1] for details and further information.) □

A global section $f \in \Gamma(X, \mathcal{M}_X^*)$ determines in the evident manner a divisor

$$D \;=\; \operatorname{div}(f) \;\in\; \operatorname{Div}(X).$$

As usual, a divisor of this form is called *principal* and the subgroup of all such is $\operatorname{Princ}(X) \subseteq \operatorname{Div}(X)$. Two divisors D_1, D_2 are *linearly equivalent*, written $D_1 \equiv_{\text{lin}} D_2$, if $D_1 - D_2$ is principal.

Let D be a divisor on X. Given a morphism $f : Y \longrightarrow X$, one would like to define a divisor f^*D on Y by pulling back the local equations for D. The following condition is sufficient to guarantee that this is meaningful:

[3] See [344] for a discussion of how one should define $\mathcal{M}_X$ on arbitrary open subsets.

Let $V \subseteq Y$ be any *associated subvariety* of Y, i.e. the subvariety defined by an associated prime of $\mathcal{O}_Y$ in the sense of primary decomposition. Then $f(V)$ should not be contained in the support of D.

If Y is reduced, the requirement is just that no component of Y map into the support of D.

A similar condition allows one to define the divisor of a section of a line bundle L on X. Specifically, let $s \in \Gamma(X, L)$ be a global section of L. Assume that s does not vanish on any associated subvariety of X — for example, if X is reduced, this just means that s shouldn't vanish identically on any component of X. Then a local equation of s determines in the natural way a divisor $\mathrm{div}(s) \in \mathrm{Div}(X)$. We leave it to the reader to formulate the analogous condition under which a "rational section" $s \in \Gamma(X, L \otimes_{\mathcal{O}_X} \mathcal{M}_X)$ gives rise to a divisor.

A Cartier divisor $D \in \mathrm{Div}(X)$ determines a line bundle $\mathcal{O}_X(D)$ on X, leading to a canonical homomorphism

$$\mathrm{Div}(X) \longrightarrow \mathrm{Pic}(X) \quad , \quad D \mapsto \mathcal{O}_X(D) \tag{1.2}$$

of abelian groups, where $\mathrm{Pic}(X)$ denotes as usual the Picard group of isomorphism classes of line bundles on X. Concretely, if D is given by data $\{(U_i, f_i)\}$ as above, then one can build $\mathcal{O}_X(D)$ by using the g_{ij} in (1.1) as transition functions. More abstractly, one can view (the isomorphism class of) $\mathcal{O}_X(D)$ as the image of D under the connecting homomorphism

$$\mathrm{Div}(X) = \Gamma(X, \mathcal{M}_X^*/\mathcal{O}_X^*) \longrightarrow H^1(X, \mathcal{O}_X^*) = \mathrm{Pic}(X)$$

determined by the exact sequence $0 \longrightarrow \mathcal{O}_X^* \longrightarrow \mathcal{M}_X^* \longrightarrow \mathcal{M}_X^*/\mathcal{O}_X^* \longrightarrow 0$ of sheaves on X. Evidently,

$$\mathcal{O}_X(D_1) \cong \mathcal{O}_X(D_2) \quad \Longleftrightarrow \quad D_1 \equiv_{\mathrm{lin}} D_2.$$

If D is effective then $\mathcal{O}_X(D)$ carries a global section $s = s_D \in \Gamma(X, \mathcal{O}_X(D))$ with $\mathrm{div}(s) = D$. In general $\mathcal{O}_X(D)$ has a rational section with the analogous property.

The question of whether every line bundle comes from a divisor is more delicate. On the positive side, there are two sufficient conditions:

Example 1.1.5. (Line bundles from divisors). There are a couple of natural hypotheses to guarantee that every line bundle arises from a divisor.

(i). If X is reduced and irreducible, or merely reduced, then the homomorphism in (1.2) is surjective.

(ii). If X is projective then the same statement holds even if it is non-reduced.

(If X is reduced and irreducible then any line bundle L has a rational section s, and one can take $D = \mathrm{div}(s)$. For the second statement — which is due to

Nakai [466] — one can use the theorem of Cartan–Serre–Grothendieck (Theorem 1.2.6) to reduce to the case in which L is globally generated (Definition 1.1.10). But then one can find a section $s \in \Gamma\big(X, L\big)$ that does not vanish on any of the associated subvarieties of X, in which case $D = \operatorname{div}(s)$ gives the required divisor.) □

On the other hand, there is also

Example 1.1.6. (Kleiman's example). Following [346] and [528] we construct a non-projective non-reduced scheme X on which the mapping (1.2) is not surjective.[4] Start by taking Y to be Hironaka's example of a smooth non-projective threefold containing two disjoint smooth rational curves A and B with $A+B \equiv_{\text{num}} 0$ as described in [280, Appendix B, Example 3.4.1]. Now fix points $a \in A$, $b \in B$ and introduce nilpotents at a and b to produce a non-reduced scheme X containing Y. Note that X has depth zero at a and b, so these points must be disjoint from the support of every Cartier divisor on X. Observe also that $\operatorname{Pic}(X) \longrightarrow \operatorname{Pic}(Y)$ is an isomorphism thanks to the fact that $\ker\big(\mathcal{O}_X^* \longrightarrow \mathcal{O}_Y^*\big)$ is supported on a finite set, and so has vanishing H^1 and H^2.

We claim that there exists a line bundle L on X with the property that

$$\int_A c_1(L) \; > \; 0. \tag{*}$$

In fact, it follows from Hironaka's construction that one can find a line bundle on Y satisfying the analogous inequality, and by what have said above this bundle extends to X. Suppose now that $L = \mathcal{O}_X(D)$ for some divisor D on X. Decompose the corresponding Weil divisor as a sum $[D] = \sum m_i[D_i]$ of prime divisors with $m_i \neq 0$. None of the D_i can pass through a or b, so each D_i is Cartier and

$$D \;=\; \sum m_i D_i$$

as Cartier divisors on X. Now $\big(D_i \cdot A\big) \geq 0$ and $\big(D_i \cdot B\big) \geq 0$ since each of the D_i — avoiding as they do the points a and b — meet A and B properly. On the other hand, it follows from (*) that there is at least one index i such that $\big(m_i D_i \cdot A\big) > 0$: in particular $m_i > 0$ and $\big(D_i \cdot A\big) > 0$. But $\big(D_i \cdot B\big) = -\big(D_i \cdot A\big)$ since $B \equiv_{\text{num}} -A$ and therefore $\big(D_i \cdot B\big) < 0$, a contradiction. (As Schröer observes, the analogous but slightly simpler example appearing in [276, I.1.3] is erroneous.) □

Later on, the canonical bundle of a smooth variety will play a particularly important role:

Notation 1.1.7. (Canonical bundle and divisor). Let X be a non-singular complete variety of dimension n. We denote by $\omega_X = \Omega_X^n$ the canonical line bundle on X, and by K_X any canonical divisor on X. Thus $\mathcal{O}_X(K_X) = \omega_X$. □

---

[4] We will use here the basic facts of intersection theory recalled later in this section.

Finally, a word about terminology. There is inevitably a certain amount of tension between the additive language of divisors and the multiplicative formalism of line bundles. Our convention is always to work additively with divisors and multiplicatively with line bundles. However, on many occasions it is natural or customary to stay in additive mode when nonetheless one has line bundles in mind. In these circumstances we will speak of divisors but use notation suggestive of bundles, as for instance in the following statement of the Kodaira vanishing theorem:

> Let L be an ample divisor on a smooth projective variety X of dimension n. Then $H^i\big(X, \mathcal{O}_X(K_X + L)\big) = 0$ for every $i > 0$.

(The mathematics here appears in Chapter 4.) While for the most part this convention seems to work well, it occasionally leads us to make extraneous projectivity or integrality hypotheses in order to be able to invoke Example 1.1.5. Specifically, we will repeatedly work with the Néron–Severi group $N^1(X)$ of X and the corresponding real vector space $N^1(X)_{\mathbf{R}} = N^1(X) \otimes \mathbf{R}$. Here additive notation seems essential, so we are led to view $N^1(X)$ as the group of divisors modulo numerical equivalence (Definition 1.1.15). On the other hand, functorial properties are most easily established by passing to $\text{Pic}(X)$. For this to work smoothly one wants to know that every line bundle comes from a divisor, and this is typically guaranteed by simply assuming that X is either a variety or a projective scheme. We try to flag this artifice when it occurs.

1.1.B Linear Series

We next review some basic facts and definitions concerning linear series. For further information the reader can consult [276, Chapter I.2], [280, Chapter II, Sections 6, 7], and [248, Chapter 1.4].

Let X be a variety (or scheme), L a line bundle on X, and $V \subseteq H^0\big(X, L\big)$ a non-zero subspace of finite dimension. We denote by $|V| = \mathbf{P}_{\text{sub}}(V)$ the projective space of one-dimensional subspaces of V. When X is a complete variety, $|V|$ is identified with the linear series of divisors of sections of V in the sense of [280, Chapter II, §7], and in general we refer to $|V|$ as a linear series.[5] Taking $V = H^0\big(X, L\big)$ — assuming that this space is finite-dimensional, as will be the case for instance if X is complete — yields the *complete linear series* $|L|$. Given a divisor D, we also write $|D|$ for the complete linear series associated to $\mathcal{O}_X(D)$.

Evaluation of sections in V gives rise to a morphism

[5] Observe that on a non-integral scheme it may happen that not every element of $|V|$ determines a divisor, since there may exist $s \in V$ for which $\text{div}(s)$ is not defined. In this case calling $|V|$ a linear series is slightly unconventional. However we trust that no confusion will result.

$$\text{eval}_V : V \otimes_{\mathbf{C}} \mathcal{O}_X \longrightarrow L$$

of vector bundles on X.

Definition 1.1.8. (Base locus and base ideal). The *base ideal* of $|V|$, written

$$\mathfrak{b}(|V|) = \mathfrak{b}(X, |V|) \subseteq \mathcal{O}_X,$$

is the image of the map $V \otimes_{\mathbf{C}} L^* \longrightarrow \mathcal{O}_X$ determined by eval_V. The *base locus*

$$\text{Bs}(|V|) \subseteq X$$

of $|V|$ is the closed subset of X cut out by the base ideal $\mathfrak{b}(|V|)$. When we wish to emphasize the scheme structure on $\text{Bs}(|V|)$ determined by $\mathfrak{b}(|V|)$ we will refer to $\text{Bs}(|V|)$ as the *base scheme* of $|V|$. When $V = H^0(X, L)$ or $V = H^0(X, \mathcal{O}_X(D))$ are finite-dimensional, we write respectively $\mathfrak{b}(|L|)$ and $\mathfrak{b}(|D|)$ for the base ideals of the indicated complete linear series. □

Very concretely, then, $\text{Bs}(|V|)$ is the set of points at which all the sections in V vanish, and $\mathfrak{b}(|V|)$ is the ideal sheaf spanned by these sections.

Example 1.1.9. (Inclusions). Assuming for the moment that X is projective (or complete), fix a Cartier divisor D on X. Then for any integers $m, \ell \geq 1$, one has an inclusion

$$\mathfrak{b}(|\ell D|) \cdot \mathfrak{b}(|mD|) \subseteq \mathfrak{b}(|(\ell + m)D|).$$

(Use the natural homomorphism

$$H^0(X, \mathcal{O}_X(\ell D)) \otimes H^0(X, \mathcal{O}_X(mD)) \longrightarrow H^0(X, \mathcal{O}_X((\ell + m)D))$$

determined by multiplication of sections.) □

The easiest linear series to deal with are those for which the base locus is empty.

Definition 1.1.10. (Free linear series). One says that $|V|$ is *free*, or *basepoint-free*, if its base locus is empty, i.e. if $\mathfrak{b}(|V|) = \mathcal{O}_X$. A divisor D or line bundle L is *free* if the corresponding complete linear series is so. In the case of line bundles one says synonymously that L is *generated by its global sections* or *globally generated.* □

In other words, $|V|$ is free if and only if for each point $x \in X$ one can find a section $s = s_x \in V$ such that $s(x) \neq 0$.

Assume now (in order to avoid trivialities) that $\dim V \geq 2$, and set $B = \text{Bs}(|V|)$. Then $|V|$ determines a morphism

$$\phi = \phi_{|V|} : X - B \longrightarrow \mathbf{P}(V)$$

from the complement of the base locus in X to the projective space of one-dimensional quotients of V. Given $x \in X$, $\phi(x)$ is the hyperplane in V consisting of those sections vanishing at x. If one chooses a basis $s_0, \ldots, s_r \in V$, this amounts to saying that ϕ is given in homogeneous coordinates by the (somewhat abusive!) expression

$$\phi(x) \; = \; [s_0(x), \ldots, s_r(x)] \; \in \; \mathbf{P}^r.$$

When X is an irreducible variety it is sometimes useful to ignore the base locus, and view $\phi_{|V|}$ as a rational mapping $\phi : X \dashrightarrow \mathbf{P}(V)$. If $|V|$ is free then $\phi_{|V|} : X \longrightarrow \mathbf{P}(V)$ is a globally defined morphism.

At least when $B = \varnothing$ these constructions can be reversed, so that a morphism to projective space gives rise to a linear series. Specifically, suppose given a morphism

$$\phi : X \longrightarrow \mathbf{P} = \mathbf{P}(V)$$

from X to the projective space of one-dimensional quotients of a vector space V, and assume that $\phi(X)$ does not lie on any hyperplanes. Then pullback of sections via ϕ realizes $V = H^0\big(\mathbf{P}, \mathcal{O}_{\mathbf{P}}(1)\big)$ as a subspace of $H^0\big(X, \phi^*\mathcal{O}_{\mathbf{P}}(1)\big)$, and $|V|$ is a free linear series on X. Moreover, ϕ is identified with the corresponding morphism $\phi_{|V|}$.

Example 1.1.11. If X is a non-singular variety and $B \subseteq X$ has codimension ≥ 2, then a similar construction works starting with a morphism

$$\phi \; : \; X - B \; \longrightarrow \; \mathbf{P} = \mathbf{P}(V).$$

(In fact, $\phi^*\mathcal{O}_{\mathbf{P}}(1)$ extends uniquely to a line bundle L on X — corresponding to the divisor obtained by taking the closure of the pullback of a hyperplane — and ϕ^* realizes V as a subspace of $H^0\big(X, L\big)$, with $\mathrm{Bs}\big(|V|\big) \subseteq B$.) □

Example 1.1.12. (Projection). Suppose that $W \subseteq V$ is a subspace (say of dimension ≥ 2). Then $\mathrm{Bs}\big(|V|\big) \subseteq \mathrm{Bs}\big(|W|\big)$, so that $\phi_{|V|}$ and $\phi_{|W|}$ are both defined on $X - \mathrm{Bs}(|W|)$. Viewed as morphisms on this set one has the relation $\phi_{|W|} = \pi \circ \phi_{|V|}$, where

$$\pi \; : \; \mathbf{P}(V) - \mathbf{P}(V/W) \; \longrightarrow \; \mathbf{P}(W)$$

is linear projection centered along the subspace $\mathbf{P}(V/W) \subseteq \mathbf{P}(V)$. Note that if $|W|$ — and hence also $|V|$ — is free, and if X is complete, then $\pi|X$ is finite (since it is affine and proper). So in this case, the two morphisms

$$\phi_{|V|} : X \longrightarrow \mathbf{P}(V) \quad , \quad \phi_{|W|} : X \longrightarrow \mathbf{P}(W)$$

differ by a finite projection of $\phi_{|V|}(X)$. □

1.1.C Intersection Numbers and Numerical Equivalence

This subsection reviews briefly some definitions and facts from intersection theory.

Intersection numbers. Let X be a complete irreducible complex variety. Given Cartier divisors $D_1, \ldots, D_k \in \mathrm{Div}(X)$ together with an irreducible subvariety $V \subseteq X$ of dimension k, the intersection number

$$\big(D_1 \cdot \ldots \cdot D_k \cdot V\big) \ \in \ \mathbf{Z} \tag{1.3}$$

can be defined in various ways. To begin with, of course, the quantity in question arises as a special case of the theory in [208]. However intersection products of divisors against subvarieties do not require the full strength of that technology: a relatively elementary direct approach based on numerical polynomials was developed in the sixties by Snapper [546] and Kleiman [341]. Extensions and modern presentations of the Snapper–Kleiman theory appear in [363, VI.2], [114, Chapter 1.2], and [22, Chapter 1]. We prefer to minimize foundational discussions by working topologically, referring to [208] for additional properties as needed.[6] Some suggestions for the novice appear in Remark 1.1.13.

Specifically, in the above situation each of the line bundles $\mathcal{O}_X(D_i)$ has a Chern class

$$c_1\big(\mathcal{O}_X(D_i)\big) \ \in \ H^2\big(X;\mathbf{Z}\big),$$

the cohomology group in question being ordinary singular cohomology of X with its classical topology. The cup product of these classes is then an element

$$c_1\big(\mathcal{O}_X(D_1)\big) \cdot \ldots \cdot c_1\big(\mathcal{O}_X(D_k)\big) \ \in \ H^{2k}\big(X;\mathbf{Z}\big) :$$

here and elsewhere we write $\alpha \cdot \beta$ or simply $\alpha\beta$ for the cup product of elements $\alpha, \beta \in H^*\big(X;\mathbf{Z}\big)$. Denoting by $[V] \in H_{2k}\big(X;\mathbf{Z}\big)$ the fundamental class of V, cap product leads finally to an integer

$$\Big(c_1\big(\mathcal{O}_X(D_1)\big) \cdot \ldots \cdot c_1\big(\mathcal{O}_X(D_k)\big)\Big) \cap [V] \ \in \ H_0\big(X;\mathbf{Z}\big) \ = \ \mathbf{Z}, \tag{1.4}$$

which of course is nothing but the quantity appearing in (1.3). We generally use one of the notations

$$(D_1 \cdot \ldots \cdot D_k \cdot V) \quad , \quad \int_V D_1 \cdot \ldots \cdot D_k$$

[6] The essential foundational savings materialize when we deal with higher-codimension intersection theory — e.g. Chern classes of vector bundles — in Part Two of this book. Working topologically allows one to bypass complications involved in specifying groups to receive the classes in question. It then seemed natural to use topologically based intersection theory throughout.

(or a small variant thereof) for the intersection product in question. By linearity one can replace V by an arbitrary k-cycle, and evidently this product depends only on the linear equivalence class of the D_i. If $D_1 = \ldots = D_k = D$ we write $(D^k \cdot V)$, and when $V = X$ is irreducible of dimension n we often use the abbreviation $(D_1 \cdot \ldots \cdot D_n) \in \mathbf{Z}$. Intersection numbers involving line bundles in place of divisors are of course defined analogously.

Similar constructions work when X or V are possibly non-reduced complete complex schemes provided only that V has pure dimension k. The homology and cohomology groups of X are those of the underlying Hausdorff space: in other words, $H^*(X;\mathbf{Z})$ and $H_*(X;\mathbf{Z})$ do not see the scheme structure of X. However, one introduces the cycle $[V]$ of V, viz. the algebraic k-cycle

$$[V] \;=\; \sum_{V_i} \big(\text{length}_{\mathcal{O}_{V_i}} \mathcal{O}_V\big) \cdot [V_i]$$

on X, where $\{V_i\}$ are the irreducible components of V (with their reduced scheme structures), and $\mathcal{O}_{V_i}$ is the local ring of V along V_i. By linearity we get a corresponding class $[V] = \sum \big(\text{length}_{\mathcal{O}_{V_i}} \mathcal{O}_V\big) \cdot [V_i] \in H_{2k}(X;\mathbf{Z})$. Then the cap product appearing in (1.4) defines the intersection number

$$(D_1 \cdot \ldots \cdot D_k \cdot [V]) \;=\; \int_{[V]} D_1 \cdot \ldots \cdot D_k \;\in\; \mathbf{Z}.$$

Somewhat abusively we often continue to write simply $(D_1 \cdot \ldots \cdot D_k \cdot V) \in \mathbf{Z}$, it being understood that one has to take into account any multiple components of V. If V has pure dimension d and $k \leq d$ then we define

$$(D_1 \cdot \ldots \cdot D_k \cdot [V]) \;=_{\text{def}}\; \Big(c_1\big(\mathcal{O}_X(D_1)\big) \cdot \ldots \cdot c_1\big(\mathcal{O}_X(D_k)\big)\Big) \cap [V] \\ \in\; H_{2d-2k}(X;\mathbf{Z}). \tag{1.5}$$

These intersection classes are compatible with the constructions in [208] and they satisfy the usual formal properties, as in [208, Chapter 2].[7] For instance if X has pure dimension n and D is an effective Cartier divisor on X, then we may view D as a subscheme of X and

$$c_1\big(\mathcal{O}_X(D)\big) \cap [X] \;=\; [D] \;\in\; H_{2(n-1)}(X;\mathbf{Z}) \tag{1.6}$$

([208, Chapter 2.5]). The same formula holds even if D is not effective provided that one interprets the right-hand side as the homology class of the Weil

[7] More precisely, the constructions of [208, Chapter 2] yield a class

$$\Big(c_1\big(\mathcal{O}_X(D_1)\big) \cdot \ldots \cdot c_1\big(\mathcal{O}_X(D_k)\big)\Big) \cap [V] \;\in\; A_{d-k}(X)$$

in the Chow group that maps to the class in (1.5) under the cycle map $A_{d-k}(X) \longrightarrow H_{2(d-k)}(X;\mathbf{Z})$.

divisor determined by D (Example 1.1.3). Similarly, if V is an irreducible variety, then

$$c_1\big(\mathcal{O}_X(D)\big) \cap [V] \;=\; c_1\big(\mathcal{O}_X(D)\,|\,V\,\big) \cap [V] \;=\; [\overline{D}],$$

where $\overline{D} \in \mathrm{Div}(V)$ is a divisor on V with $\mathcal{O}_V(\overline{D}) = \mathcal{O}_X(D)\,|\,V$. This inductively leads to the important fact that the intersection class $\big(D_1 \cdot \ldots \cdot D_k \cdot [V]\big)$ in (1.5) is represented by an algebraic $(d-k)$-cycle on X. In fact it is even represented by a $(d-k)$-cycle on $\mathrm{Supp}(D_1) \cap \ldots \cap \mathrm{Supp}(D_k) \cap V$.

Remark 1.1.13. (Advice for the novice). The use of topological definitions as our "official" foundation for intersection theory might not be the most accessible approach for a novice. So we say here a few words about what we actually require, and where one can learn it. In the present volume, all one needs for the most part is to be able to define the intersection number

$$\big(D_1 \cdot \ldots \cdot D_n\big) \;=\; \int_X D_1 \cdot \ldots \cdot D_n \;\in\; \mathbf{Z}$$

of n Cartier divisors $D_1, \ldots, D_n$ on an n-dimensional irreducible projective (or complete) variety X. The most important features of this product (which in fact characterize it in the projective case) are:

(i). The integer $\big(D_1 \cdot \ldots \cdot D_n\big)$ is symmetric and multilinear as a function of its arguments;

(ii). $\big(D_1 \cdot \ldots \cdot D_n\big)$ depends only on the linear equivalence classes of the D_i;

(iii). If $D_1, \ldots, D_n$ are effective divisors that meet transversely at smooth points of X, then

$$\big(D_1 \cdot \ldots \cdot D_n\big) \;=\; \#\{\, D_1 \cap \ldots \cap D_n \,\}.$$

Given an irreducible subvariety $V \subseteq X$ of dimension k, the intersection number

$$\big(D_1 \cdot \ldots \cdot D_k \cdot V\big) \;\in\; \mathbf{Z} \tag{*}$$

is then defined by replacing each divisor D_i with a linearly equivalent divisor D_i' whose support does not contain V, and intersecting the restrictions of the D_i' on V.[8] It is also important to know that if D_n is reduced, irreducible and effective, then one can compute $(D_1 \cdot \ldots \cdot D_n)$ by taking $V = D_n$ in (*). The intersection product satisfies the projection formula: if $f : Y \longrightarrow X$ is a generically finite surjective proper map, then

$$\int_Y f^*D_1 \cdot \ldots \cdot f^*D_n \;=\; \big(\deg f\,\big) \cdot \int_X D_1 \cdot \ldots \cdot D_n.$$

By linearity, one can replace V in (*) by an arbitrary k-cycle, and the analogous constructions when X or V carries a possibly non-reduced scheme structure are handled as above by passing to cycles.

[8] This is independent of the choice of D_i' thanks to (ii).

The case $\dim X = 2$ is treated very clearly in Chapter 5, Section 1, of [280], and this is certainly the place for a beginner to start. The extension to higher dimensions might to some extent be taken on faith. Alternatively, as noted above the theory is developed in detail via the method of Snapper and Kleiman in [363, Chapter 6.2], [114, Chapter 1.2] or [22, Chapter 1]. In this approach the crucial Theorem 1.1.24 is established along the way to defining intersection products. A more elementary presentation appears in [532, Chapter 4] provided that one is willing to grant 1.1.24. □

Numerical equivalence. We continue to assume that X is a complete algebraic scheme over $\mathbf{C}$. Of the various natural equivalence relations defined on $\mathrm{Div}(X)$, we will generally deal with the weakest:

Definition 1.1.14. (Numerical equivalence). Two Cartier divisors

$$D_1\,,\,D_2\ \in\ \mathrm{Div}(X)$$

are *numerically equivalent*, written $D_1 \equiv_{\mathrm{num}} D_2$, if

$$\big(D_1 \cdot C\big)\ =\ \big(D_2 \cdot C\big)\ \text{ for every irreducible curve } C \subseteq X,$$

or equivalently if $(D_1 \cdot \gamma) = (D_2 \cdot \gamma)$ for all one-cycles γ on X. Numerical equivalence of line bundles is defined in the analogous manner. A divisor or line bundle is *numerically trivial* if it is numerically equivalent to zero, and $\mathrm{Num}(X) \subseteq \mathrm{Div}(X)$ is the subgroup consisting of all numerically trivial divisors. □

Definition 1.1.15. (Néron–Severi group). The *Néron–Severi group* of X is the group

$$N^1(X) = \mathrm{Div}(X)/\mathrm{Num}(X)$$

of numerical equivalence classes of divisors on X. □

The first basic fact is that this group is finitely generated:

Proposition 1.1.16. (Theorem of the base). *The Néron–Severi group* $N^1(X)$ *is a free abelian group of finite rank.*

Definition 1.1.17. (Picard number). The rank of $N^1(X)$ is called the *Picard number* of X, written $\rho(X)$. □

Proof of Proposition 1.1.16. A divisor D on X determines a cohomology class

$$[D]_{\mathrm{hom}}\ =\ c_1(\mathcal{O}_X(D))\ \in\ H^2\big(X;\mathbf{Z}\big),$$

and if $[D]_{\mathrm{hom}} = 0$ then evidently D is numerically trivial. Therefore the group $\mathrm{Hom}(X)$ of cohomologically trivial Cartier divisors is a subgroup of $\mathrm{Num}(X)$. It follows that $N^1(X)$ is a quotient of a subgroup of $H^2\big(X;\mathbf{Z}\big)$, and in particular is finitely generated. It is torsion-free by construction. □

The next point is that intersection numbers respect numerical equivalence:

Lemma 1.1.18. *Let X be a complete variety or scheme, and let*

$$D_1, \dots, D_k, D_1', \dots, D_k' \in \mathrm{Div}(X)$$

be Cartier divisors on X. If $D_i \equiv_{\mathrm{num}} D_i'$ for each i, then

$$\big(D_1 \cdot D_2 \cdot \ldots \cdot D_k \cdot [V]\big) \ = \ \big(D_1' \cdot D_2' \cdot \ldots \cdot D_k' \cdot [V]\big)$$

for every subscheme $V \subseteq X$ of pure dimension k.

The lemma allows one to discuss intersection numbers among numerical equivalence classes:

Definition 1.1.19. (Intersection of numerical equivalence classes). Given classes $\delta_1, \dots, \delta_k \in N^1(X)$, we denote by

$$\int_{[V]} \delta_1 \cdot \ldots \cdot \delta_k \qquad \text{or} \qquad \big(\delta_1 \cdot \ldots \cdot \delta_k \cdot [V]\big)$$

the intersection number of any representatives of the classes in question. □

Proof of Lemma 1.1.18. We assert first that if $E \equiv_{\mathrm{num}} 0$ is a numerically trivial Cartier divisor on X, then $\big(E \cdot D_2 \cdot \ldots \cdot D_k \cdot [V]\big) = 0$ for any Cartier divisors $D_2, \dots, D_k$. In fact, $c_1\big(\mathcal{O}_X(D_2)\big) \cdot \ldots \cdot c_1\big(\mathcal{O}_X(D_k)\big) \cap [V]$ is represented by a one-cycle γ on X, and so $\big(E \cdot D_2 \cdot \ldots \cdot D_k \cdot [V]\big) = (E \cdot \gamma) = 0$ by definition of numerical equivalence. This shows that $\big(D_1 \cdot D_2 \cdot \ldots \cdot D_k \cdot [V]\big) = \big(D_1' \cdot D_2 \cdot \ldots \cdot D_k \cdot [V]\big)$ provided that $D_1 \equiv_{\mathrm{num}} D_1'$, and the lemma follows by induction on k. □

Remark 1.1.20. (Characterization of numerically trivial line bundles). A useful characterization of numerically trivial line bundles is established in Kleiman's exposé [52, XIII, Theorem 4.6] in SGA6. Specifically, consider a line bundle L on a complete scheme X. Then L is numerically trivial if and only if there is an integer $m \neq 0$ such that $L^{\otimes m} \in \mathrm{Pic}^0(X)$, i.e. such that $L^{\otimes m}$ is a deformation of the trivial line bundle. We sketch a proof in Section 1.4.D in the projective case based on a vanishing theorem of Fujita and Grothendieck's Quot schemes. (We also give there a fuller explanation of the statement.) □

Remark 1.1.21. (Lefschetz (1 , 1)-theorem). When X is a non-singular projective variety, Hodge theory gives an alternative description of $N^1(X)$. Set

$$H^2\big(X;\mathbf{Z}\big)_{\text{t.f.}} \ = \ H^2\big(X;\mathbf{Z}\big) / \text{ (torsion) }.$$

It follows from the result quoted in the previous remark that if D is a numerically trivial divisor then $[D]_{\mathrm{hom}} \in H^2\big(X;\mathbf{Z}\big)$ is a torsion class. Therefore

$N^1(X)$ embeds into $H^2(X;\mathbf{Z})_{\text{t.f.}}$. On the other hand, the Lefschetz $(1,1)$-theorem asserts that a class $\alpha \in H^2(X;\mathbf{Z})$ is algebraic if and only if α has type $(1,1)$ under the Hodge decomposition of $H^2(X;\mathbf{C})$ (cf. [248, Chapter 1, §2]). Therefore

$$N^1(X) \;=\; H^2(X;\mathbf{Z})_{\text{t.f.}} \cap H^{1,1}(X;\mathbf{C}). \quad \square$$

Finally we say a word about functoriality. Let $f : Y \longrightarrow X$ be a morphism of complete varieties or projective schemes. If $\alpha \in \text{Pic}(X)$ is a class mapping to zero in $N^1(X)$, then it follows from the projection formula that $f^*(\alpha)$ is numerically trivial on Y. Therefore the pullback mapping on Picard groups determines thanks to Example 1.1.5 a functorial induced homomorphism $f^* : N^1(X) \longrightarrow N^1(Y)$.

Remark 1.1.22. (Non-projective schemes). As indicated at the end of Section 1.1.A, the integrality and projectivity hypotheses in the previous paragraph arise only in order to use the functorial properties of line bundles to discuss divisors. To have a theory that runs smoothly for possibly non-projective schemes, it would be better — as in [341] — to take $N^1(X)$ to be the additive group of numerical equivalence classes of line bundles: we leave this modification to the interested reader. As explained above we prefer to stick with the classical language of divisors. $\square$

1.1.D Riemann–Roch

We will often have occasion to draw on asymptotic forms of the Riemann–Roch theorem, and we give a first formulation here. More detailed treatments appear in [363, VI.2], [114, Chapter 1.2], and [22, Chapter 1], to which we will refer for proofs.

We start with a definition:

Definition 1.1.23. (Rank and cycle of a coherent sheaf). Let X be an irreducible variety (or scheme) of dimension n, and $\mathcal{F}$ a coherent sheaf on X. The *rank* $\text{rank}(\mathcal{F})$ of $\mathcal{F}$ is the length of the stalk of $\mathcal{F}$ at the generic point of X. If X is reduced, then

$$\text{rank}(\mathcal{F}) \;=\; \dim_{\mathbf{C}(X)} \mathcal{F} \otimes \mathbf{C}(X).$$

If X is reducible (but still of dimension n), then one defines similarly the rank of $\mathcal{F}$ along any n-dimensional irreducible component V of X: $\text{rank}_V(\mathcal{F}) = \text{length}_{\mathcal{O}_v}\mathcal{F}_v$, where $\mathcal{F}_v$ is the stalk of $\mathcal{F}$ at the generic point v of V. The *cycle* of $\mathcal{F}$ is the n-cycle

$$Z_n(\mathcal{F}) \;=\; \sum_V \text{rank}_V(\mathcal{F}) \cdot [V],$$

the sum being taken over all n-dimensional components of X. $\square$

One then has

Theorem 1.1.24. (Asymptotic Riemann–Roch, I). *Let X be an irreducible projective variety of dimension n, and let D be a divisor on X. Then the Euler characteristic $\chi\big(X, \mathcal{O}_X(mD)\big)$ is a polynomial of degree $\leq n$ in m, with*

$$\chi\big(X, \mathcal{O}_X(mD)\big) \;=\; \frac{\big(D^n\big)}{n!}\, m^n \;+\; O(m^{n-1}). \tag{1.7}$$

More generally, for any coherent sheaf $\mathcal{F}$ on X,

$$\chi\big(X, \mathcal{F} \otimes \mathcal{O}_X(mD)\big) \;=\; \mathrm{rank}(\mathcal{F}) \cdot \frac{\big(D^n\big)}{n!} \cdot m^n \;+\; O(m^{n-1}). \tag{1.8}$$

Corollary 1.1.25. *In the setting of the theorem, if $H^i\big(X, \mathcal{F} \otimes \mathcal{O}_X(mD)\big) = 0$ for $i > 0$ and $m \gg 0$ then*

$$h^0\big(X, \mathcal{F} \otimes \mathcal{O}_X(mD)\big) \;=\; \mathrm{rank}(\mathcal{F}) \cdot \frac{\big(D^n\big)}{n!}\, m^n \;+\; O(m^{n-1}) \tag{1.9}$$

for large m. More generally, (1.9) *holds provided that*

$$h^i\big(X, \mathcal{F} \otimes \mathcal{O}_X(mD)\big) \;=\; O(m^{n-1})$$

for $i > 0$. □

Remark 1.1.26. (Reducible schemes). The formula (1.7) remains valid if X is a possibly reducible complete scheme of pure dimension n provided as usual that we interpret $\big(D^n\big)$ as the intersection number $\int_{[X]} D^n$. The same is true of (1.8) provided that the first term on the right is replaced by

$$\Big(\int_{Z_n(\mathcal{F})} D^n \Big) \cdot \frac{m^n}{n!}.$$

With the analogous modifications, the corollary likewise extends to possibly reducible complete schemes. □

We do not prove Theorem 1.1.24 here. The result is established in a relatively elementary fashion via the approach of Snapper–Kleiman in [363, Corollary VI.2.14] (see also [22, Chapter 1]). Debarre [114, Theorem 1.5] gives a very accessible account of the main case $\mathcal{F} = \mathcal{O}_X$. However, one can quickly obtain 1.1.24 and 1.1.26 as special cases of powerful general results: see the next example.

Example 1.1.27. (Theorem 1.1.24 via Hirzebruch–Riemann–Roch). Theorem 1.1.24 and the extension in 1.1.26 yield easily to heavier machinery. In fact, if X is a non-singular variety then the Euler characteristic in question is computed by the Hirzebruch–Riemann–Roch theorem:

$$\chi\big(X, \mathcal{F} \otimes \mathcal{O}_X(mD)\big) \;=\; \int_X \mathrm{ch}(\mathcal{F} \otimes \mathcal{O}_X(mD)) \cdot \mathrm{Td}(X).$$

Viewing $\mathrm{ch}(\mathcal{F} \otimes \mathcal{O}_X(mD))$ as a polyomial in m, one has

$$\begin{aligned}\mathrm{ch}(\mathcal{F} \otimes \mathcal{O}_X(mD)) &= \mathrm{ch}(\mathcal{F}) \cdot \mathrm{ch}\big(\mathcal{O}_X(mD)\big) \\ &= \mathrm{rank}(\mathcal{F}) \cdot \frac{c_1\big(\mathcal{O}_X(D)\big)^n}{n!} m^n + \text{ lower-degree terms,}\end{aligned}$$

which gives (1.8) in this case. On an arbitrary complete scheme one can invoke similarly the general Riemann–Roch theorem for singular varieties [208, Corollary 18.3.11 and Example 18.3.6].[9] □

Finally we record two additional results for later reference. The first asserts that Euler characteristics are multiplicative under étale covers:

Proposition 1.1.28. (Étale multiplicativity of Euler characteristics). *Let $f : Y \longrightarrow X$ be a finite étale covering of complete schemes, and let $\mathcal{F}$ be any coherent sheaf on X. Then*

$$\chi\big(Y\,,\,f^*\mathcal{F}\big) \;=\; \deg\left(Y \longrightarrow X\right) \cdot \chi\big(X\,,\,\mathcal{F}\big).$$

This follows for example from the Riemann–Roch theorem and [208, Example 18.3.9]. An elementary direct approach — communicated by Kleiman — is outlined in Example 1.1.30.

Example 1.1.29. The Riemann–Hurwitz formula for branched coverings of curves shows that Proposition 1.1.28 fails in general if f is not étale. □

Example 1.1.30. (Kleiman's proof of Proposition 1.1.28). We sketch a proof of 1.1.28 when X is projective. Set $d = \deg(f)$. Arguing as in Example 1.4.42 one reduces to the case in which X is an integral variety and $\mathcal{F} = \mathcal{O}_X$, and by induction on dimension one can assume that the result is known for all sheaves supported on a proper subset of X. Since f is finite one has $\chi(Y, \mathcal{O}_Y) = \chi(X, f_*\mathcal{O}_Y)$, so the issue is to show that

$$\chi(X, f_*\mathcal{O}_Y) \;=\; d \cdot \chi(X, \mathcal{O}_X). \tag{*}$$

For this, choose an ample divisor H on X. Then for $p \gg 0$ one can construct exact sequences

$$\begin{gathered}0 \longrightarrow \mathcal{O}_X(-pH)^d \longrightarrow f_*\mathcal{O}_Y \longrightarrow \mathcal{G}_1 \longrightarrow 0, \\ 0 \longrightarrow \mathcal{O}_X(-pH)^d \longrightarrow \mathcal{O}_X^d \longrightarrow \mathcal{G}_2 \longrightarrow 0,\end{gathered} \tag{**}$$

where $\mathcal{G}_1$, $\mathcal{G}_2$ are supported on proper subsets of X. Now suppose one knew that 1.1.28 held for $\mathcal{F} = f_*\mathcal{O}_Y$, i.e. suppose that one knows

$$\chi(Y, f^*f_*\mathcal{O}_Y) \;=\; d \cdot \chi(X, f_*\mathcal{O}_Y). \tag{***}$$

[9] Note however that the term $\tau_{X,n}(\mathcal{F})$ is missing from the last displayed formula in [208, 18.3.6].

Since we can assume that 1.1.28 holds for $\mathcal{G}_1$ and $\mathcal{G}_2$, the exact sequences (**) will then yield (*). So it remains to prove (***).

To this end, consider the fibre square

$$\begin{array}{ccc} W & \xrightarrow{g} & Y \\ {\scriptstyle g}\downarrow & & \downarrow{\scriptstyle f} \\ Y & \xrightarrow[f]{} & X \end{array}$$

where $W = Y \times_X Y$. Since f is étale, W splits as the disjoint union of a copy of Y and another scheme W' étale of degree $d-1$ over Y. So by induction on d, we can assume that $\chi(W, \mathcal{O}_W) = \chi(Y, g_*\mathcal{O}_W) = d \cdot \chi(Y, \mathcal{O}_Y)$. On the other hand, $f^*f_*\mathcal{O}_Y = g_*g^*\mathcal{O}_Y = g_*\mathcal{O}_W$ since f is flat ([280, III.9.3]), and then (***) follows. □

The second result, allowing one to produce very singular divisors, will be useful in Chapters 5 and 10.

Proposition 1.1.31. (Constructing singular divisors). *Let X be an irreducible projective (or complete) variety of dimension n, and let D be a divisor on X with the property that $h^i\big(X, \mathcal{O}_X(mD)\big) = O(m^{n-1})$ for $i > 0$. Fix a positive rational number α with*

$$0 \ < \ \alpha^n \ < \ \big(D^n\big).$$

Then when $m \gg 0$ there exists for any smooth point $x \in X$ a divisor $E = E_x \in |mD|$ with

$$\operatorname{mult}_x(E) \ \geq \ m \cdot \alpha. \tag{1.10}$$

Here $\operatorname{mult}_x(E)$ denotes as usual the multiplicity of the divisor E at x, i.e. the order of vanishing at x of a local equation for E. The proof will show that there is one large value of m that works simultaneously at all smooth points $x \in X$.

Proof. Producing a divisor with prescribed multiplicity at a given point involves solving the system of linear equations determined by the vanishing of an appropriate number of partial derivatives of a defining equation. To prove the Proposition we simply observe that under the stated assumptions there are more variables than equations. Specifically, the number of sections of $\mathcal{O}_X(mD)$ is estimated by 1.1.25:

$$h^0\big(X, \mathcal{O}_X(mD)\big) \ = \ \frac{(D^n)}{n!} m^n + O(m^{n-1}).$$

On the other hand, it is at most

$$\binom{n+c-1}{n} \ = \ \frac{c^n}{n!} + O(c^{n-1})$$

conditions for a section of $\mathcal{O}_X(mD)$ to vanish to order $\geq c$ at a smooth point $x \in X$. Taking $m \gg 0$ and suitable $m \cdot \alpha < c < m \cdot \left(D^n\right)^{1/n}$, we get the required divisor. □

Remark 1.1.32. (Other ground fields). The discussion in this section goes through with only minor changes if X is an algebraic variety or scheme defined over an algebraically closed field of any characteristic. (In Section 1.1.C one would use the algebraic definition of intersection numbers, and a different argument is required to prove that $N^1(X)$ has finite rank: see [341, Chapter IV]). □

1.2 The Classical Theory

Given a divisor D on a projective variety X, what should it mean for D to be positive? The most appealing idea from an intuitive point of view is to ask that D be a hyperplane section under some projective embedding of X — one says then that D is *very ample.* However this turns out to be rather difficult to work with technically: already on curves it can be quite subtle to decide whether or not a given divisor is very ample. It is found to be much more convenient to focus instead on the condition that some positive multiple of D be very ample; in this case D is *ample.* This definition leads to a very satisfying theory, which was largely worked out in the fifties and early sixties. The fundamental conclusion is that on a projective variety, amplitude can be characterized geometrically (which we take as the definition), cohomologically (theorem of Cartan–Serre–Grothendieck) or numerically (Nakai–Moishezon–Kleiman criterion).

This section is devoted to an overview of the classical theory of ample line bundles. One of our purposes is to set down the basic facts in a form convenient for later reference. The cohomological material in particular is covered (in greater generality and detail) in Hartshorne's text [280], to which we will refer where convenient. Chapter I of Hartshorne's earlier book [276] contains a nice exposition of the theory, and in several places we have drawn on his discussion quite closely.

We begin with the basic definition.

Definition 1.2.1. (Ample and very ample line bundles and divisors on a complete scheme). Let X be a complete scheme, and L a line bundle on X.

(i). L is *very ample* if there exists a closed embedding $X \subseteq \mathbf{P}$ of X into some projective space $\mathbf{P} = \mathbf{P}^N$ such that

$$L = \mathcal{O}_X(1) =_{\text{def}} \mathcal{O}_{\mathbf{P}^N}(1) \mid X.$$

(ii). L is *ample* if $L^{\otimes m}$ is very ample for some $m > 0$.

A Cartier divisor D on X is ample or very ample if the corresponding line bundle $\mathcal{O}_X(D)$ is so. □

Remark 1.2.2. (Amplitude). We will interchangeably use "ampleness" and "amplitude" to describe the property of being ample. We feel that the euphonious quality of the latter term compensates for the fact that it may not be completely standard in the present context. (Confusion with other meanings of amplitude seems very unlikely.) □

Example 1.2.3. (Ample line bundle on curves). If X is an irreducible curve, and L is a line bundle on X, then L is ample if and only if $\deg(L) > 0$. □

Example 1.2.4. (Varieties with Pic = Z). If X is a projective variety with $\mathrm{Pic}(X) = \mathbf{Z}$, then any non-zero effective divisor on X is ample. (Use 1.2.6 (iv).) This applies, for instance, when X is a projective space or a Grassmannian. □

Example 1.2.5. (Intersection products). If $D_1, \ldots, D_n$ are ample divisors on an n-dimensional projective variety X, then $\big(D_1 \cdot \ldots \cdot D_n\big) > 0$. (One can assume that each D_i is very ample, and then the inequality reduces to Example 1.2.3.) □

1.2.A Cohomological Properties

The first basic fact is that amplitude can be detected cohomologically:

Theorem 1.2.6. (Cartan–Serre–Grothendieck theorem). *Let L be a line bundle on a complete scheme X. The following are equivalent:*

(i). *L is ample.*

(ii). *Given any coherent sheaf $\mathcal{F}$ on X, there exists a positive integer $m_1 = m_1(\mathcal{F})$ having the property that*

$$H^i\big(X, \mathcal{F} \otimes L^{\otimes m}\big) = 0 \quad \textit{for all} \ \ i > 0 \ , \ m \geq m_1(\mathcal{F}).$$

(iii). *Given any coherent sheaf $\mathcal{F}$ on X, there exists a positive integer $m_2 = m_2(\mathcal{F})$ such that $\mathcal{F} \otimes L^{\otimes m}$ is generated by its global sections for all $m \geq m_2(\mathcal{F})$.*

(iv). *There is a positive integer $m_3 > 0$ such that $L^{\otimes m}$ is very ample for every $m \geq m_3$.*

Remark 1.2.7. (Serre vanishing). The conclusion in (ii) is often referred to as Serre's vanishing theorem. □

Outline of Proof of Theorem 1.2.6. (i) ⇒ (ii). We assume to begin with that L is very ample, defining an embedding of X into some projective space $\mathbf{P}$.

In this case, extending $\mathcal{F}$ by zero to a coherent sheaf on $\mathbf{P}$, we are reduced to the vanishing of $H^i\big(\mathbf{P}, \mathcal{F}(m)\big)$ for $m \gg 0$, which is the content of [280, Theorem III.5.2]. In general, when L is merely ample, fix m_0 such that $L^{\otimes m_0}$ is very ample. Then apply the case already treated to each of the sheaves $\mathcal{F}, \mathcal{F} \otimes L, \ldots, \mathcal{F} \otimes L^{\otimes m_0 - 1}$.

(ii) $\Rightarrow$ (iii). Fix a point $x \in X$, and denote by $\mathfrak{m}_x \subset \mathcal{O}_X$ the maximal ideal sheaf of x. By (ii) there is an integer $m_2(\mathcal{F}, x)$ such that

$$H^1\big(X, \mathfrak{m}_x \cdot \mathcal{F} \otimes L^{\otimes m}\big) = 0 \quad \text{for} \quad m \geq m_2(\mathcal{F}, x).$$

It then follows from the exact sequence

$$0 \longrightarrow \mathfrak{m}_x \cdot \mathcal{F} \longrightarrow \mathcal{F} \longrightarrow \mathcal{F} \ / \ \mathfrak{m}_x \cdot \mathcal{F} \longrightarrow 0$$

upon twisting by $L^{\otimes m}$ and taking cohomology that $\mathcal{F} \otimes L^{\otimes m}$ is globally generated in a neighborhood of x for every $m \geq m_2(\mathcal{F}, x)$. By quasi-compactness we can then choose a single natural number $m_2(\mathcal{F})$ that works for all $x \in X$.

(iii) $\Rightarrow$ (iv). It follows first of all from (iii) that there exists a positive integer p_1 such that $L^{\otimes m}$ is globally generated for all $m \geq p_1$. Denote by

$$\phi_m : X \longrightarrow \mathbf{P}H^0\big(X, L^{\otimes m}\big)$$

the corresponding map to projective space. We need to show that we can arrange for ϕ_m to be an embedding by taking $m \gg 0$, for which it is sufficient to prove that ϕ_m is one-to-one and unramified ([280, II.7.3]). To this end, consider the set

$$U_m = \big\{ \, y \in X \mid L^{\otimes m} \otimes \mathfrak{m}_y \text{ is globally generated} \, \big\}.$$

This is an open set (Example 1.2.9), and $U_m \subset U_{m+p}$ for $p \geq p_1$ thanks to the fact that $L^{\otimes p}$ is generated by its global sections. Given any point $x \in X$ we can find by (iii) an integer $m_2(x)$ such that $x \in U_m$ for all $m \geq m_2(x)$, and therefore $X = \cup U_m$. By quasi-compactness there is a single integer $m_3 \geq p_1$ such that $L^{\otimes m} \otimes \mathfrak{m}_x$ is generated by its global sections for every $x \in X$ whenever $m \geq m_3$. But the global generation of $L^{\otimes m} \otimes \mathfrak{m}_x$ implies that $\phi_m(x) \neq \phi_m(x')$ for all $x' \neq x$, and that ϕ_m is unramified at x. Thus ϕ_m is an embedding for all $m \geq m_3$, as required.

(iv) $\Rightarrow$ (i): Definitional. □

Remark 1.2.8. (Amplitude on non-complete schemes). One can also discuss ample line bundles on possibly non-complete schemes. In this more general setting, property (iii) from 1.2.6 is taken as the definition of amplitude. □

Example 1.2.9. Let B be a globally generated line bundle on a complete variety or scheme X. Then

$$U =_{\text{def}} \{ y \in X \mid B \otimes \mathfrak{m}_y \text{ is globally generated} \}$$

is an open subset of X. (Since B is globally generated, it suffices by Nakayama's lemma to prove the openness of the set

$$V =_{\text{def}} \{ y \in X \mid H^0(X, B) \longrightarrow B \otimes \mathcal{O}_X/\mathfrak{m}_y^2 \text{ is surjective} \}.$$

But this follows from the existence of a coherent sheaf $\mathcal{P}$ on X, whose fibre at y is canonically $\mathcal{P}(y) = B \otimes \mathcal{O}_X/\mathfrak{m}_y^2$, together with a map $u : H^0(X, B) \otimes \mathcal{O}_X \longrightarrow \mathcal{P}$ that fibre by fibre is given by evaluation of sections. In fact if $u(y)$ is surjective at one point y then it is surjective in a neighborhood of y by the coherence of coker(u). As for $\mathcal{P}$, it is the sheaf $\mathcal{P} = P_X^2(B)$ of second-order principal parts of B: starting with the ideal sheaf $\mathcal{I}_\Delta$ of the diagonal on $X \times X$ one takes

$$P_X^2(B) = \mathrm{pr}_{2,*}\big(\mathrm{pr}_1^* B \otimes (\mathcal{O}_{X \times X}/\mathcal{I}_\Delta^2)\big).$$

See [257, Chapter 16] for details on these sheaves.) □

Example 1.2.10. (Sums of divisors). Let D and E be (Cartier) divisors on a projective scheme X. If D is ample, then so too is $mD + E$ for all $m \gg 0$. In fact, $mD + E$ is very ample if $m \gg 0$. (For the second assertion choose positive integers m_1, m_2 such that mD is very ample for $m \geq m_1$ and $mD + E$ is free when $m \geq m_2$. Then $mD + E$ is very ample once $m \geq m_1 + m_2$.) □

Example 1.2.11. (Ample line bundles on a product). If L and M are ample line bundles on projective schemes X and Y respectively, then $\mathrm{pr}_1^* L \otimes \mathrm{pr}_2^* M$ is ample on $X \times Y$. □

Remark 1.2.12. (Matsusaka's theorem). According to Theorem 1.2.6, if L is an ample line bundle on a projective variety X then there is an integer $m(L)$ such that $L^{\otimes m}$ is very ample for $m \geq m(L)$. However, the proof of the theorem fails to give any concrete information about the value of this integer. So it is interesting to ask what geometric information $m(L)$ depends on, and whether one can give effective estimates. A theorem of Matsusaka [420] and Kollár–Matsusaka [366] states that if X is a smooth projective variety of dimension n, then one can find $m(L)$ depending only on the intersection numbers $\int c_1(L)^n$ and $\int c_1(L)^{n-1} c_1(X)$. Siu [537] used the theory of multiplier ideals to give an effective statement, which was subsequently improved and clarified by Demailly [126]. A proof of the theorem of Kollár–Matsusaka via the approach of Siu–Demailly appears in Section 10.2. An example due to Kollár, showing that in general one cannot take $m(L)$ independent of L, is presented in Example 1.5.7. □

Proposition 1.2.13. (Finite pullbacks, I). *Let $f : Y \longrightarrow X$ be a finite mapping of complete schemes, and L an ample line bundle on X. Then $f^* L$ is an ample line bundle on Y. In particular, if $Y \subseteq X$ is a subscheme of X, then the restriction $L \mid Y$ of L to Y is ample.*

Remark 1.2.14. See Corollary 1.2.28 for a partial converse. □

Proof of Proposition 1.2.13. Let $\mathcal{F}$ be a coherent sheaf on Y. Then $f_*\big(\mathcal{F}\otimes f^*L^{\otimes m}\big) = f_*\mathcal{F}\otimes L^{\otimes m}$ by the projection formula, and $R^jf_*\big(\mathcal{F}\otimes f^*L^{\otimes m}\big) = 0$ for $j > 0$ thanks to the finiteness of f. Therefore

$$H^i\big(Y, \mathcal{F}\otimes f^*L^{\otimes m}\big) = H^i\big(X, f_*\mathcal{F}\otimes L^{\otimes m}\big)$$

for all i, and the statement then follows from the characterization (ii) of amplitude in Theorem 1.2.6. □

Corollary 1.2.15. (Globally generated line bundles). *Suppose that L is globally generated, and let*

$$\phi = \phi_{|L|} : X \longrightarrow \mathbf{P} = \mathbf{P}H^0\big(X, L\big)$$

be the resulting map to projective space defined by the complete linear system $|L|$. Then L is ample if and only if ϕ is a finite mapping, or equivalently if and only if

$$\int_C c_1(L) > 0$$

for every irreducible curve $C \subseteq X$.

Proof. The preceding proposition shows that if ϕ is finite, then L is ample. In this case evidently $\int_C c_1(L) > 0$ for every irreducible curve $C \subseteq X$. Conversely, if ϕ is not finite then there is a subvariety $Z \subseteq X$ of positive dimension that is contracted by ϕ to a point. Since $L = \phi^*\mathcal{O}_{\mathbf{P}}(1)$, we see that L restricts to a trivial line bundle on Z. In particular, $L \mid Z$ is not ample, and so thanks again to the previous proposition, neither is L. Moreover, if $C \subseteq Z$ is any irreducible curve, then $\int_C c_1(L) = 0$. □

The next result allows one in practice to restrict attention to reduced and irreducible varieties.

Proposition 1.2.16. *Let X be a complete scheme, and L a line bundle on X.*

(i). *L is ample on X if and only if L_{red} is ample on X_{red}.*

(ii). *L is ample on X if and only if the restriction of L to each irreducible component of X is ample.*

Proof. In each case the "only if" statement is a consequence of the previous proposition. So for (i) we need to show that if L_{red} is ample on X_{red}, then L itself is already ample. To this end we again use characterization (ii) of Theorem 1.2.6. Fix a coherent sheaf $\mathcal{F}$ on X, and let $\mathcal{N}$ be the nilradical of $\mathcal{O}_X$, so that $\mathcal{N}^r = 0$ for some r. Consider the filtration

$$\mathcal{F} \supset \mathcal{N}\cdot\mathcal{F} \supset \mathcal{N}^2\cdot\mathcal{F} \supset \cdots \supset \mathcal{N}^r\cdot\mathcal{F} = 0.$$

The quotients $\mathcal{N}^i\mathcal{F}/\mathcal{N}^{i+1}\mathcal{F}$ are coherent $\mathcal{O}_{X_{\mathrm{red}}}$-modules, and therefore

$$H^j\big(X, (\mathcal{N}^i\mathcal{F}/\mathcal{N}^{i+1}\mathcal{F}) \otimes L^{\otimes m}\big) \ = \ 0 \quad \text{for } \ j > 0 \text{ and } m \gg 0$$

thanks to the amplitude of $L \mid X_{\mathrm{red}}$. Twisting the exact sequences

$$0 \longrightarrow \mathcal{N}^{i+1}\mathcal{F} \longrightarrow \mathcal{N}^i\mathcal{F} \longrightarrow \mathcal{N}^i\mathcal{F}/\mathcal{N}^{i+1}\mathcal{F} \longrightarrow 0$$

by $L^{\otimes m}$ and taking cohomology, we then find by decreasing induction on i that

$$H^j\big(X, \mathcal{N}^i\mathcal{F} \otimes L^{\otimes m}\big) = 0 \quad \text{for } \ j > 0 \text{ and } m \gg 0.$$

When $i = 0$ this gives the vanishings required for 1.2.6 (ii). The proof of (ii) is similar. Specifically, supposing as we may that X is reduced, let $X = X_1 \cup \cdots \cup X_r$ be its decomposition into irreducible components, and assume that $L \mid X_i$ is ample for every i. Fix a coherent sheaf $\mathcal{F}$ on X, let $\mathcal{I}$ be the ideal sheaf of X_1 in X, and consider the exact sequence

$$0 \longrightarrow \mathcal{I} \cdot \mathcal{F} \longrightarrow \mathcal{F} \longrightarrow \mathcal{F} \mathbin{/} \mathcal{I} \cdot \mathcal{F} \longrightarrow 0. \tag{*}$$

The outer terms of (*) are supported on $X_2 \cup \cdots \cup X_r$ and X_1 respectively. So by induction on the number of irreducible components, we may assume that

$$H^j\big(X, \mathcal{I}\mathcal{F} \otimes L^{\otimes m}\big) \ = \ H^j\big(X, (\mathcal{F}/\mathcal{I}\mathcal{F}) \otimes L^{\otimes m}\big) \ = \ 0$$

for $j > 0$ and $m \gg 0$. It then follows from (*) that $H^j\big(X, \mathcal{F} \otimes L^{\otimes m}\big) = 0$ when $j > 0$ and $m \gg 0$, as required. □

A theorem of Grothendieck [256, III.4.7.1] shows that — in an extremely strong sense — amplitude is an open condition in families.

Theorem 1.2.17. (Amplitude in families). *Let $f : X \longrightarrow T$ be a proper morphism of schemes, and L a line bundle on X. Given $t \in T$, write*

$$X_t = f^{-1}(t) \quad , \quad L_t = L \mid X_t.$$

Assume that L_0 is ample on X_0 for some point $0 \in T$. Then there is an open neighborhood U of 0 in T such that L_t is ample on X_t for all $t \in U$.

Observe that we do not assume that f is flat.

Proof of Theorem 1.2.17. We follow a proof given by Kollár and Mori [368, Proposition 1.41]. The statement being local on T, we suppose that $T = \mathrm{Spec}(A)$ is affine.

We assert to begin with that for any coherent sheaf $\mathcal{F}$ on X, there is a positive integer $m(\mathcal{F}, L)$ such that

$$R^i f_*\big(\mathcal{F} \otimes L^{\otimes m}\big) = 0 \ \text{ in a neighborhood } U_m \subseteq T \text{ of } 0 \tag{*}$$

for all $i \geq 1$ and $m \geq m(\mathcal{F}, L)$. In fact, this is certainly true for large i — e.g. $i > \dim X_0$ — and we proceed by decreasing induction on i. Assuming then that (*) is known for given $i \geq 2$ and all $\mathcal{F}$, we need to show that it holds also for $i-1$.

To this end, consider the maximal ideal $\mathfrak{m}_0 \subset A$ of 0 in T, and choose generators $u_1, \ldots, u_p \in \mathfrak{m}_0$. This gives rise to a presentation

$$A^{\oplus p} \xrightarrow{\ u\ } A \longrightarrow A/\mathfrak{m}_0 \longrightarrow 0$$

of $A/\mathfrak{m}_0$, where $u(a_1, \ldots, a_p) = \sum a_i u_i$. Pulling back by f and tensoring by $\mathcal{F}$ we arrive at an exact diagram:

$$\begin{array}{ccccccccc}
0 \to \ker(f^*u \otimes 1) \to \mathcal{O}_X^p \otimes \mathcal{F} & & \xrightarrow{\quad f^*u\otimes 1 \quad} & & \mathcal{F} \to \mathcal{F} \otimes \mathcal{O}_{X_0} \to 0 \\
& \searrow & & \nearrow & \\
& & \operatorname{im}(f^*u \otimes 1) & & \\
& \nearrow & & \searrow & \\
0 & & & & 0\,.
\end{array}$$

By the induction hypothesis applied to the kernel sheaf,

$$R^i f_*\big(\ker(f^*u \otimes 1) \otimes L^{\otimes m}\big) \ = \ 0 \ \text{ near } 0$$

for $m \gg 0$. Furthermore, since L_0 is ample, the higher direct images of $\mathcal{F} \otimes \mathcal{O}_{X_0} \otimes L^{\otimes m}$ — which are just the cohomology groups on X_0 of the sheaves in question — vanish when m is large. It then follows upon tensoring by $L^{\otimes m}$ and chasing through the above diagram that the map

$$R^{i-1} f_*(\mathcal{F} \otimes L^{\otimes m}) \otimes_{\mathcal{O}_T} \mathcal{O}_T^p \xrightarrow{\ 1\otimes u\ } R^{i-1} f_*(\mathcal{F} \otimes L^{\otimes m})$$

on direct images is surjective in a neighborhood U'_m of 0 for $m \gg 0$. On the other hand, by construction $1 \otimes u$ factors through the inclusion $\mathfrak{m}_0 \cdot R^{i-1} f_*(\mathcal{F} \otimes L^{\otimes m}) \subset R^{i-1} f_*(\mathcal{F} \otimes L^{\otimes m})$. In other words, if m is sufficiently large, then

$$R^{i-1} f_*(\mathcal{F} \otimes L^{\otimes m}) \ = \ \mathfrak{m}_0 \cdot R^{i-1} f_*(\mathcal{F} \otimes L^{\otimes m})$$

in a neighborhood of 0. But by Nakayama's lemma, this implies that

$$R^{i-1} f_*(\mathcal{F} \otimes L^{\otimes m}) \ = \ 0$$

near 0, as required. Thus we have verified (*).

We assert next that the canonical mapping

$$\rho_m : f^* f_* L^{\otimes m} \longrightarrow L^{\otimes m}$$

is surjective along X_t for all t in a neighborhood U''_m of 0 provided that m is sufficiently large. To see this, apply (*) to the ideal sheaf $\mathcal{I}_{X_0/X}$ of X_0 in X. One finds that

$$f_*(L^{\otimes m}) \longrightarrow f_*(L^{\otimes m} \otimes \mathcal{O}_{X_0}) \ = \ H^0\big(X_0, L_0^{\otimes m}\big) \qquad (**)$$

is surjective when $m \gg 0$. But all sufficiently large powers of the ample line bundle L_0 are globally generated. Composing (**) with the evaluation $H^0\big(X_0, L_0^{\otimes m}\big) \otimes \mathcal{O}_{X_0} \longrightarrow L_0^{\otimes m}$, then shows that ρ_m is surjective along X_0 for $m \gg 0$. By the coherence of $\operatorname{coker} \rho_m$, it follows that ρ_m is also surjective along X_t for t near 0, as claimed.

Shrinking T we can suppose that ρ_m is globally surjective for some fixed large integer m. Now $f_*(L^{\otimes m})$ is itself globally generated since T is affine. Choosing finitely many sections generating $f_*(L^{\otimes m})$ and pulling back to X, we arrive at a surjective homomorphism $f^*\mathcal{O}_T^{r+1} \twoheadrightarrow L^{\otimes m}$ of sheaves on X. This defines a mapping

$$\phi : X \longrightarrow \mathbf{P}(\mathcal{O}_T^{r+1}) = \mathbf{P}^r \times T$$

over T. The amplitude of L_0 implies that ϕ is finite on X_0, and hence $\phi_t = \phi | X_t : X_t \longrightarrow \mathbf{P}^r$ is likewise finite for t in a neighborhood of 0. Thus $L_t^{\otimes m} = \phi_t^* \mathcal{O}_{\mathbf{P}^r}(1)$ is indeed ample. □

Remark 1.2.18. Observe for later reference that the final step of the proof just completed shows that there is a neighborhood $U = U_m$ of 0 in T such that the mapping

$$\phi : X_U =_{\text{def}} f^{-1}(U) \longrightarrow \mathbf{P}^r \times U$$

over U determined by $L^{\otimes m}$ is finite. □

We close this discussion by presenting some useful applications of Serre vanishing.

Example 1.2.19. (Asymptotic Riemann–Roch, II). Let D be an ample Cartier divisor on an irreducible projective variety X of dimension n. Then

$$h^0\big(X, \mathcal{O}_X(mD)\big) \ = \ \frac{(D^n)}{n!} \cdot m^n + O(m^{n-1}).$$

More generally, if $\mathcal{F}$ is any coherent sheaf on X then

$$h^0\big(X, \mathcal{F} \otimes \mathcal{O}_X(mD)\big) \ = \ \operatorname{rank}(\mathcal{F}) \frac{(D^n)}{n!} \cdot m^n + O(m^{n-1}).$$

(This follows immediately from Theorem 1.1.24 by virtue of the vanishing

$$H^i\big(X, \mathcal{F} \otimes \mathcal{O}_X(mD)\big) = 0$$

for $i > 0$ and $m \gg 0$. In fact, in the case at hand the dimensions in question are given for $m \gg 0$ by polynomials with the indicated leading terms.) This extends to reducible or non-reduced schemes X as in Remark 1.1.26; we leave the statement to the reader. □

Example 1.2.20. (Upper bounds on $\mathbf{h^0}$). If E is any divisor on an irreducible projective variety X of dimension n, then there is a constant $C > 0$ such that

$$h^0\big(X, \mathcal{O}_X(mE)\big) \ \leq\ Cm^n \quad \text{for all } \ m.$$

(Fix an ample divisor D on X. Then $aD - E$ is effective for some $a \gg 0$, and consequently $h^0\big(X, \mathcal{O}_X(mE)\big) \leq h^0\big(X, \mathcal{O}_X(maD)\big)$. The assertion then follows from Example 1.2.19.) See Example 1.2.33 for a generalization. □

Example 1.2.21. (Resolutions of a sheaf). Let X be a projective variety, and D an ample divisor on X. Then any coherent sheaf $\mathcal{F}$ on X admits a (possibly non-terminating) resolution of the form

$$\ldots \longrightarrow \oplus\, \mathcal{O}_X(-p_1 D) \longrightarrow \oplus\, \mathcal{O}_X(-p_0 D) \longrightarrow \mathcal{F} \longrightarrow 0$$

for suitable integers $0 \ll p_0 \ll p_1 \ll \ldots$. (Choose $p_0 \gg 0$ such that $\mathcal{F} \otimes \mathcal{O}_X(p_0 D)$ is globally generated. Fixing a collection of generating sections determines a surjective map $\oplus\, \mathcal{O}_X \longrightarrow \mathcal{F} \otimes \mathcal{O}_X(p_0 D)$. Twisting by $\mathcal{O}_X(-p_0 D)$ then gives rise to a surjection $\oplus\, \mathcal{O}_X(-p_0 D) \longrightarrow \mathcal{F}$, and one continues by applying the same argument to the kernel of this map.) Even though they may be infinite, one can sometimes use such resolutions to reduce cohomological questions about coherent sheaves to the case of line bundles. The next example provides an illustration. □

Example 1.2.22. (Surjectivity of multiplication maps). Let X be a projective variety or scheme, and let D and E be ample Cartier divisors on X. Then there is a positive integer $m_0 = m_0(D, E)$ such that the natural maps

$$H^0\big(X, \mathcal{O}_X(aD)\big) \otimes H^0\big(X, \mathcal{O}_X(bE)\big) \longrightarrow H^0\big(X, \mathcal{O}_X(aD + bE)\big)$$

are surjective whenever $a, b \geq m_0$. More generally, for any coherent sheaves $\mathcal{F}, \mathcal{G}$ on X, there is an integer $m_1 = m_1(D, E, \mathcal{F}, \mathcal{G})$ such that

$$H^0\Big(X\,,\, \mathcal{F} \otimes \mathcal{O}_X(aD)\Big) \otimes H^0\Big(X\,,\, \mathcal{G} \otimes \mathcal{O}_X(bE)\Big) \longrightarrow \\ H^0\Big(X\,,\, \mathcal{F} \otimes \mathcal{G} \otimes \mathcal{O}_X(aD + bE)\Big)$$

is surjective for $a, b \geq m_1$. (For the first statement consider on $X \times X$ the exact sequence

$$0 \longrightarrow \mathcal{I}_\Delta \longrightarrow \mathcal{O}_{X \times X} \longrightarrow \mathcal{O}_\Delta \longrightarrow 0, \tag{*}$$

$\Delta \subset X \times X$ being the diagonal. Writing (aD, bE) for the divisor $\mathrm{pr}_1^*(aD) + \mathrm{pr}_2^*(bE)$ on $X \times X$, the displayed sequence (*) shows that it suffices to verify that

$$H^1\big(X \times X, \mathcal{I}_\Delta\big((aD, bE)\big)\big) = 0 \tag{**}$$

for $a, b \geq m_0$. To this end, apply 1.2.21 to the ample divisor (D, E) to construct a resolution

$$\ldots \to \oplus \mathcal{O}_{X\times X}\big((-p_1 D, -p_1 E)\big) \longrightarrow \oplus \mathcal{O}_{X\times X}\big((-p_0 D, -p_0 E)\big) \longrightarrow \mathcal{I}_\Delta \to 0.$$

By Proposition B.1.2 from Appendix B, it is enough for (**) to produce an integer m_0 such that

$$H^i\Big(X \times X\,,\, \mathcal{O}_{X\times X}\big((a-p_{i-1})D, (b-p_{i-1})E\big)\Big) = 0$$

whenever $i > 0$ and $a, b \geq m_0$. But this is non-trivial only when $i \leq \dim X \times X$, and the cohomology group in question is computed by the Künneth formula. So the existence of the required integer m_0 follows immediately from Serre vanishing. The second statement is similar, except that one works with $\mathcal{I}_\Delta \otimes \mathrm{pr}_1^*(\mathcal{F}) \otimes \mathrm{pr}_2^*(\mathcal{G})$ in place of $\mathcal{I}_\Delta$, observing that (*) remains exact after tensoring through by $\mathrm{pr}_1^*(\mathcal{F}) \otimes \mathrm{pr}_2^*(\mathcal{G})$ thanks to flatness. Alternatively, one could use Fujita's vanishing theorem (Theorem 1.4.35) to bypass 1.2.21.) □

1.2.B Numerical Properties

A second very fundamental fact is that amplitude is characterized numerically:

Theorem 1.2.23. (Nakai–Moishezon–Kleiman criterion). *Let L be a line bundle on a projective scheme X. Then L is ample if and only if*

$$\int_V c_1(L)^{\dim(V)} \; > \; 0 \tag{1.11}$$

for every positive-dimensional irreducible subvariety $V \subseteq X$ (including the irreducible components of X).

Kleiman's paper [340] contains an illuminating discussion of the history of this basic result. In brief, it was originally established by Nakai [464] for smooth surfaces. Moishezon [433] proved 1.2.23 for non-singular varieties of higher dimension, and suggested in [434] a definition of intersection numbers that led to its validity on singular varieties as well. Nakai [465] subsequently extended the statement to arbitrary projective algebraic schemes, and finally Kleiman [341, Chapter III] treated the case of arbitrary complete schemes. The (now standard) proof we will give is due to Kleiman. In spite of the collaborative nature of Theorem 1.2.23, we will generally refer to it in the interests of brevity simply as Nakai's criterion.

Before giving the proof we mention two important consequences. First, it follows from the theorem that the amplitude of a divisor depends only on its numerical equivalence class:

Corollary 1.2.24. (Numerical nature of amplitude). *If $D_1, D_2 \in \mathrm{Div}(X)$ are numerically equivalent Cartier divisors on a projective variety or scheme X, then D_1 is ample if and only if D_2 is.* □

In particular it makes sense to discuss the amplitude of a class $\delta \in N^1(X)$:

Definition 1.2.25. (Ample classes). A numerical equivalence class $\delta \in N^1(X)$ is *ample* if it is the class of an ample divisor. Ample (algebraic) classes in $H^2(X, \mathbf{Z})$ or $H^2(X, \mathbf{Q})$ are defined in the same way. □

Example 1.2.26. (Varieties with Picard number 1). If X is a projective variety having Picard number $\rho(X) = 1$, then any non-zero effective divisor on X is ample. This extends Example 1.2.4, and applies for example to a very general abelian variety having a polarization of fixed type. □

Remark 1.2.27. The structure of the cone of all ample classes on a fixed projective variety is discussed further in Section 1.4. Several examples are worked out in Section 1.5. □

The second corollary shows that amplitude can be tested after pulling back by a finite surjective morphism:

Corollary 1.2.28. (Finite pullbacks, II). *Let $f : Y \longrightarrow X$ be a finite and surjective mapping of projective schemes, and let L be a line bundle on X. If f^*L is ample on Y, then L is ample on X.*

Proof. Let $V \subseteq X$ be an irreducible variety. Since f is surjective, there is an irreducible variety $W \subseteq Y$ mapping (finitely) onto V: starting with $f^{-1}(V)$, one constructs W by taking irreducible components and cutting down by general hyperplanes. Then by the projection formula

$$\int_W c_1(f^*L)^{\dim W} = \deg(W \longrightarrow V) \cdot \int_V c_1(L)^{\dim V},$$

so the assertion follows from the theorem. □

We now turn to the very interesting proof of the Nakai–Moishezon–Kleiman criterion. The argument will lead to several other results as well.

Proof of Theorem 1.2.23. Suppose first that L is ample. Then $L^{\otimes m}$ is very ample for some $m \gg 0$, and

$$m^{\dim V} \cdot \int_V \big(c_1(L)\big)^{\dim V} = \int_V \big(c_1(L^{\otimes m})\big)^{\dim V}$$

is the degree of V in the corresponding projective embedding of X. Consequently, this integral is strictly positive. (Alternatively, one could invoke Example 1.2.5.)

Conversely, assuming the positivity of the intersection numbers appearing in the theorem, we prove that L is ample. By Proposition 1.2.16 we are free to suppose that X is reduced and irreducible. The result being clear if $\dim X = 1$, we put $n = \dim X$ and assume inductively that the theorem is known for all schemes of dimension $\leq n - 1$. It is convenient at this point to switch to additive notation, so write $L = \mathcal{O}_X(D)$ for some divisor D on X.

We assert first that

$$H^0\big(X, \mathcal{O}_X(mD)\big) \neq 0 \quad \text{for} \quad m \gg 0.$$

In fact, asymptotic Riemann–Roch (Theorem 1.1.24) gives to begin with that

$$\chi\big(X, \mathcal{O}_X(mD)\big) \;=\; m^n \,\frac{(D^n)}{n!} \;+\; O(m^{n-1}), \tag{*}$$

and $(D^n) = \int_X c_1(L)^n > 0$ by assumption. Now write $D \equiv_{\text{lin}} A - B$ as a difference of very ample effective divisors A and B (using e.g. Example 1.2.10). We have two exact sequences:

$$0 \longrightarrow \mathcal{O}_X(mD - B) \xrightarrow{\cdot A} \mathcal{O}_X((m+1)D) \longrightarrow \mathcal{O}_A((m+1)D) \longrightarrow 0,$$

$$0 \longrightarrow \mathcal{O}_X(mD - B) \xrightarrow{\cdot B} \mathcal{O}_X(mD) \longrightarrow \mathcal{O}_B(mD) \longrightarrow 0.$$

By induction, $\mathcal{O}_A(D)$ and $\mathcal{O}_B(D)$ are ample. Consequently the higher cohomology of each of the two sheaves on the right vanishes when $m \gg 0$. So we find that if $m \gg 0$, then

$$H^i\big(X, \mathcal{O}_X(mD)\big) \;=\; H^i\big(X, \mathcal{O}_X(mD - B)\big) \;=\; H^i\big(X, \mathcal{O}_X((m+1)D)\big)$$

for $i \geq 2$. In other words, if $i \geq 2$ then the dimensions $h^i\big(X, \mathcal{O}_X(mD)\big)$ are eventually constant. Therefore

$$\chi\big(X, \mathcal{O}_X(mD)\big) \;=\; h^0\big(X, \mathcal{O}_X(mD)\big) - h^1\big(X, \mathcal{O}_X(mD)\big) \;+\; C$$

for some constant C and $m \gg 0$. So it follows from (*) that $H^0\big(X, \mathcal{O}_X(mD)\big)$ is non-vanishing when m is sufficiently large, as asserted. Since D is ample if and only if mD is, there is no loss in generality in replacing D by mD. Therefore we henceforth suppose that D is effective.

We next show that $\mathcal{O}_X(mD)$ is generated by its global sections if $m \gg 0$. Since D is assumed to be effective this is evidently true away from Supp(D), so the issue is to show that no point of D is a base point of the linear series $|\mathcal{O}_X(mD)|$. Consider to this end the exact sequence

$$0 \longrightarrow \mathcal{O}_X\big((m-1)D\big) \xrightarrow{\cdot D} \mathcal{O}_X\big(mD\big) \longrightarrow \mathcal{O}_D\big(mD\big) \longrightarrow 0. \tag{*}$$

As before, $\mathcal{O}_D(D)$ is ample by induction. Consequently $\mathcal{O}_D(mD)$ is globally generated and $H^1\big(X, \mathcal{O}_D(mD)\big) = 0$ for $m \gg 0$. It then follows first of all from (*) that the natural homomorphism

$$H^1\big(X, \mathcal{O}_X\big((m-1)D\big)\big) \longrightarrow H^1\big(X, \mathcal{O}_X\big(mD\big)\big) \tag{**}$$

is surjective for every $m \gg 0$. The spaces in question being finite-dimensional, the maps in (**) must actually be isomorphisms for sufficiently large m. Therefore the restriction mappings

$$H^0\big(X, \mathcal{O}_X(mD)\big) \longrightarrow H^0\big(X, \mathcal{O}_D(mD)\big)$$

are surjective for $m \gg 0$. But since $\mathcal{O}_D(mD)$ is globally generated, it follows that no point of $\mathrm{Supp}(D)$ is a basepoint of $|mD|$, as required.

Finally, the amplitude of $\mathcal{O}_X(mD)$ — and hence also of $\mathcal{O}_X(D)$ — now follows from Corollary 1.2.15 since by assumption $\big(mD \cdot C\big) > 0$ for every irreducible curve $C \subseteq X$. □

Remark 1.2.29. (Nakai's criterion on proper schemes). The statement of Theorem 1.2.23 remains true for any complete scheme X, without assuming at the outset that X is projective. The projectivity hypothesis was used in the previous proof to write the given divisor D as a difference of two very ample divisors, but with a little more care one can modify this step to work on any complete X. See [341, Chapter III] or [276, p. 31] for details. □

We conclude this subsection with several other applications of the line of reasoning that led to the Nakai criterion.

Example 1.2.30. (Divisors with ample normal bundle). We outline a result due to Hartshorne [276, III.4.2] concerning divisors having ample normal bundles. Let X be a projective variety, and let $D \subset X$ be an effective Cartier divisor on X whose normal bundle $\mathcal{O}_D(D)$ is ample. Then:

(i). For $m \gg 0$, $\mathcal{O}_X(mD)$ is globally generated.

(ii). For $m \gg 0$, the restriction

$$H^0\big(X, \mathcal{O}_X(mD)\big) \longrightarrow H^0\big(D, \mathcal{O}_D(mD)\big)$$

is surjective.

(iii). There is a proper birational morphism

$$f : X \longrightarrow \overline{X}$$

from X to a projective variety $\overline{X}$ such that f is an isomorphism in a neighborhood of D, and $\overline{D} =_{\mathrm{def}} f(D)$ is an ample effective divisor on $\overline{X}$.

(For (i) and (ii), argue as in the proof of Theorem 1.2.23. For (iii) assume that m is sufficiently large so that (i) and (ii) hold, and in addition $\mathcal{O}_D(mD)$ is very ample. Then take $\overline{X}$ to be the image of the Stein factorization of the mapping

$$\phi : X \longrightarrow \mathbf{P} = \mathbf{P}H^0\big(X, \mathcal{O}_X(mD)\big),$$

defined by $|\mathcal{O}_X(mD)|$, with $f : X \longrightarrow \overline{X}$ the evident morphism. By (ii) and the assumption on $\mathcal{O}_D(mD)$, f is finite over a neighborhood of $f(D)$. Then since $f_*\mathcal{O}_X = \mathcal{O}_{\overline{X}}$ it follows that f maps a neighborhood of D isomorphically to its image in the variety $\overline{X}$. In particular, f is birational.) □

Example 1.2.31. (Irreducible curves of positive self-intersection on a surface). Let X be a smooth projective surface, and let $C \subseteq X$ be an irreducible curve with $\big(C^2\big) > 0$. Then $\mathcal{O}_X(mC)$ is free for $m \gg 0$. □

Example 1.2.32. (Further characterizations of amplitude). Kleiman [341, Chapter 3, §1] gives some additional characterizations of amplitude. Let D be a Cartier divisor on a projective algebraic scheme X. Then D is ample if and only if it satisfies either of the following properties:

(i). For every irreducible subvariety $V \subseteq X$ of positive dimension, there is a positive integer $m = m(V)$, together with a non-zero section $0 \neq s = s_V \in H^0\big(V, \mathcal{O}_V(mD)\big)$, such that s vanishes at some point of V.

(ii). For every irreducible subvariety $V \subseteq X$ of positive dimension,

$$\chi\big(V, \mathcal{O}_V(mD)\big) \to \infty \quad \text{as } m \to \infty.$$

(It is enough to prove this when X is reduced and irreducible. Suppose that (i) holds. Taking first $V = X$ and replacing D by a multiple, we can assume that D is effective. By induction on dimension $\mathcal{O}_D(D)$ is ample, and so by 1.2.30 (i) $\mathcal{O}_X(mD)$ is free for $m \gg 0$. But the hypothesis implies that $\mathcal{O}_X(D)$ is non-trivial on every curve, and hence Corollary 1.2.15 applies. Supposing (ii) holds, one can again assume inductively that $\mathcal{O}_E(D)$ is ample for every effective divisor E on X. Then the proof of Theorem 1.2.23 goes through with little change, except that one uses (ii) rather than Riemann–Roch to control $\chi(X, \mathcal{O}_X(mD))$, while in order to apply 1.2.15 one notes that (ii) implies that $\mathcal{O}_X(D)$ restricts to an ample bundle on every irreducible curve in X.) □

The next example shows that on a scheme of dimension n, dimensions of cohomology groups can grow at most like a polynomial of degree n:

Example 1.2.33. (Growth of cohomology). Let X be a projective scheme of dimension n and D a divisor on X. If $\mathcal{F}$ is any coherent sheaf on X then

$$h^i\big(X, \mathcal{F}(mD)\big) \ = \ O(m^n)$$

for every i. (Write $D = A - B$ as the difference of very ample divisors having the property that neither A nor B contains any of the subvarieties of X defined by the associated primes of $\mathcal{F}$. Then the two sequences

$$0 \longrightarrow \mathcal{F}\big(mD - B\big) \xrightarrow{\cdot A} \mathcal{F}\big((m+1)D\big) \longrightarrow \mathcal{F} \otimes \mathcal{O}_A\big((m+1)D\big) \longrightarrow 0,$$

$$0 \longrightarrow \mathcal{F}\big(mD - B\big) \xrightarrow{\cdot B} \mathcal{F}\big(mD\big) \longrightarrow \mathcal{F} \otimes \mathcal{O}_B\big(mD\big) \longrightarrow 0$$

are exact. By induction on dimension one finds that

$$\big|\ h^i\big(X, \mathcal{F}((m+1)D)\big) - h^i\big(X, \mathcal{F}(mD)\big)\ \big| \ = \ O(m^{n-1}),$$

and the assertion follows.) □

Example 1.2.34. In the setting of 1.2.33, it can easily happen that the higher cohomology groups have maximal growth. For instance, if X is smooth and $-D$ is ample, then $h^n\big(X, \mathcal{O}_X(mD)\big) = h^0\big(X, \mathcal{O}_X(K_X - mD)\big)$ by Serre duality, and the latter group grows like m^n. For a more interesting example, let X be the blowing-up $\mathrm{Bl}_p(\mathbf{P}^2)$ of $\mathbf{P}^2$ at a point, with exceptional divisor E. Then $h^1\big(X, \mathcal{O}_X(mE)\big) = \binom{m}{2}$ has quadratic growth. □

Example 1.2.35. (Growth of cohomology of pullbacks). Let

$$\mu : X' \longrightarrow X$$

be a surjective and generically finite mapping of projective varieties or schemes of dimension n. Fix a divisor D on X and put $D' = \mu^* D$. Then for every $i \geq 0$,

$$h^i\big(X', \mathcal{O}_{X'}(mD')\big) \;=\; h^i\big(X, (\mu_*\mathcal{O}_{X'}) \otimes \mathcal{O}_X(mD)\big) + O(m^{n-1}).$$

(This follows from the Leray spectral sequence and Example 1.2.33 in view of the fact that the higher direct images $R^j\mu_*\mathcal{O}_{X'}$ $(j > 0)$ are supported on proper subschemes of X.) □

Example 1.2.36. (Higher cohomology of nef divisors, I). Kollár [363, V.2.15] shows that one can adapt the proof of Nakai's criterion to establish the following

> **Theorem.** *Let X be a projective scheme of dimension n, and D a divisor on X having the property that*
>
> $$\big(D^{\dim V} \cdot V\big) \;\geq\; 0 \quad \textit{for all irreducible subvarieties } V \subseteq X.$$
>
> *Then*
>
> $$h^i\big(X, \mathcal{O}_X(mD)\big) \;=\; O(m^{n-1}) \quad \text{for } i \geq 1. \tag{1.12}$$

We outline Kollár's argument here. As we will see in Section 1.4, the hypothesis on D is equivalent to the assumption that it be nef. A stronger statement is established in Theorem 1.4.40 using a vanishing theorem of Fujita.

(i). Arguing as in the proof of 1.2.23, one shows by induction on dimension that

$$\big|\, h^i\big(X, \mathcal{O}_X((m+1)D)\big) \;-\; h^i\big(X, \mathcal{O}_X(mD)\big) \,\big| \;=\; O(m^{n-2})$$

provided that $i \geq 2$. This yields (1.12) for $i \geq 2$.

(ii). Combining (i) with asymptotic Riemann–Roch (Theorem 1.1.24), it follows that

$$h^0\big(X, \mathcal{O}_X(mD)\big) \;-\; h^1\big(X, \mathcal{O}_X(mD)\big) \;=\; \frac{\big(D^n\big)}{n!} \cdot m^n \;+\; O(m^{n-1}).$$

If $h^0\big(X, \mathcal{O}_X(mD)\big) = 0$ for all $m > 0$, then the left-hand side is non-positive. But since by assumption $\big(D^n\big) \geq 0$, this forces $\big(D^n\big) = 0$ and we get the required estimate on $h^1\big(X, \mathcal{O}_X(mD)\big)$.

(iii). In view of (ii), we can assume that $H^0\big(X, \mathcal{O}_X(m_0 D)\big) \neq 0$ for some $m_0 > 0$. Fix $E \in |m_0 D|$ and consider the exact sequence

$$0 \longrightarrow \mathcal{O}_X\big((m - m_0)D\big) \longrightarrow \mathcal{O}_X\big(mD\big) \longrightarrow \mathcal{O}_E\big(mD\big) \longrightarrow 0.$$

Applying the induction hypothesis to E we find that

$$\begin{aligned} h^1\big(X, \mathcal{O}_X(mD)\big) - h^1\big(X, \mathcal{O}_X((m - m_0)D)\big) \ &\leq \ h^1\big(E, \mathcal{O}_E(mD)\big) \\ &= \ O(m^{n-2}), \end{aligned}$$

and the case $i = 1$ of (1.12) follows. □

Remark 1.2.37. (Complete schemes). With a little more care, one can show that 1.2.33 and 1.2.36 remain valid for arbitrary complete schemes. See [114, Proposition 1.31]. □

Remark 1.2.38. (Other ground fields). Except for Matsusaka's theorem (Example 1.2.12) all of the results and arguments appearing so far in this section remain valid without change for varieties defined over an algebraically closed field of arbitrary characteristic. □

1.2.C Metric Characterizations of Amplitude

The final basic result we recall here is that when X is smooth — and so may be considered as a complex manifold — amplitude can be detected analytically. The discussion will be rather brief, and we refer for instance to [604] or [248, Chapter 1, Sections 1, 2, and 4], for background and details.

We start with some remarks on positivity of differential forms. Let X be a complex manifold. For $x \in X$, write $T_x X_{\mathbf{R}}$ for the tangent space to the underlying $\mathcal{C}^\infty$ real manifold, and let $J : T_x X_{\mathbf{R}} \longrightarrow T_x X_{\mathbf{R}}$ be the endomorphism determined by the complex structure on X, so that $J^2 = -\mathrm{Id}$. Given a $\mathcal{C}^\infty$ 2-form ω on X, we denote by $\omega(v, w) = \omega_x(v, w) \in \mathbf{R}$ the result of evaluating ω on a pair of real tangent vectors $v, w \in T_x X_{\mathbf{R}}$.

Definition 1.2.39. (Positive, (1, 1) and Kähler forms). The 2-form ω has *type* $(1, 1)$ if

$$\omega_x(Jv, Jw) \ = \ \omega_x(v, w)$$

for every $v, w \in T_x X_{\mathbf{R}}$ and every $x \in X$. A $(1, 1)$-form is *positive* if

$$\omega_x(v, Jv) > 0$$

for every $x \in X$ and all non-zero tangent vectors $0 \neq v \in T_x X_{\mathbf{R}}$. A *Kähler form* is a closed positive $(1, 1)$-form, i.e. a positive form ω of type $(1, 1)$ such that $d\omega = 0$. □

Example 1.2.40. (Local description). Let $z_1, \ldots, z_n$ be local holomorphic coordinates on the complex manifold X. A two-form ω is of type $(1,1)$ if and only if it can be written locally as

$$\omega \;=\; \frac{i}{2} \sum h_{\alpha\beta}\, dz_\alpha \wedge d\bar{z}_\beta,$$

where $\big(h_{\alpha\beta}\big)$ is a Hermitian matrix of $\mathbf{C}$-valued $\mathcal{C}^\infty$ functions on X. It is positive if and only if $\big(h_{\alpha\beta}\big)$ is positive definite at each point $x \in X$. □

Example 1.2.41. (Hermitian metrics). Let H be a Hermitian metric on X, given by Hermitian forms $H_x(\ ,\ )$ on $T_x X_{\mathbf{R}}$ varying smoothly with $x \in X$.[10] Then the negative imaginary part

$$\omega \;=\; -\mathrm{Im}\, H$$

of H is a $(1,1)$-form on X, and if H is positive definite then ω is positive. Conversely a positive $(1,1)$-form ω determines a positive definite Hermitian metric by the rule

$$H_x(v,w) \;=\; \omega_x(v, Jw) - i\omega_x(v,w). \quad \square$$

Fix a Kähler form ω on X. Let Δ be the unit disk, with complex coordinate $z = x + iy$, and suppose that $\mu : \Delta \longrightarrow X$ is a holomorphic mapping. Then $\mu^*\omega$ is likewise positive of type $(1,1)$, and hence

$$\mu^*\omega \;=\; \phi(x,y) \cdot dx \wedge dy,$$

where $\phi(x,y)$ is a positive $\mathcal{C}^\infty$ function on Δ. Therefore if $C \subseteq X$ is a one-dimensional complex submanifold, then $\int_C \omega > 0$ (provided that the integral is finite). Similarly, for any complex submanifold $V \subseteq X$, the integrand computing $\int_V \omega^{\dim V}$ is everywhere positive. So it is suggestive to think of a positive $(1,1)$-form as one satisfying "pointwise" inequalities of Nakai–Moishezon–Kleiman type.

Example 1.2.42. (Fubini–Study form on $\mathbf{P}^n$, I). Complex projective space $\mathbf{P}^n$ carries a very beautiful $\mathrm{SU}(n+1)$-invariant Kähler form ω_{FS}. Following [17, Appendix 3], we construct it by first building an $\mathrm{SU}(n{+}1)$-invariant Hermitian metric H_{FS} on $\mathbf{P}^n$. The Fubini–Study form ω_{FS} will then arise as the negative imaginary part $\omega_{\mathrm{FS}} = -\mathrm{Im}\, H_{\mathrm{FS}}$ of this Fubini–Study metric.

Consider the standard Hermitian inner product $\langle v, w\rangle = {}^t v \cdot \overline{w}$ on $V = \mathbf{C}^{n+1}$. Set $V^0 = V - \{0\}$ and denote by

$$\rho : V^0 \longrightarrow \mathbf{P}^n \;=\; \mathbf{P}_{\mathrm{sub}}(V)$$

the canonical map. Define to begin with a Hermitian metric H' on V^0 by associating to $x \in V^0$ the Hermitian inner product

[10] Our convention is that H_x should be complex linear in the first argument and conjugate linear in the second.

$$H'_x(v,w) = \left\langle \frac{v}{|x|}, \frac{w}{|x|} \right\rangle \qquad \text{for } v, w \in T_x V^0 = V$$

and $|x| = \sqrt{\langle x, x\rangle}$. The metric H' is constructed so as to be invariant under the natural $\mathbf{C}^*$-action on V^0. Now $\rho^* T\mathbf{P}^n$ is canonically a quotient of TV^0, and so H' induces in the usual manner a $\mathbf{C}^*$-invariant metric on $\rho^* T\mathbf{P}^n$, which then descends to a Hermitian metric H_{FS} on $T\mathbf{P}^n$.

More explicitly, write $W_x \subseteq V$ for the H'_x-orthogonal complement to $\mathbf{C}{\cdot}x \subseteq V$, and let $\pi_x : V \longrightarrow W_x$ be orthogonal projection:

$$\pi_x(v) = v - \frac{\langle v, x\rangle}{\langle x, x\rangle} \cdot x.$$

Then W_x is identified with $T_{\rho(x)}\mathbf{P}^n$ and π_x with $d\rho_x$, and

$$\begin{aligned} H_{\rho(x)}\big(d\rho_x v, d\rho_x w\big)_{\mathrm{FS}} &= H'_x(\pi_x v, \pi_x w) \\ &= \frac{\langle v, w\rangle\langle x, x\rangle - \langle v, x\rangle\langle x, w\rangle}{\langle x, x\rangle^2}. \end{aligned}$$

If we take the usual affine local coordinates $z_1, \ldots, z_n$ on $\mathbf{P}^n$ — corresponding to $x = (1, z_1, \ldots, z_n) \in V^0$ — then one finds[11] that

$$\begin{aligned} \omega_{\mathrm{FS}} &=_{\mathrm{def}} -\mathrm{Im}\, H_{\mathrm{FS}} \\ &=_{\mathrm{locally}} \frac{i}{2} \cdot \left(\frac{\sum dz_\alpha \wedge d\bar{z}_\alpha}{1 + \sum |z_\alpha|^2} - \frac{\left(\sum \bar{z}_\alpha dz_\alpha\right) \wedge \left(\sum z_\alpha d\bar{z}_\alpha\right)}{\left(1 + \sum |z_\alpha|^2\right)^2} \right). \end{aligned}$$

By construction H_{FS} is invariant under the natural $\mathrm{SU}(n+1)$-action on $\mathbf{P}^n$, and hence so too is ω_{FS}.

We next verify that ω_{FS} is indeed a Kähler form. The positivity of ω_{FS} follows using Example 1.2.41 from the fact that H_{FS} is positive definite. Alternatively, since ω_{FS} is $\mathrm{SU}(n+1)$-invariant it is enough to prove positivity at any one point $p \in \mathbf{P}^n$, and when $p = [1, 0, \ldots, 0]$ this is clear from the local description. Following [453, Lemma 5.20], $\mathrm{SU}(n+1)$-invariance also leads to a quick proof that ω_{FS} is closed. In fact, given $p \in \mathbf{P}^n$ choose an element $\gamma \in \mathrm{SU}(n+1)$ such that $\gamma(p) = p$ while $d\gamma_p = -\mathrm{Id}$. Then for any three tangent vectors $u, v, w \in T_p\mathbf{P}^n$ one has

---

[11] It may be useful to consider here an n-dimensional vector space $W = \mathbf{C}^n$, with its standard Hermitian product $\langle u, v\rangle = {}^t u \cdot \overline{v}$. Then as w varies over W the expressions

$$\eta_w(u,v) = -\mathrm{Im}\big(\langle u, v\rangle\big) \;,\; \eta'_w(u,v) = -\mathrm{Im}\big(\langle u, w\rangle\langle w, v\rangle\big)$$

define $(1,1)$-forms η and η' on W, which in terms of standard linear coordinates $w_1, \ldots, w_n$ on W are given by $\eta = \frac{i}{2} \cdot \sum dw_\alpha \wedge d\overline{w}_\alpha$ and $\eta' = \frac{i}{2} \cdot \left(\sum \overline{w}_\alpha dw_\alpha\right) \wedge \left(\sum w_\alpha d\overline{w}_\alpha\right)$.

$$d\omega_{\mathrm{FS}}(u,v,w) \;=\; \gamma^*(d\omega_{\mathrm{FS}})(u,v,w) \;=\; d\omega_{\mathrm{FS}}(-u,-v-w),$$

and hence $d\omega_{\mathrm{FS}} = 0$. □

Example 1.2.43. (Fubini–Study form on $\mathbf{P^n}$, II). Another approach to the Fubini–Study form involves the Hopf map. Keeping the notation of the previous example, consider the unit sphere

$$\mathbf{C}^{n+1} \;\supseteq\; S^{2n+1} \;=\; S$$

with respect to the standard inner product $\langle\ ,\ \rangle$, with

$$p : S \longrightarrow \mathbf{P}^n$$

the Hopf mapping. Denote by ω_{std} the standard symplectic form on $\mathbf{C}^{n+1}$, i.e.

$$\omega_{\mathrm{std}} \;=\; \sum dx_\alpha \wedge dy_\alpha,$$

where $z_\alpha = x_\alpha + iy_\alpha$ are the usual complex coordinates on $\mathbf{C}^{n+1}$. Then ω_{FS} is characterized as the unique symplectic form on $\mathbf{P}^n$ having the property that

$$p^*\omega_{\mathrm{FS}} \;=\; \omega_{\mathrm{std}} \mid S.$$

(This follows from the construction in the previous example.) □

Suppose now given a holomorphic line bundle L on X on which a Hermitian metric h has been fixed. We write $|\ \ |_h$ for the corresponding length function on the fibres of L. The Hermitian line bundle (L,h) determines a *curvature form*

$$\Theta(L,h) \;\in\; C^\infty(X, \Lambda^{1,1}T^*X_{\mathbf{R}}) :$$

this is a closed $(1,1)$-form on X having the property that $\frac{i}{2\pi}\Theta(L,h)$ represents $c_1(L)$. If $s \in \Gamma(U,L)$ is a local holomorphic section of L that doesn't vanish at any point of an open set U, then for instance one can define $\Theta(L,h)$ locally by the formula

$$\Theta \;=\; -\partial\bar{\partial}\log|s|_h^2 \ ,$$

this being independent of the choice of s.

The analytic approach to positivity is to ask that the form $\frac{i}{2\pi}\Theta(L,h)$ representing $c_1(L)$ be positive:

Definition 1.2.44. (Positive line bundles). The line bundle L is *positive* (in the sense of Kodaira) if it carries a Hermitian metric h such that $\frac{i}{2\pi}\Theta(L,h)$ is a Kähler form. □

Example 1.2.45. (Fubini–Study metric on the hyperplane bundle). The hyperplane bundle $\mathcal{O}_{\mathbf{P}^n}(1)$ on $\mathbf{P}^n$ carries a Hermitian metric h whose curvature form is a multiple of the Fubini–Study form ω_{FS} (Example 1.2.42). In fact, the standard Hermitian product $\langle v,w\rangle = {}^tv\cdot\overline{w}$ on $V = \mathbf{C}^{n+1}$ gives rise to a Hermitian metric on the trivial bundle $V_{\mathbf{P}^n}$ on $\mathbf{P}^n = \mathbf{P}_{\text{sub}}(V)$. Then $\mathcal{O}_{\mathbf{P}^n}(-1)$ inherits a metric as a sub-bundle of $V_{\mathbf{P}^n}$, which in turn determines a metric h on $\mathcal{O}_{\mathbf{P}^n}(1)$. Very explicitly, write $[x] \in \mathbf{P}^n$ for the point corresponding to a vector $x \in V - \{0\}$ and consider a section $s \in V^* = H^0\big(\mathbf{P}^n, \mathcal{O}_{\mathbf{P}^n}(1)\big)$. Then h is determined by the rule

$$\big|s\big([x]\big)\big|_h^2 = \frac{|s(x)|^2}{\langle x, x\rangle},$$

where the numerator on the right is the squared modulus of the result of evaluating the linear functional s on the vector x.

If we work with the usual affine coordinates $z = (z_1, \dots, z_n)$ on $\mathbf{P}^n$ corresponding to the point $x = (1, z_1, \dots, z_n) \in V$ and take $s \in V^*$ to be the functional given by projection onto the zeroth coordinate, then

$$\big|s\big([x]\big)\big|_h^2 = \frac{1}{1 + \sum |z_\alpha|^2}.$$

An explicit calculation [248, p. 30] shows that

$$\begin{aligned}
\frac{i}{2\pi}\Theta(\mathcal{O}_{\mathbf{P}^n}(1), h) &=_{\text{locally}} -\frac{i}{2\pi} \cdot \partial\bar{\partial} \log \left(\frac{1}{1 + \sum |z_\alpha|^2} \right) \\
&= \frac{i}{2\pi} \cdot \left(\frac{\sum dz_\alpha \wedge d\bar{z}_\alpha}{1 + \sum |z_\alpha|^2} - \frac{\left(\sum \bar{z}_\alpha dz_\alpha\right) \wedge \left(\sum z_\alpha d\bar{z}_\alpha\right)}{\left(1 + \sum |z_\alpha|^2\right)^2} \right) \\
&= \frac{1}{\pi} \cdot \omega_{\text{FS}}.
\end{aligned}$$

In particular, $\mathcal{O}_{\mathbf{P}^n}(1)$ is positive in the sense of Kodaira. □

The beautiful fact is that if L is a positive line bundle on a compact Kähler manifold X, then X is algebraic and L is ample:

Theorem. (Kodaira embedding theorem). *Let X be a compact Kähler manifold, and L a holomorphic line bundle on X. Then L is positive if and only if there is a holomorphic embedding*

$$\phi : X \hookrightarrow \mathbf{P}$$

of X into some projective space such that $\phi^*\mathcal{O}_{\mathbf{P}}(1) = L^{\otimes m}$ for some $m > 0$.

One direction is elementary: if such an embedding exists, then the pullback of the standard Fubini–Study metric on $\mathcal{O}_{\mathbf{P}}(1)$ determines a positive metric on $L^{\otimes m}$ and hence also on L. Conversely, if one assumes that X is already a projective variety — in which case it follows by the GAGA theorems that L is an algebraic line bundle — then the amplitude of a positive line bundle is a consequence of the Nakai criterion. Indeed, $\frac{i}{2\pi}\Theta(L, h)$ represents $c_1(L)$, and as we have noted, the positivity of this form implies the positivity of the intersection numbers appearing in 1.2.23.[12]

The deeper assertion is that as soon as L is a positive line bundle on a compact Kähler manifold, some power of L has enough sections to define a projective embedding of X. This is traditionally proved by establishing for positive line bundles an analogue of the sort of vanishings appearing in Theorem 1.2.6. We refer to [248, Chapter 2, §4] for details.

1.3 Q-Divisors and R-Divisors

For questions of positivity, it is very useful to be able to discuss small perturbations of a given divisor class. The natural way to do so is through the formalism of **Q**- and **R**-divisors, which we develop in this section. As an application, we establish that amplitude is an open condition on numerical equivalence classes.

1.3.A Definitions for Q-Divisors

As one would expect, a **Q**-divisor is simply a **Q**-linear combination of integral Cartier divisors:

Definition 1.3.1. (Q-divisors). Let X be an algebraic variety or scheme. A **Q**-*divisor* on X is an element of the **Q**-vector space

$$\mathrm{Div}_{\mathbf{Q}}(X) =_{\mathrm{def}} \mathrm{Div}(X) \otimes_{\mathbf{Z}} \mathbf{Q}.$$

We represent a **Q**-divisor $D \in \mathrm{Div}_{\mathbf{Q}}(X)$ as a finite sum

$$D = \sum c_i \cdot A_i, \tag{1.13}$$

where $c_i \in \mathbf{Q}$ and $A_i \in \mathrm{Div}(X)$. By clearing denominators we can also write $D = cA$ for a single rational number c and integral divisor A, and if $c \neq 0$ then $cA = 0$ if and only if A is a torsion element of $\mathrm{Div}(X)$. A **Q**-divisor D is *integral* if it lies in the image of the natural map $\mathrm{Div}(X) \longrightarrow \mathrm{Div}_{\mathbf{Q}}(X)$. The **Q**-divisor D is *effective* if it is of the form $D = \sum c_i A_i$ with $c_i \geq 0$ and A_i effective. □

Definition 1.3.2. (Supports). Let $D \in \mathrm{Div}_{\mathbf{Q}}(X)$ be a **Q**-divisor. A codimension one subset $E \subseteq X$ *supports* D, or *is a support* of D, if D admits a representation (1.13) in which the union of the supports of the A_i is contained in E. □

---

[12] Recall ([248, p. 32]) that if V is singular, one computes $\int_V c_1(L)^{\dim V}$ by integrating the appropriate power of $\frac{i}{2\pi}\Theta(L,h)$ over the smooth locus of V.

Since the expression (1.13) may not be unique, E is not canonically determined. But this does not cause any problems.

All the usual operations and properties of Cartier divisors extend naturally to this setting simply by tensoring with $\mathbf{Q}$:

Definition 1.3.3. (Equivalences and operations on Q-divisors). Assume henceforth that X is complete.

(i). Given a subvariety or subscheme $V \subseteq X$ of pure dimension k, a $\mathbf{Q}$-valued intersection product

$$\mathrm{Div}_{\mathbf{Q}}(X) \times \ldots \times \mathrm{Div}_{\mathbf{Q}}(X) \longrightarrow \mathbf{Q},$$
$$(D_1, \ldots, D_k) \mapsto \int_{[V]} D_1 \cdot \ldots \cdot D_k \;=\; (D_1 \cdot \ldots \cdot D_k \cdot [V])$$

is defined via extension of scalars from the analogous product on $\mathrm{Div}(X)$.

(ii). Two $\mathbf{Q}$-divisors $D_1, D_2 \in \mathrm{Div}_{\mathbf{Q}}(X)$ are *numerically equivalent*, written

$$D_1 \equiv_{\mathrm{num}} D_2,$$

(or $D_1 \equiv_{\mathrm{num},\mathbf{Q}} D_2$ when confusion seems possible) if

$$(D_1 \cdot C) \;=\; (D_2 \cdot C)$$

for every curve $C \subseteq X$. We denote by $N^1(X)_{\mathbf{Q}}$ the resulting finite-dimensional $\mathbf{Q}$-vector space of numerical equivalence classes of $\mathbf{Q}$-divisors.

(iii). Two $\mathbf{Q}$-divisors $D_1, D_2 \in \mathrm{Div}_{\mathbf{Q}}(X)$ are *linearly equivalent*, written

$$D_1 \equiv_{\mathrm{lin},\mathbf{Q}} D_2 \quad (\text{or simply } D_1 \equiv_{\mathrm{lin}} D_2)$$

if there is an integer r such that rD_1 and rD_2 are integral and linearly equivalent in the usual sense, i.e. if $r\big(D_1 - D_2\big)$ is the image of a principal divisor in $\mathrm{Div}(X)$.

(iv). If $f : Y \longrightarrow X$ is a morphism such that the image of every associated subvariety of Y meets a support of $D \in \mathrm{Div}_{\mathbf{Q}}(X)$ properly, then $f^*D \in \mathrm{Div}_{\mathbf{Q}}(Y)$ is defined by extension of scalars from the corresponding pullback on integral divisors (this being independent of the representation of D in (1.13)).

(v). If $f : Y \longrightarrow X$ is an arbitrary morphism of complete varieties or projective schemes, extension of scalars gives rise to a functorial induced homomorphism $f^* : N^1(X)_{\mathbf{Q}} \longrightarrow N^1(Y)_{\mathbf{Q}}$ compatible with the divisor-level pullback defined in (iv). □

Remark 1.3.4. More concretely, these operations and equivalences are determined from those on integral divisors by writing $D = \sum c_i A_i$ — or, after clearing denominators, $D = cA$ — and expanding by linearity. So for instance

$D \equiv_{\mathrm{num},\mathbf{Q}} 0$ if and only if $\sum c_i(A_i \cdot C) = 0$ for every curve $C \subseteq X$. Note also that there is an isomorphism

$$N^1(X)_{\mathbf{Q}} = N^1(X) \otimes_{\mathbf{Z}} \mathbf{Q}. \quad \square$$

Remark 1.3.5. It can happen that two integral divisors in distinct linear equivalence classes become linearly equivalent in the sense of (iii) when considered as $\mathbf{Q}$-divisors. For this reason one has to be careful when dealing with $\mathbf{Q}$-linear equivalence. For the most part we will work with numerical equivalence, where this problem does not arise. □

Continue to assume that X is complete. The definition of ampleness for $\mathbf{Q}$-divisors likewise presents no problems:

Definition 1.3.6. (Amplitude for Q-divisors). A $\mathbf{Q}$-divisor

$$D \in \mathrm{Div}_{\mathbf{Q}}(X)$$

is *ample* if any one of the following three equivalent conditions is satisfied:

(i). D is of the form $D = \sum c_i A_i$ where $c_i > 0$ is a positive rational number and A_i is an ample Cartier divisor.

(ii). There is a positive integer $r > 0$ such that $r \cdot D$ is integral and ample.

(iii). D satisfies the statement of Nakai's criterion, i.e.

$$\left(D^{\dim V} \cdot V\right) > 0$$

for every irreducible subvariety $V \subseteq X$ of positive dimension. □

(The equivalence of (i)–(iii) is immediate.) As before, amplitude is preserved by numerical equivalence, and we speak of ample classes in $N^1(X)_{\mathbf{Q}}$.

As an illustration, we prove that amplitude is an open condition under small perturbations of a divisor:

Proposition 1.3.7. *Let X be a projective variety, H an ample $\mathbf{Q}$-divisor on X, and E an arbitrary $\mathbf{Q}$-divisor. Then $H + \varepsilon E$ is ample for all sufficiently small rational numbers $0 \leq |\varepsilon| \ll 1$. More generally, given finitely many $\mathbf{Q}$-divisors $E_1, \ldots, E_r$ on X,*

$$H + \varepsilon_1 E_1 + \ldots + \varepsilon_r E_r$$

is ample for all sufficiently small rational numbers $0 \leq |\varepsilon_i| \ll 1$.

Proof. Clearing denominators, we may assume that H and each E_i are integral. By taking $m \gg 0$ we can arrange for each of the $2r$ divisors $mH \pm E_1, \ldots, mH \pm E_r$ to be ample (Example 1.2.10). Now provided that $|\varepsilon_i| \ll 1$ we can write any divisor of the form

$$H + \varepsilon_1 E_2 + \ldots + \varepsilon_r E_r$$

as a positive **Q**-linear combination of H and some of the **Q**-divisors $H \pm \frac{1}{m} E_i$. But a positive linear combination of ample **Q**-divisors is ample. □

Remark 1.3.8. (Weil Q-divisors). As in Example 1.1.3, write $\mathrm{WDiv}(X)$ for the additive group of Weil divisors on an irreducible variety X. The group of Weil **Q**-divisors is defined to be

$$\mathrm{WDiv}_{\mathbf{Q}}(X) \ =_{\mathrm{def}} \ \mathrm{WDiv}(X) \otimes_{\mathbf{Z}} \mathbf{Q}.$$

So a Weil **Q**-divisor is just a **Q**-linear combination of codimension-one subvarieties. As before, there is a cycle class map $[\ \ ] : \mathrm{Div}_{\mathbf{Q}}(X) \longrightarrow \mathrm{WDiv}_{\mathbf{Q}}(X)$.

Now assume that X is normal. Then the cycle mapping is injective, and in this case one can identify $\mathrm{Div}_{\mathbf{Q}}(X)$ with a subgroup of $\mathrm{WDiv}_{\mathbf{Q}}(X)$. This provides a convenient and concrete way of manipulating Cartier **Q**-divisors on a normal variety. A Weil **Q**-divisor $E \in \mathrm{WDiv}_{\mathbf{Q}}(X)$ is said to be **Q**-Cartier if it lies in $\mathrm{Div}_{\mathbf{Q}}(X)$. Thus all the operations and equivalence relations defined in Definition 1.3.3 make sense for **Q**-Cartier Weil **Q**-divisors provided that X is normal. (However we do not attempt to pass to Weil divisors when X fails to be normal.) □

Example 1.3.9. To illustrate the preceding remark, consider the quadric cone $X \subset \mathbf{P}^3$ with vertex O, and let $E \subset X$ be a ruling of X, i.e. a line through O (Figure 1.1). Viewed as a Weil divisor, E is not Cartier. But if A

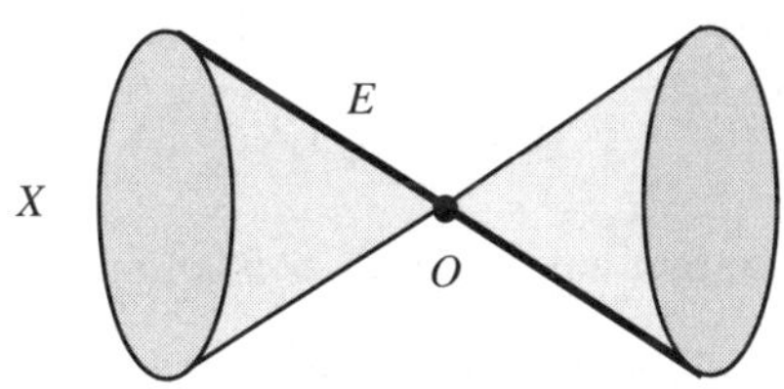

Figure 1.1. Ruling of quadric cone

is the Cartier divisor obtained by intersecting X with the hyperplane in $\mathbf{P}^3$ tangent to X along E, then $A = 2 \cdot E$. Thus E is **Q**-Cartier, and in particular we can compute its self-intersection:

$$\begin{aligned} (E \cdot E) \ &= \left(\tfrac{1}{2}A \cdot \tfrac{1}{2}A\right) \\ &= \tfrac{1}{4}(A \cdot A) \\ &= \tfrac{1}{4} \cdot 2 \\ &= \tfrac{1}{2}. \quad \square \end{aligned}$$

1.3.B R-Divisors and Their Amplitude

The definition of **R**-divisors proceeds in an exactly analogous fashion. Thus one defines the real vector space

$$\mathrm{Div}_{\mathbf{R}}(X) = \mathrm{Div}(X) \otimes \mathbf{R}$$

of **R**-divisors on X. Supposing X is complete, there is an associated **R**-valued intersection theory, giving rise in particular to the notion of numerical equivalence. Very concretely, an **R**-divisor D is represented by a finite sum $D = \sum c_i A_i$ where $c_i \in \mathbf{R}$ and $A_i \in \mathrm{Div}(X)$. It is numerically trivial if and only if $\sum c_i\big(A_i \cdot C\big) = 0$ for every curve $C \subseteq X$. The resulting vector space of equivalence classes is denoted by $N^1(X)_{\mathbf{R}}$. We say that D is *effective* if $D = \sum c_i A_i$ with $c_i \geq 0$ and A_i effective. Pullbacks and supports of **R**-divisors are likewise defined as before.

Example 1.3.10. One has an isomorphism

$$N^1(X)_{\mathbf{R}} \ = \ N^1(X) \otimes_{\mathbf{Z}} \mathbf{R}.$$

(Use the fact — to be established shortly in the proof of Proposition 1.3.13 — that a numerically trivial **R**-divisor is an **R**-linear combination of numerically trivial integral divisors.) □

For ampleness of **R**-divisors, however, the situation is slightly more subtle (Remark 1.3.12). We take as our definition the analogue of (i) in 1.3.6:

Definition 1.3.11. (Amplitude for R-divisors). Assume that X is complete. An **R**-divisor D on X is *ample* if it can be expressed as a finite sum

$$D = \sum c_i A_i$$

where $c_i > 0$ is a positive real number and A_i is an ample Cartier divisor. □

Observe that a finite positive **R**-linear combination of ample **R**-divisors is therefore ample.

Remark 1.3.12. (Nakai inequalities for R-divisors). If D is an ample **R**-divisor then certainly

$$\big(D^{\dim V} \cdot V\big) \ > \ 0 \qquad (*)$$

for every irreducible $V \subseteq X$ of positive dimension. However, now it is no longer clear that these inequalities characterize amplitude. For instance, if $D = \sum c_i A_i$ with $c_i > 0$ and A_i integral and ample, then

$$\big(D^{\dim V} \cdot V\big) \ \geq \ \big(\sum c_i\big)^{\dim V}$$

thanks to the fact that the intersection product on integral Cartier divisors is **Z**-valued. In particular, for fixed ample D the intersection numbers in (*) are bounded away from zero. On the other hand, one could imagine the existence of an **R**-divisor D satisfying (*) but for which the intersection numbers in question cluster toward 0 as V varies over all subvarieties of a given dimension. Surprisingly enough, however, these difficulties don't actually occur: a theorem of Campana and Peternell [78] states that the Nakai inequalities (*) do in fact imply that an **R**-divisor D on a projective variety is ample. This appears as Theorem 2.3.18 below. However we prefer to develop the general theory without appealing to this result. □

As before, amplitude depends only on numerical equivalence classes:

Proposition 1.3.13. (Ample classes for R-divisors). *The amplitude of an* **R***-divisor depends only upon its numerical equivalence class.*

Proof. It is sufficient to show that if D and B are **R**-divisors, with D ample and $B \equiv_{\text{num}} 0$, then $D + B$ is again ample. To this end, observe first that B is an **R**-linear combination of numerically trivial integral divisors. Indeed, the condition that an **R**-divisor

$$B = \sum r_i B_i \quad (r_i \in \mathbf{R}\ ,\ B_i \in \mathrm{Div}(X)\,)$$

be numerically trivial is given by finitely many integer linear equations on the r_i, determined by integrating over a set of generators of the subgroup of $H_2\big(X, \mathbf{Z}\big)$ spanned by algebraic 1-cycles on X. The assertion then follows from the fact that any real solution to these equations is an **R**-linear combination of integral ones.

We are now reduced to showing that if A and B are integral divisors, with A ample and $B \equiv_{\text{num}} 0$, then $A + rB$ is ample for any $r \in \mathbf{R}$. If r is rational we already know this. In general, we can fix rational numbers $r_1 < r < r_2$, together with a real number $t \in [0,1]$, such that $r = tr_1 + (1-t)r_2$. Then

$$A + rB \;=\; t\big(A + r_1 B\big) \;+\; (1-t)\big(A + r_2 B\big),$$

exhibiting $A + rB$ as a positive **R**-linear combination of ample **Q**-divisors. □

Example 1.3.14. (Openness of amplitude for R-divisors). The statement of Proposition 1.3.7 remains valid for **R**-divisors. In other words:

> Let X be a projective variety and H an ample **R**-divisor on X. Given finitely many **R**-divisors $E_1, \ldots, E_r$, the **R**-divisor
>
> $$H + \varepsilon_1 E_1 + \ldots + \varepsilon_r E_r$$
>
> is ample for all sufficiently small real numbers $0 \leq |\varepsilon_i| \ll 1$.

(When H and each E_i are rational this follows from the proof of Proposition 1.3.7, and one reduces the general case to this one. To begin with, since each

E_j is a finite **R**-linear combination of integral divisors, there is no loss of generality in assuming at the outset that all of the E_j are integral. Now write $H = \sum c_i A_i$ with $c_i > 0$ and A_i ample and integral, and fix a rational number $0 < c < c_1$. Then

$$H + \sum \varepsilon_j E_j = \left(cA_1 + \sum \varepsilon_j E_j\right) + (c_1 - c)A_1 + \sum_{i \geq 2} c_i A_i.$$

The first term on the right is governed by the case already treated, and the remaining summands are ample.) □

Example 1.3.15. If X is projective, then the finite-dimensional vector space $N^1(X)_{\mathbf{R}}$ is spanned by the classes of ample divisors on X. (Use 1.3.14.) □

Remark 1.3.16. (More general ground fields). The discussion in this section again goes through with only minor changes if X is a projective scheme over an arbitrary algebraically closed ground field. (In the proof of 1.3.13 one would replace $H_2(X;\mathbf{Z})$ with the group $N_1(X)$ of numerical equivalence classes of curves (Definition 1.4.25).) □

1.4 Nef Line Bundles and Divisors

We have seen that if X is a projective variety, then a class $\delta \in N^1(X)_{\mathbf{Q}}$ is ample if and only if it satisfies the Nakai inequalities:

$$\int_V \delta^{\dim V} > 0 \text{ for all irreducible } V \subseteq X \text{ with } \dim V > 0.$$

This suggests that limits of ample classes should be characterized by the corresponding weak inequalities

$$\int_V \delta^{\dim V} \geq 0 \quad \text{for all } V \subseteq X. \tag{*}$$

It is a basic and remarkable fact (Kleiman's theorem) that it suffices to test (*) when V is a curve. For this reason, it turns out to be very profitable to work systematically with such limits of ample classes. From the contemporary viewpoint these so-called nef divisors lie at the heart of the theory of positivity for line bundles.

We start in Section 1.4.A with the definition and basic properties. The most important material appears in Section 1.4.B, which contains Kleiman's theorem and its consequences. It is reinterpreted in the following subsection, where we introduce the ample and nef cones. Finally we discuss in Section 1.4.D an extremely useful vanishing theorem due to Fujita.

1.4.A Definitions and Formal Properties

We begin with the definition.

Definition 1.4.1. (Nef line bundles and divisors). Let X be a complete variety or scheme. A line bundle L on X is *numerically effective*, or *nef*, if

$$\int_C c_1(L) \geq 0$$

for every irreducible curve $C \subseteq X$. Similarly, a Cartier divisor D on X (with $\mathbf{Z}$, $\mathbf{Q}$, or $\mathbf{R}$ coefficients) is *nef* if

$$\big(D \cdot C\big) \geq 0$$

for all irreducible curves $C \subset X$. □

The definition evidently depends only on the numerical equivalence class of L or D, and so one has a notion of nef classes in $N^1(X), N^1(X)_{\mathbf{Q}}$, and $N^1(X)_{\mathbf{R}}$. Note that any ample class is nef, as is the sum of two nef classes.

Remark 1.4.2. The terminology "nef," although now standard, did not come into use until the mid 1980s: it was introduced by Reid.[13] The concept previously appeared in the literature under various different names. For example, in his paper [623] Zariski speaks of "arithmetically effective" divisors. Kleiman used "numerically effective" in [341]. In that paper, a divisor satisfying the conclusion of Theorem 1.4.9 was called "pseudoample." □

Remark 1.4.3. (Chow's lemma). In dealing with nefness, one can often use Chow's Lemma to reduce statements about complete varieties or schemes to the projective case. Specifically, suppose that X is a complete variety (or scheme). Then there exists a projective variety (or scheme) X', together with a surjective morphism $f : X' \longrightarrow X$ that is an isomorphism over a dense open subset of X. In the relative setting, an analogous statement holds starting from a proper morphism $X \longrightarrow T$. See [280, Exercise II.4.10] for details. □

The formal properties of nefness are fairly immediate:

Example 1.4.4. (Formal properties of nefness). Let X be a complete variety or scheme, and L a line bundle on X.

(i). Let $f : Y \longrightarrow X$ be a proper mapping. If L is nef, then $f^*(L)$ is a nef line bundle on Y. In particular, restrictions of nef bundles to subschemes remain nef.

(ii). In the situation of (i), if f is surjective and $f^*(L)$ is nef on Y, then L itself is nef.

[13] Reid was motivated by the fact that nef is also an abbreviation for "numerically eventually free."

(iii). L is nef if and only if L_{red} is nef on X_{red}.

(iv). L is nef if and only if its restriction to each irreducible component of X is nef.

(For (ii) one needs to check that if $f : Y \longrightarrow X$ is a surjective morphism of (possibly non-projective) complete varieties, and if $C \subset X$ is an irreducible curve, then there is a curve $C' \subset Y$ mapping onto C. To this end, one can use Chow's lemma to reduce to the case in which Y is projective, where the assertion is clear. See [341, Chapter I, Section 4, Lemma 1].) □

Example 1.4.5. Let X be a complete variety (or scheme) and L a globally generated line bundle on X. Then L is nef. □

Example 1.4.6. (Divisors with nef normal bundle). Let X be a complete variety, and $D \subseteq X$ an effective divisor on X. If the normal bundle $N_{D/X} = \mathcal{O}_D(D)$ to D in X is nef, then D itself is a nef divisor. In particular, if X is a surface and $C \subseteq X$ is an irreducible curve with $(C^2) \geq 0$, then C is nef. (This generalizes the previous example.) □

Example 1.4.7. (Nefness on homogeneous varieties). Let X be a complete variety, and suppose that a connected algebraic group G acts transitively on X. Then any effective divisor D on X is nef. This applies for instance to arbitrary flag manifolds and abelian varieties. (Fix an irreducible curve $C \subset X$. Then the translate gD of D by a general element $g \in G$ meets C properly. Moreover, $gD \equiv_{\text{num}} D$ since G is connected. Therefore

$$\big(D \cdot C\big) \;=\; \big(gD \cdot C\big) \;\geq\; 0,$$

and hence D is nef.) □

Remark 1.4.8. (Metric characterizations of nefness). Let X be a complex projective manifold, and L a line bundle on X. In the spirit of Section 1.2.C, it is natural to ask whether one can recognize the nefness of L metrically. If L carries a Hermitian metric h such that the corresponding Chern form $c_1(L) = \frac{i}{2\pi}\Theta(L,h)$ is non-negative, then certainly L is nef. However, there are examples [133, Example 1.7] of nef bundles that do not admit such metrics.

On the other hand, one can in effect use Corollary 1.4.10 below to reduce to the case of positive bundles. Specifically, fix a Kähler form ω on X. Then L is nef if and only if for every $\varepsilon > 0$ there exists a Hermitian metric h_ε on L such that

$$\frac{i}{2\pi}\,\Theta(L,h_\varepsilon) \;>\; -\varepsilon\cdot\omega$$

in the sense that $\frac{i}{2\pi}\Theta(L,h_\varepsilon) + \varepsilon\,\omega$ is a Kähler form. This also gives a way of defining nefness on arbitrary compact Kähler manifolds (which might not contain any curves). We refer to [133, §1.A] for details, and to [126, §6] for a survey. □

1.4.B Kleiman's Theorem

The fundamental result concerning nef divisors is due to Kleiman [341]:

Theorem 1.4.9. (Kleiman's theorem). *Let X be a complete variety (or scheme). If D is a nef $\mathbf{R}$-divisor on X, then*

$$\big(D^k \cdot V\big) \ \geq \ 0$$

for every irreducible subvariety $V \subseteq X$ of dimension k. Similarly,

$$\int_V c_1(L)^{\dim V} \ \geq \ 0$$

for every nef line bundle L on X.

Before giving the proof, we present several applications and corollaries.

The essential content of Kleiman's theorem is to characterize nef divisors as limits of ample ones. The next statement gives a first illustration of this principle; another formulation appears in Theorem 1.4.23.

Corollary 1.4.10. *Let X be a projective variety or scheme, and D a nef $\mathbf{R}$-divisor on X. If H is any ample $\mathbf{R}$-divisor on X, then*

$$D \ + \ \varepsilon \cdot H$$

is ample for every $\varepsilon > 0$. Conversely, if D and H are any two divisors such that $D + \varepsilon H$ is ample for all sufficiently small $\varepsilon > 0$, then D is nef.

Proof. If $D + \varepsilon H$ is ample for $\varepsilon > 0$, then

$$\big(D \cdot C\big) + \varepsilon\big(H \cdot C\big) \ = \ \big((D + \varepsilon H) \cdot C\big) \ > \ 0$$

for every irreducible curve C. Letting $\varepsilon \to 0$ it follows that $(D \cdot C) \geq 0$, and hence that D is nef.

Assume conversely that D is nef and H is ample. Replacing εH by H, it suffices to show that $D + H$ is ample. To this end, the main point is to verify that $D + H$ satisfies the Nakai-type inequalities appearing in Definition 1.3.6 (iii). Provided that $D+H$ is (numerically equivalent to) a rational divisor, this will establish that it is ample; the general case will follow by an approximation argument.

So fix an irreducible subvariety $V \subseteq X$ of dimension $k > 0$. Then

$$\Big((D + H)^k \cdot V\Big) \ = \ \sum_{s=0}^{k} \binom{k}{s} \Big(H^s \cdot D^{k-s} \cdot V\Big). \qquad (*)$$

Since H is a positive $\mathbf{R}$-linear combination of integral ample divisors, the intersection $(H^s \cdot V)$ is represented by an effective real $(k-s)$-cycle. Applying

Kleiman's theorem to each of the components of this cycle, it follows that $(H^s \cdot D^{k-s} \cdot V) \geq 0$. Thus each of the terms in (*) is non-negative, and the last intersection number $(H^k \cdot V)$ is strictly positive. Therefore $\big((D+H)^k \cdot V\big) > 0$ for every V, and in particular if $D + H$ is rational then it is ample.

It remains to prove that $D+H$ is ample even when it is irrational. To this end, choose ample divisors $H_1, \ldots, H_r$ whose classes span $N_1(X)_{\mathbf{R}}$. By the open nature of amplitude (Example 1.3.14), the $\mathbf{R}$-divisor

$$H(\varepsilon_1, \ldots, \varepsilon_r) = H - \varepsilon_1 H_1 - \ldots - \varepsilon_r H_r$$

remains ample for all $0 \leq \varepsilon_i \ll 1$. But the classes of these divisors fill up an open[14] subset of $N^1(X)_{\mathbf{R}}$, and consequently there exist $0 < \varepsilon_i \ll 1$ such that $D' = D + H(\varepsilon_1, \ldots, \varepsilon_r)$ represents a rational class in $N^1(X)_{\mathbf{R}}$. The case of the corollary already treated shows that D' is ample. Consequently so too is

$$D + H \;=\; D' + \varepsilon_1 H_1 + \ldots + \varepsilon_r H_r,$$

as required. □

The corollary in turn gives rise to a test for amplitude involving only intersections with curves:

Corollary 1.4.11. *Let X be a projective variety or scheme, and H an ample $\mathbf{R}$-divisor on X. Fix an $\mathbf{R}$-divisor D on X. Then D is ample if and only if there exists a positive number $\varepsilon > 0$ such that*

$$\frac{(D \cdot C)}{(H \cdot C)} \geq \varepsilon \tag{1.14}$$

for every irreducible curve $C \subset X$.

In other words, the amplitude of a divisor D is characterized by the requirement that the degree of any curve C with respect to D be uniformly bounded below in terms of the degree C with respect to a known ample divisor. (See Example 1.5.3 for a concrete illustration of how this can fail.)

Proof of Corollary 1.4.11. The inequality (1.14) is equivalent to the condition that $D - \varepsilon H$ be nef. So assuming (1.14) holds, it follows from the previous Corollary 1.4.10 that

$$D \;=\; \big(D - \varepsilon H\big) + \varepsilon H$$

is ample. Conversely, if D is ample then $D - \varepsilon H$ is even ample for $0 \leq \varepsilon \ll 1$ (Example 1.3.14). □

---

[14] Openness refers here to the usual topology on the finite-dimensional real vector space $N^1(X)_{\mathbf{R}}$.

Example 1.4.12. If H_1 and H_2 are ample divisors on a projective variety X, then there are positive rational numbers $M, m > 0$ such that

$$m \cdot \big(H_1 \cdot C\big) \ \le\ \big(H_2 \cdot C\big) \ \le\ M \cdot \big(H_1 \cdot C\big)$$

for every irreducible curve $C \subset X$. (Choose M and m such that $M \cdot H_1 - H_2$ and $H_2 - m \cdot H_1$ are both ample.) □

Seshadri's criterion for amplitude is another application. Aside from its intrinsic interest, this result forms the basis for our discussion of local positivity in Chapter 5. As a matter of notation, given an irreducible curve C, $\mathrm{mult}_x C$ denotes the multiplicity of C at a point $x \in C$.

Theorem 1.4.13. (Seshadri's criterion). *Let X be a projective variety and D a divisor on X. Then D is ample if and only if there exists a positive number $\varepsilon > 0$ such that*

$$\frac{(D \cdot C)}{\mathrm{mult}_x C} \ \ge\ \varepsilon \tag{1.15}$$

for every point $x \in X$ and every irreducible curve $C \subseteq X$ passing through x.

In other words, we ask that the degree of every curve be uniformly bounded below in terms of its singularities.

Proof of Theorem 1.4.13. We first show that (1.15) holds when D is ample. To this end note that if E_x is an effective divisor which passes through x and meets an irreducible curve C properly, then the local intersection number $i(E_x, C; x)$ of E_x and C at x is bounded below by $\mathrm{mult}_x C$. In particular,

$$\big(E_x \cdot C\big) \ \ge\ \mathrm{mult}_x C.$$

But if D is ample, so that mD is very ample for some $m \gg 0$, then for every x and C one can find an effective divisor $E_x \equiv_{\mathrm{lin}} mD$ with the stated properties. Therefore $(D \cdot C) \ge \frac{1}{m}\mathrm{mult}_x C$ for all x and C.

Conversely, suppose that (1.15) holds for some $\varepsilon > 0$. Arguing by induction on $n = \dim X$, we can assume that $\mathcal{O}_V(D)$ is ample for every irreducible proper subvariety $V \subset X$. In particular, $\big(D^{\dim V} \cdot V\big) > 0$ for every proper $V \subset X$ of positive dimension. By Nakai's criterion, it therefore suffices to show that $(D^n) > 0$.

To this end, fix any smooth point $x \in X$, and consider the blowing-up

$$\mu : X' \ =\ \mathrm{Bl}_x(X) \longrightarrow X$$

of X at x, with exceptional divisor $E = \mu^{-1}(x)$. We claim that the $\mathbf{R}$-divisor

$$\mu^* D - \varepsilon \cdot E$$

is nef on X'. Granting this, Theorem 1.4.9 implies:

$$\begin{aligned}(D^n)_X - \varepsilon^n &= \Big(\big(\mu^*D - \varepsilon \cdot E\big)^n\Big)_{X'} \\ &\geq 0\end{aligned}$$

(where we indicate with a subscript the variety on which intersection numbers are being computed). Therefore $(D^n) > 0$, as required. For the nefness of $\big(\mu^*D - \varepsilon \cdot E\big)$, fix an irreducible curve $C' \subset X'$ not contained in E and set $C = \mu(C')$, so that C' is the proper transform of C. Then

$$\big(C' \cdot E\big) = \mathrm{mult}_x C$$

thanks to [208, p. 79] (see Lemma 5.1.10). On the other hand,

$$\big(C' \cdot \mu^*D\big)_{X'} = \big(C \cdot D\big)_X$$

by the projection formula. So the hypothesis (1.15) implies that $\big((\mu^*D - \varepsilon \cdot E) \cdot C'\big) \geq 0$. Since $\mathcal{O}_E(E)$ is a negative line bundle on the projective space E the same inequality certainly holds if $C' \subset E$. Therefore $\big(\mu^*D - \varepsilon E\big)$ is nef and the proof is complete. □

As a final application, we prove a result about the variation of nefness in families:

Proposition 1.4.14. (Nefness in families). *Let $f : X \longrightarrow T$ be a surjective proper morphism of varieties, and let L be a line bundle on X. For $t \in T$ put*

$$X_t = f^{-1}(t) \quad , \quad L_t = L \mid X_t.$$

If L_0 is nef for some given $0 \in T$, then there is a countable union $B \subset T$ of proper subvarieties of T, not containing 0, such that L_t is nef for all $t \in T - B$.

Proof. First, we can assume by Chow's lemma that f is projective. Next, after possibly shrinking T, we can write $L = \mathcal{O}_X(D)$ where D is a Cartier divisor on X whose support does not contain any of the fibres X_t. Fix also a Cartier divisor A on X such that $A_t = A|X_t$ is ample for all t. According to Corollary 1.4.10, D_t is nef if and only if $D_t + \frac{1}{m}A_t$ is ample for every positive integer $m > 0$. By assumption this holds when $t = 0$, and it follows from Theorem 1.2.17 that the locus on T where $D_t + \frac{1}{m}A_t$ fails to be ample is contained in a proper algebraic subset $B_m \subset T$ not containing 0. Then take $B = \cup B_m$. □

Remark 1.4.15. It seems to be unknown whether one actually needs a countable union of subvarieties in 1.4.14. □

We now turn to the proof of Kleiman's theorem.

Proof of Theorem 1.4.9. One can assume that X is irreducible and reduced, and by Chow's Lemma one can assume in addition that X is projective. We proceed by induction on $n = \dim X$, the assertion being evident if X is a curve. We therefore suppose that

$$(D^k \cdot V) \geq 0 \quad \text{for all irreducible } V \subset X \text{ of dimension } \leq n-1, \qquad (*)$$

and the issue is to show that $(D^n) \geq 0$. Until further notice we suppose that D is a $\mathbf{Q}$-divisor: the argument reducing the general case to this one appears at the end of the proof.

Fix an ample divisor H on X, and consider for $t \in \mathbf{R}$ the self-intersection number

$$P(t) =_{\text{def}} \big(D + tH\big)^n \in \mathbf{R}.$$

Expanding out the right-hand side, we can view $P(t)$ as a polynomial in t, and we are required to verify that $P(0) \geq 0$. Aiming for a contradiction, we assume to the contrary that $P(0) < 0$.

Note first that if $k < n$, then

$$\big(D^k \cdot H^{n-k}\big) \geq 0. \qquad (**)$$

In fact, H being ample, H^{n-k} is represented by an effective rational k-cycle. So (**) follows by applying the induction hypothesis (*) to the components of this cycle. In particular, for $k < n$ the coefficient of t^{n-k} in $P(t)$ is non-negative. Since by assumption $P(0) < 0$, it follows that $P(t)$ has a single real root $t_0 > 0$.

We claim next that for any rational number $t > t_0$, the $\mathbf{Q}$-divisor $D + tH$ is ample. To verify this, it is equivalent to check that

$$\Big(\big(D + tH\big)^k \cdot V\Big) > 0$$

for every irreducible $V \subseteq X$ of dimension k. When $V = X$ this follows from the fact that $P(t) > P(t_0) = 0$. If $V \subsetneq X$ one expands out the intersection number in question as a polynomial in t. As in (*) and (**) all the coefficients are non-negative, while the leading coefficient $\big(H^k \cdot V\big)$ is strictly positive. The claim is established.

Now write $P(t) = Q(t) + R(t)$, where

$$Q(t) = \Big(D \cdot \big(D + tH\big)^{n-1}\Big),$$
$$R(t) = \Big(tH \cdot \big(D + tH\big)^{n-1}\Big).$$

If $t > t_0$ then $(D+tH)$ is ample, and hence $\big(D \cdot (D+tH)^{n-1}\big)$ is the intersection of a nef divisor with an effective 1-cycle. Therefore $Q(t) \geq 0$ for all rational $t > t_0$, and consequently $Q(t_0) \geq 0$ by continuity. On the other hand, thanks to (**) all the coefficients of $R(t)$ are non-negative, and the highest one (H^n) is strictly positive. Therefore $R(t_0) > 0$. But then $P(t_0) > 0$, a contradiction. Thus we have proven the theorem in the case that D is rational.

It remains only to check that the theorem holds when D is an arbitrary nef **R**-divisor. To this end, choose ample divisors $H_1, \ldots, H_r$ whose classes span $N^1(X)_{\mathbf{R}}$. Then $\varepsilon_1 H_1 + \ldots + \varepsilon_r H_r$ is ample for all $\varepsilon_i > 0$. In particular,

$$D(\varepsilon_1, \ldots, \varepsilon_r) \;=\; D + \varepsilon_1 H_1 + \ldots + \varepsilon_r H_r,$$

being the sum of a nef and an ample **R**-divisor, is (evidently) nef. But the classes of these divisors fill up an open subset in $N^1(X)_{\mathbf{R}}$, and therefore we can find arbitrarily small $0 < \varepsilon_i \ll 1$ such that $D(\varepsilon_1, \ldots, \varepsilon_r)$ is (numerically equivalent to) a rational divisor. For such divisors, the case of the Theorem already treated shows that

$$\Big(D(\varepsilon_1, \ldots, \varepsilon_r)^k \cdot V\Big) \;\geq\; 0$$

for all irreducible V of dimension k. Letting the $\varepsilon_i \to 0$, it follows that $(D^k \cdot V) \geq 0$. □

Example 1.4.16. (Intersection products of nef classes). Let

$$\delta_1, \ldots, \delta_n \;\in\; N^1(X)_{\mathbf{R}}$$

be nef classes on a complete variety or scheme X. Then

$$\int_X \delta_1 \cdot \ldots \cdot \delta_n \;\geq\; 0.$$

(By Chow's lemma, one can assume that X is projective. Then using Corollary 1.4.10 one can perturb the δ_i slightly so that they become ample classes. But in this case the assertion is clear.) □

Remark 1.4.17. Let X be a projective variety and L a line bundle on X. By analogy with Kleiman's theorem, one might be tempted to wonder whether the strict positivity of the degrees $\int_C c_1(L)$ of L on every curve $C \subseteq X$ implies that $\int_V c_1(L)^{\dim V} > 0$ for every $V \subseteq X$. However, this is not the case. A counter-example due to Mumford is outlined in Example 1.5.2 below. □

Example 1.4.18. (Minimal surfaces). Let X be a smooth projective surface of non-negative Kodaira dimension, i.e. with the property that $|mK_X| \neq \varnothing$ for some $m > 0$. Then X is minimal — in other words, it contains no smooth rational curves having self-intersection (-1) — if and only if the canonical divisor K_X is nef. (Fix $D \in |mK_X|$, and write $D = \sum a_i C_i$ with $a_i > 0$ and C_i irreducible. If $C \subseteq X$ is an irreducible curve with $\big(K_X \cdot C\big) < 0$, then evidently C must appear as one of the C_i, say $C = C_1$. Then $\big(a_1 C \cdot C\big) \leq \big(D \cdot C\big) < 0$, and it follows from the adjunction formula that C is a (-1)-curve.) □

Remark 1.4.19. (Higher-dimensional minimal varieties). The previous example points to a notion of minimality for varieties of higher dimension.

A non-singular projective variety is *minimal* if its canonical divisor K_X is nef. More generally, a *minimal variety* is a normal projective variety, having only canonical singularities, with K_X nef.[15] Kawamata, Shokurov, Reid, and others have shown that minimal varieties share many of the excellent properties of minimal surfaces (cf. [326] or [368] for an overview). By analogy with the case of surfaces, it is natural to ask whether every smooth projective variety — say of general type, to fix ideas — is birationally equivalent to a minimal variety. Mori [440] proved that this is so in dimension three, but in arbitrary dimensions the question remains open as of this writing. However the *minimal model program* of trying to construct such models has led to many important developments. We again refer to [368] or [114] for a survey. Section 1.5.F describes some related work. □

1.4.C Cones

The meaning of Theorem 1.4.9 is clarified by introducing some natural and important cones in the Néron–Severi space $N^1(X)_{\mathbf{R}}$ and its dual. This viewpoint was pioneered by Kleiman in [341, Chapter 4].

Let X be a complete complex variety or scheme. We start by defining the nef and ample cones. As a matter of terminology, if V is a finite-dimensional real vector space, by a *cone* in V we understand a set $K \subseteq V$ stable under multiplication by positive scalars.[16]

Definition 1.4.20. (Ample and nef cones). The *ample cone*

$$\mathrm{Amp}(X) \subset N^1(X)_{\mathbf{R}}$$

of X is the convex cone of all ample $\mathbf{R}$-divisor classes on X. The *nef cone*

$$\mathrm{Nef}(X) \subset N^1(X)_{\mathbf{R}}$$

is the convex cone of all nef $\mathbf{R}$-divisor classes. □

It follows from the definitions that one could equivalently define $\mathrm{Amp}(X)$ to be the convex cone in $N^1(X)_{\mathbf{R}}$ spanned by the classes of all ample integral (or rational) divisors, i.e. the convex hull of all positive real multiples of such classes.

Remark 1.4.21. As soon as $\rho(X) = \dim N^1(X)_{\mathbf{R}} \geq 3$ the structure of these cones can become quite complicated. For example, they may or may not be polyhedral. Several concrete examples are worked out in the next section. □

---

15 See [516], [364], [368], or [114] for the relevant definitions. One of the requirements is that K_X exist as a $\mathbf{Q}$-Cartier divisor, so that the intersection products $(K_X \cdot C)$ defining nefness make sense.

16 In order that the ample classes form a cone in $N^1(X)_{\mathbf{R}}$, we do not require that cones contain the origin.

Remark 1.4.22. (Visualization). It is sometimes convenient to represent a cone by drawing its intersection with a hyperplane not passing through the origin. For example, the pentagonal cone shown in Figure 1.2 would be drawn as a pentagon in the plane. □

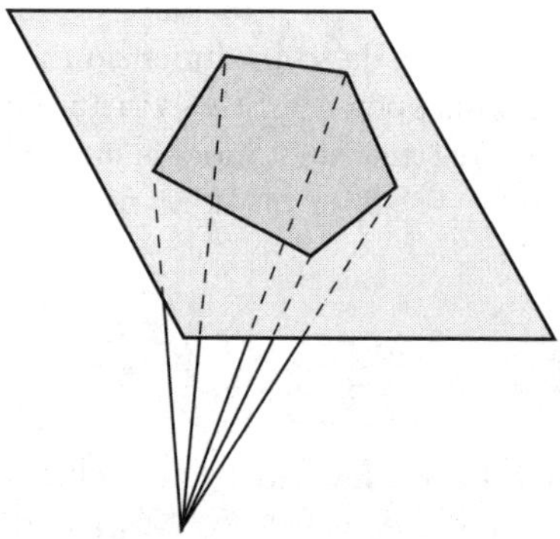

Figure 1.2. Representing cones

We view $N^1(X)_{\mathbf{R}}$ as a finite-dimensional vector space with its standard Euclidean topology. This allows one in particular to discuss closures and interiors of sets of numerical equivalence classes of **R**-divisors.

At least in the projective case, Kleiman's theorem is equivalent to the fact that the nef cone is the closure of the ample cone.

Theorem 1.4.23. (Kleiman, [341]). *Let X be any projective variety or scheme.*

(i). *The nef cone is the closure of the ample cone:*

$$\mathrm{Nef}(X) \;=\; \overline{\mathrm{Amp}(X)}.$$

(ii). *The ample cone is the interior of the nef cone:*

$$\mathrm{Amp}(X) \;=\; \mathrm{int}\Big(\mathrm{Nef}(X)\Big).$$

Proof. It is evident that the nef cone is closed, and it follows from Example 1.3.14 that Amp(X) is open. This gives the inclusions

$$\overline{\mathrm{Amp}(X)} \subseteq \mathrm{Nef}(X) \qquad \text{and} \qquad \mathrm{Amp}(X) \subseteq \mathrm{int}\big(\mathrm{Nef}(X)\big).$$

The remaining two inclusions

$$\mathrm{Nef}(X) \subseteq \overline{\mathrm{Amp}(X)} \qquad \text{and} \qquad \mathrm{int}\big(\mathrm{Nef}(X)\big) \subseteq \mathrm{Amp}(X) \qquad (*)$$

are consequences of Corollary 1.4.10. In fact let H be an ample divisor on X. If D is any nef **R**-divisor then 1.4.10 shows that $D+\varepsilon H$ is ample for all $\varepsilon > 0$. Therefore D is a limit of ample divisors, establishing the first inclusion in (*). For the second, observe that if the class of D lies in the interior of $\text{Nef}(X)$, then $D - \varepsilon H$ remains nef for $0 < \varepsilon \ll 1$. Consequently

$$D \;=\; \big(D - \varepsilon H\big) + \varepsilon H$$

is ample thanks again to Corollary 1.4.10. □

Remark 1.4.24. (Non-projective complete varieties). Kleiman [341] shows that 1.4.23 (ii) holds on a possibly non-projective complete variety X assuming only that the Zariski topology on X is generated by the complements of effective Cartier divisors: this is automatic, for instance, if X is smooth or **Q**-factorial. In this setting, $\text{Nef}(X)$ has non-empty interior if and only if X is actually projective. □

Another perspective is provided by introducing the vector space of curves dual to $N^1(X)_{\mathbf{R}}$:

Definition 1.4.25. (Numerical equivalence classes of curves). Let X be a complete variety. We denote by $Z_1(X)_{\mathbf{R}}$ the **R**-vector space of *real one-cycles* on X, consisting of all finite **R**-linear combinations of irreducible curves on X. An element $\gamma \in Z_1(X)_{\mathbf{R}}$ is thus a formal sum

$$\gamma \;=\; \sum a_i \cdot C_i$$

where $a_i \in \mathbf{R}$ and $C_i \subset X$ is an irreducible curve. Two one-cycles $\gamma_1, \gamma_2 \in Z_1(X)_{\mathbf{R}}$ are *numerically equivalent* if

$$\big(D \cdot \gamma_1\big) \;=\; \big(D \cdot \gamma_2\big)$$

for every $D \in \text{Div}_{\mathbf{R}}(X)$. The corresponding vector space of *numerical equivalence classes* of one-cycles is written $N_1(X)_{\mathbf{R}}$. Thus by construction one has a perfect pairing

$$N^1(X)_{\mathbf{R}} \times N_1(X)_{\mathbf{R}} \longrightarrow \mathbf{R} \quad , \quad (\delta, \gamma) \mapsto \big(\delta \cdot \gamma\big) \in \mathbf{R}.$$

In particular, $N_1(X)_{\mathbf{R}}$ is a finite dimensional real vector space on which we put the standard Euclidean topology. (Of course one defines numerical equivalence of integral and rational one-cycles similarly.) □

The relevant cones in $N_1(X)_{\mathbf{R}}$ are those spanned by effective curves:

Definition 1.4.26. (Cone of curves). Let X be a complete variety. The *cone of curves*

$$\text{NE}(X) \;\subseteq\; N_1(X)_{\mathbf{R}}$$

is the cone spanned by the classes of all effective one-cycles on X. Concretely,

$$\mathrm{NE}(X) = \Big\{ \sum a_i [C_i] \;\Big|\; C_i \subset X \text{ an irreducible curve, } a_i \geq 0 \Big\}.$$

Its closure

$$\overline{\mathrm{NE}}(X) \subseteq N_1(X)_{\mathbf{R}}$$

is the *closed cone of curves* on X. □

The notation $\mathrm{NE}(X)$ seems to have been introduced by Mori in his fundamental paper [438]. The abbreviation is suggested by the observation — to be established momentarily — that $\overline{\mathrm{NE}}(X)$ is dual to the cone of numerically effective divisors. □

Remark 1.4.27. An example in which $\mathrm{NE}(X)$ is not itself closed is given in 1.5.1. □

A basic fact is that $\overline{\mathrm{NE}}(X)$ and $\mathrm{Nef}(X)$ are dual:

Proposition 1.4.28. *In the situation of Definition 1.4.26,* $\overline{\mathrm{NE}}(X)$ *is the closed cone dual to* $\mathrm{Nef}(X)$, *i.e.*

$$\overline{\mathrm{NE}}(X) = \{ \gamma \in N_1(X)_{\mathbf{R}} \mid (\delta \cdot \gamma) \geq 0 \quad \textit{for all} \;\; \delta \in \mathrm{Nef}(X) \}.$$

Proof. This is a consequence of the theory of duality for cones. Specifically, suppose that $K \subseteq V$ is a closed convex cone in a finite-dimensional real vector space. Recall that the dual of K is defined to be the cone in V^* given by

$$K^* = \{ \phi \in V^* \mid \phi(x) \geq 0 \; \forall \; x \in K \}.$$

The duality theorem for cones (cf. [35, p. 162]) states that under the natural identification of V^{**} with V, one has $K^{**} = K$. In the situation at hand, take

$$V = N_1(X)_{\mathbf{R}} \quad , \quad K = \overline{\mathrm{NE}}(X).$$

Then $\mathrm{Nef}(X) = \overline{\mathrm{NE}}(X)^*$ by definition. Consequently

$$\overline{\mathrm{NE}}(X) = \mathrm{Nef}(X)^*,$$

which is the assertion of the proposition. □

Continue to assume that X is complete, and fix a divisor $D \in \mathrm{Div}_{\mathbf{R}}(X)$, not numerically trivial. We denote by

$$\phi_D : N_1(X)_{\mathbf{R}} \longrightarrow \mathbf{R}$$

the linear functional determined by intersection with D, and we set

$$\begin{aligned} D^{\perp} &= \{ \gamma \in N_1(X)_{\mathbf{R}} \mid (D \cdot \gamma) = 0 \}, \\ D_{>0} &= \{ \gamma \in N_1(X)_{\mathbf{R}} \mid (D \cdot \gamma) > 0 \}. \end{aligned}$$

Thus $D^{\perp} = \ker \phi_D$ is a hyperplane and $D_{>0}$ an open half-space in $N_1(X)_{\mathbf{R}}$. One defines $D_{\geq 0}$, $D_{\leq 0}$, $D_{<0} \subset N_1(X)_{\mathbf{R}}$ similarly.

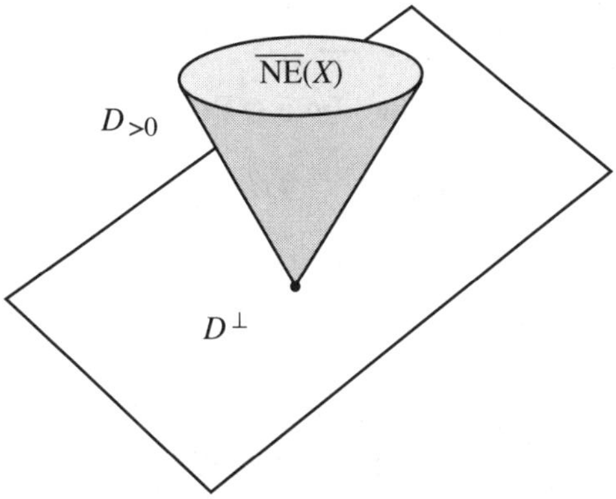

Figure 1.3. Test for amplitude via the cone of curves

Theorem 1.4.29. (Amplitude via cones). *Let X be a projective variety (or scheme), and let D be an $\mathbf{R}$-divisor on X. Then D is ample if and only if*

$$\overline{\mathrm{NE}}(X) - \{0\} \ \subseteq \ D_{>0}.$$

Equivalently, choose any norm $\| \ \|$ on $N_1(X)_{\mathbf{R}}$, and denote by

$$S \ = \ \{ \ \gamma \in N_1(X)_{\mathbf{R}} \ \mid \ \|\gamma\| = 1 \ \}$$

the "unit sphere" of classes in $\mathrm{N}_1(X)_{\mathbf{R}}$ of length 1. Then D is ample if and only if

$$\Big(\overline{\mathrm{NE}}(X) \ \cap \ S\Big) \ \subseteq \ \Big(D_{>0} \ \cap S \ \Big). \tag{1.16}$$

The theorem is illustrated in Figure 1.3: D is ample if and only if the closed cone $\overline{\mathrm{NE}}(X)$ (except the origin) lies entirely in the positive halfspace determined by D. The result is sometimes known as Kleiman's criterion for amplitude.

Proof of Theorem 1.4.29. We assume that (1.16) holds, and show that D is ample. To this end, consider the linear functional $\phi_D : N_1(X)_{\mathbf{R}} \longrightarrow \mathbf{R}$ determined by intersection with D. Then $\phi_D(\gamma) > 0$ for all $\gamma \in \big(\overline{\mathrm{NE}}(X) \cap S\big)$. But $\overline{\mathrm{NE}}(X) \ \cap \ S$ is compact, and therefore ϕ_D is bounded away from zero on this set. In other words, there exists a positive real number $\varepsilon > 0$ such that

$$\phi_D(\gamma) \geq \varepsilon \quad \text{for all} \quad \gamma \in \overline{\mathrm{NE}}(X) \cap S.$$

Thus

$$(D \cdot C) \ \geq \ \varepsilon \cdot \|C\| \tag{*}$$

for every irreducible curve $C \subseteq X$. On the other hand, choose ample divisors $H_1, \ldots, H_r$ on X whose classes form a basis of $N^1(X)_{\mathbf{R}}$. Then $\| \ \|$ is equivalent to the "taxicab" norm

$$\|\gamma\|_{\text{taxi}} = \sum \big|(H_i \cdot \gamma)\big|.$$

Setting $H = \sum H_i$ it therefore follows from (*) that for suitable $\varepsilon' > 0$,

$$(D \cdot C) \geq \varepsilon' \cdot (H \cdot C)$$

for every irreducible curve $C \subseteq X$. But then the amplitude of D is a consequence of Corollary 1.4.11. We leave the converse to the reader (cf. [363, Proposition II.4.8]). □

Example 1.4.30. Let X be a projective variety. Then the closed cone of curves $\overline{\mathrm{NE}}(X) \subset N_1(X)_{\mathbf{R}}$ does not contain any infinite straight lines. In other words, if $\gamma \in N_1(X)_{\mathbf{R}}$ is a class such that both $\gamma, -\gamma \in \overline{\mathrm{NE}}(X)$, then $\gamma = 0$. □

Example 1.4.31. (Finiteness of integral classes of bounded degree). Let X be a projective variety, and H an ample divisor on X. Denote by $N_1(X) = N_1(X)_{\mathbf{Z}}$ the group of numerical equivalence classes of integral one-cycles, and put

$$\overline{\mathrm{NE}}(X)_{\mathbf{Z}} = \overline{\mathrm{NE}}(X) \cap N_1(X)_{\mathbf{Z}}.$$

Then for any positive number $M > 0$, the set

$$\Big\{\gamma \in \overline{\mathrm{NE}}(X)_{\mathbf{Z}} \;\Big|\; (H \cdot \gamma) \leq M \Big\}$$

is finite. (One can choose ample **R**-divisors $H_1, \ldots, H_r$ forming a basis of $N^1(X)_{\mathbf{R}}$ such that $H = \sum H_i$. If $\gamma \in \overline{\mathrm{NE}}(X)_{\mathbf{Z}}$, then

$$(H \cdot \gamma) = \sum (H_i \cdot \gamma) = \sum \big|(H_i \cdot \gamma)\big|$$

is the norm $\|\gamma\|_{\text{taxi}}$ of γ in the "taxi-cab" norm determined by the H_i. So the set in question is contained in the closed ball of radius M with respect to this norm. Being compact, this ball contains only finitely many integer points.) □

Definition 1.4.32. (Extremal rays). Let $K \subseteq V$ be a closed convex cone in a finite-dimensional real vector space. An *extremal ray* $\mathbf{r} \subseteq K$ is a one-dimensional subcone having the property that if $v + w \in \mathbf{r}$ for some vectors $v, w \in K$, then necessarily $v, w \in \mathbf{r}$. □

An extremal ray is contained in the boundary of K.

Example 1.4.33. (Curves on a surface). If X is a smooth projective surface, then a one-cycle is the same thing as a divisor. Hence

$$N^1(X)_{\mathbf{R}} = N_1(X)_{\mathbf{R}},$$

and in particular the various cones we have defined all live in the same finite-dimensional vector space.

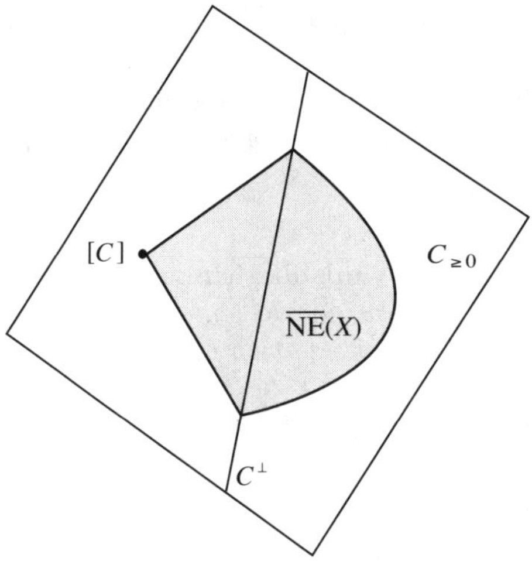

Figure 1.4. Curve of negative self-intersection on a surface

(i). One has the inclusion

$$\mathrm{Nef}(X) \ \subseteq \ \overline{\mathrm{NE}}(X),$$

with equality if and only if $\big(C^2\big) \geq 0$ for every irreducible curve $C \subset X$.

(ii). If $C \subset X$ is an irreducible curve with $\big(C^2\big) \leq 0$, then $\overline{\mathrm{NE}}(X)$ is spanned by $[C]$ and the subcone

$$\overline{\mathrm{NE}}(X)_{C \geq 0} \ =_{\mathrm{def}} \ C_{\geq 0} \cap \overline{\mathrm{NE}}(X).$$

(iii). In the situation of (ii), $[C]$ lies on the boundary of $\overline{\mathrm{NE}}(X)$. If in addition $\big(C^2\big) < 0$ then $[C]$ spans an extremal ray in that cone.

The conclusion of (iii) when $\big(C^2\big) < 0$ is illustrated in Figure 1.4 (drawn according to the convention of Remark 1.4.22). (It is evident that $\mathrm{Amp}(X) \subseteq \mathrm{NE}(X)$, and the inclusion in (i) follows by passing to closures. For (ii), observe that if $C' \subset X$ is any effective curve not containing C as a component, then $\big(C \cdot C'\big) \geq 0$. If $\big(C^2\big) < 0$, then $[C]$ does not lie in $\overline{\mathrm{NE}}(X)_{C \geq 0}$, and the second assertion of (iii) follows.) Kollár analyzes the cone of curves on an algebraic surface in more detail in [363, II.4]. □

Remark 1.4.34. (Positive characteristics). The material so far in this section goes through for varieties defined over an algebraically closed field of arbitrary characteristic. □

1.4.D Fujita's Vanishing Theorem

We now discuss a theorem of Fujita [195] showing that Serre-type vanishings can be made to operate uniformly with respect to twists by nef divisors.

Fujita's result is very useful in applications. The proof will call on vanishing theorems for big and nef line bundles to be established in Section 4.3, so from a strictly logical point of view Fujita's statement is somewhat out of sequence here.[17] We felt however that an early presentation is justified by the insight it provides.

Here is Fujita's theorem:

Theorem 1.4.35. (Fujita's vanishing theorem). *Let X be a complex projective scheme and let H be an ample (integral) divisor on X. Given any coherent sheaf $\mathcal{F}$ on X, there exists an integer $m(\mathcal{F}, H)$ such that*

$$H^i\big(X, \mathcal{F} \otimes \mathcal{O}_X(mH + D)\big) \;=\; 0 \quad \textit{for all} \quad i > 0\ ,\ m \geq m(\mathcal{F}, H),$$

and any nef divisor D on X.

The essential point here is that the integer $m(\mathcal{F}, H)$ is independent of the nef divisor D.

Proof of Theorem 1.4.35. Arguing as in the proof of Proposition 1.2.16, it suffices to prove the theorem under the additional assumption that X is irreducible and reduced. Moreover, by induction on $\dim X$ one can assume that the theorem is known for all sheaves $\mathcal{F}$ supported on a proper subscheme of X. To streamline the discussion, we will henceforth say that the theorem holds for a given coherent sheaf $\mathcal{G}$ if the statement is true when $\mathcal{F} = \mathcal{G}$.

We claim next that it is enough to exhibit any one integer $a \in \mathbf{Z}$ such that the theorem holds for $\mathcal{O}_X(aH)$. In fact, according to Example 1.2.21 an arbitrary coherent sheaf $\mathcal{F}$ admits a (possibly infinite) resolution by bundles of the form $\oplus \mathcal{O}_X(-pH)$. Using Proposition B.1.2 and Remark B.1.4 from Appendix B this reduces one to proving the stated vanishing for finitely many such bundles. On the other hand, if the theorem holds for $\mathcal{O}_X(aH)$ for any one integer a, then it follows formally (by suitably adjusting $m(\mathcal{F}, H)$) that it holds for any finite collection of the line bundles $\mathcal{O}_X(bH)$.

Let $\mu : X' \longrightarrow X$ be a resolution of singularities, and consider the torsion-free sheaf

$$\mathcal{K}_X = \mu_* \mathcal{O}_{X'}(K_{X'})$$

where $K_{X'}$ is a canonical divisor on X'.[18] If $a \gg 0$ there is an injective homomorphism

$$u \,:\, \mathcal{K}_X \longrightarrow \mathcal{O}_X(aH)$$

of coherent sheaves on X, deduced from a non-zero section of $\mathcal{O}_{X'}\big(\mu^*(aH) - K_{X'}\big)$. The cokernel of u is supported on a proper subscheme of X, so one can

[17] Needless to say, the proof of vanishing in 4.3 does not draw on Fujita's result. The interested reader can go directly to Chapter 4 after skimming Sections 2.1.A, 2.2.A, and 3.1.

[18] This is the Grauert–Riemenschneider canonical sheaf on X: see Example 4.3.12.

assume that the theorem holds for coker(u). Therefore it is enough to show that the theorem holds for $\mathcal{F} = \mathcal{K}_X$, for this then implies the statement for $\mathcal{O}_X(aH)$.

Finally, take $\mathcal{F} = \mathcal{K}_X$. Here Example 4.3.12 applies, but we recall the argument. The vanishing theorem of Grauert–Riemenschneider (Theorem 4.3.9) guarantees that

$$R^j\mu_*\mathcal{O}_{X'}(K_{X'}) \;=\; 0 \quad \text{for } \; j > 0.$$

Thus Proposition B.1.1 in Appendix B yields

$$H^i\Big(X\,,\mathcal{K}_X \otimes \mathcal{O}_X(aH+D)\Big) \;=\; H^i\Big(X'\,,\mathcal{O}_{X'}(K_{X'} + \mu^*(aH+D))\Big) \quad (*)$$

for all i. On the other hand, if $a > 0$ then $\mu^*(aH+D)$ is a nef divisor on X' whose top self-intersection number is positive (i.e. it is "big" in the sense of 2.2.1: see 2.2.16). But then the group on the right in (*) vanishes thanks to Theorem 4.3.1. □

Remark 1.4.36. (Positive characteristics). Fujita uses an argument with the Frobenius to show that the theorem also holds over algebraically closed ground fields of positive characteristic. □

We next indicate some applications of Fujita's result. The first shows that the set of all numerically trivial line bundles on a projective variety forms a bounded family.

Proposition 1.4.37. (Boundedness of numerically trivial line bundles). *Let X be a projective variety or scheme. Then there is a scheme T (of finite type!) together with a line bundle $\mathcal{L}$ on $X \times T$ having the property that any numerically trivial line bundle L on X arises as the restriction*

$$\mathcal{L}_t \;=\; \mathcal{L}\,|\,X_t \quad \textit{for some } \; t \,\in\, T,$$

where $X_t = X \times \{t\}$.

Proof. It is equivalent to prove the boundedness of the bundles $L \otimes B$ for any fixed line bundle B independent of L. With this in mind, choose a very ample line bundle $\mathcal{O}_X(1)$ on X. Since any numerically trivial line bundle is nef, 1.4.35 shows that there exists an integer $m_0 \gg 0$ such that

$$H^i\big(X, L \otimes \mathcal{O}_X(m_0 - i)\big) \;=\; 0$$

for $i > 0$ and every numerically trivial bundle L. By an elementary result of Mumford — appearing below as Theorem 1.8.3 — this implies first that

$$A_L \;=_{\text{def}}\; L \otimes \mathcal{O}_X(m_0)$$

is globally generated. Theorem 1.8.3 also gives $H^i\big(X, L \otimes \mathcal{O}_X(m)\big) = 0$ for $i > 0$ and $m \geq m_0$, so that

$$\begin{aligned} h^0\big(X, L \otimes \mathcal{O}_X(m)\big) &= \chi\big(X, L \otimes \mathcal{O}_X(m)\big) \\ &= \chi\big(X, \mathcal{O}_X(m)\big) \end{aligned}$$

for $m \geq m_0$ and every numerically trivial L: in the second equality we are using Riemann–Roch to know that twisting by L does not affect the Euler characteristic. In particular, all the bundles A_L have the same Hilbert polynomial, and they can each be written as a quotient of the trivial bundle $\mathcal{O}_X^N$ for $N = \chi\big(X, \mathcal{O}_X(m_0)\big)$. But Grothendieck's theory of Quot schemes implies that the set of all quotients of $\mathcal{O}_X^N$ having fixed Hilbert polynomial is parametrized by a scheme H of finite type, and there exists moreover a universal quotient sheaf $\mathcal{O}_{X \times H}^N \twoheadrightarrow \mathcal{F}$ flat over H. (See [300, Chapter 2.2] for a nice account of Grothendieck's theory.) We then obtain T as the open subscheme of H consisting of points $t \in H$ for which $\mathcal{F}_t$ is locally free on $X = X_t$ (cf. [300, Lemma 2.18]). □

Proposition 1.4.37 has as a consequence the characterization of numerically trivial bundles mentioned in Remark 1.1.20, at least for projective schemes:

Corollary 1.4.38. (Characterization of numerically trivial line bundles). *Let X be a projective variety or scheme, and L a line bundle on X. Then L is numerically trivial if and only if there is an integer $m > 0$ such that $L^{\otimes m}$ is a deformation of the trivial line bundle.*

Keeping notation as in 1.4.37, the conclusion means that there exists an irreducible scheme T, points $0, 1 \in T$, and a line bundle $\mathcal{L}$ on $X \times T$ such that

$$\mathcal{L}_1 = L^{\otimes m} \text{ and } \mathcal{L}_0 = \mathcal{O}_X.$$

The proof will show that one could even take T to be a smooth connected curve.

Proof of Corollary 1.4.38. If $L^{\otimes m}$ is a deformation of $\mathcal{O}_X$ then evidently L is numerically trivial. Conversely, according to the previous result all numerically trivial line bundles fall into finitely many irreducible families. Therefore there must be two distinct integers $p \neq q$ such that $L^{\otimes p}$ and $L^{\otimes q}$ lie in the same family. But then $L^{\otimes (p-q)}$ is a deformation of the trivial bundle. The statement immediately before the proof follows from the fact that any two points on an irreducible variety can be joined by a map from a smooth irreducible curve to the variety (Example 3.3.5). □

Remark 1.4.39. It is established in [52, XIII, Theorem 4.6] that the corollary continues to hold if X is complete but possibly non-projective. □

We conclude with a result that strengthens the statement of Example 1.2.36. It shows that the growth of the cohomology of a nef line bundle is bounded in terms of the degree of the cohomology:

Theorem 1.4.40. (Higher cohomology of nef divisors, II). *Let X be a projective variety or scheme of dimension n, and D a nef divisor on X. Then for any coherent sheaf $\mathcal{F}$ on X,*

$$h^i\big(X, \mathcal{F}(mD)\big) \;=\; O(m^{n-i}). \tag{1.17}$$

Proof. We may suppose by induction that the statement is known for all schemes of dimension $\leq n-1$. Thanks to Fujita's theorem, there exists a very ample divisor H having the property that $H^i\big(X, \mathcal{F}(mD+H)\big) = 0$ for $i > 0$ and every $m \geq 0$. Assuming as we may that H doesn't contain any of the subvarieties of X defined by the associated primes of $\mathcal{F}$, we have the exact sequence

$$0 \longrightarrow \mathcal{F}(mD) \xrightarrow{\cdot H} \mathcal{F}(mD+H) \longrightarrow \mathcal{F}(mD+H) \otimes \mathcal{O}_H \longrightarrow 0.$$

Therefore when $i \geq 1$,

$$h^i\big(X, \mathcal{F}(mD)\big) \;\leq\; h^{i-1}\big(H, \mathcal{F}(mD+H) \otimes \mathcal{O}_H\big) \;=\; O\big(m^{(n-1)-(i-1)}\big)$$

as required. □

Combining the Theorem with 1.1.25 one has:

Corollary 1.4.41. (Asymptotic Riemann–Roch, III). *Let X be an irreducible projective variety or scheme of dimension n, and let D be a nef divisor on X. Then*

$$h^0\big(X, \mathcal{O}_X(mD)\big) \;=\; \frac{(D^n)}{n!} \cdot m^n \,+\, O(m^{n-1}).$$

More generally,

$$h^0\big(X, \mathcal{F} \otimes \mathcal{O}_X(mD)\big) \;=\; \mathrm{rank}(\mathcal{F}) \cdot \frac{(D^n)}{n!} \cdot m^n \;+\; O(m^{n-1}) \tag{1.18}$$

for any coherent sheaf $\mathcal{F}$ on X. □

Example 1.4.42. Kollár [363, VI.2.15] shows that in the situation of Theorem 1.4.40 one can prove a slightly weaker statement without using Fujita's vanishing theorem. Specifically, with X, $\mathcal{F}$, and D as in 1.4.40 he establishes the bound

$$h^i\big(X, \mathcal{F} \otimes \mathcal{O}_X(mD)\big) \;=\; O(m^{n-1}) \quad \text{for } \; i \geq 1. \tag{1.19}$$

In particular, one can avoid Fujita's result for Corollary 1.4.41. (When $\mathcal{F} = \mathcal{O}_X$, Kollár's argument was outlined in Example 1.2.36. In general, arguing as in Proposition 1.2.16, one first reduces to the case in which X is reduced and irreducible. If $\mathrm{rank}(\mathcal{F}) = 0$ then $\mathcal{F}$ is supported on a proper subscheme, so the statement follows by induction on dimension. Assuming $\mathrm{rank}(\mathcal{F}) = r > 0$ fix a very ample divisor H. Then $\mathcal{F}(pH)$ is globally generated for $p \gg 0$, so there is an injective homomorphism $u : \mathcal{O}_X^r(-pH) \longrightarrow \mathcal{F}$ whose cokernel is supported on a proper subscheme of X. Then (1.19) for the given sheaf $\mathcal{F}$ is implied by the analogous assertion for the line bundle $\mathcal{O}_X(-pH)$. This in turn follows from the previously treated case of $\mathcal{O}_X$ by choosing a general divisor $A \in |pH|$ and arguing by induction on dimension from the exact sequence

$$0 \longrightarrow \mathcal{O}_X(-A) \longrightarrow \mathcal{O}_X \longrightarrow \mathcal{O}_A \longrightarrow 0.$$

We refer to [363, VI.2.15] for details.) □

1.5 Examples and Complements

This section gives some concrete examples of ample and nef cones, and presents some further information about their structure. We begin with ruled surfaces in Section 1.5.A. The product of a curve with itself is discussed in 1.5.B, where in particular we prove an interesting theorem of Kouvidakis. Abelian varieties are treated in Section 1.5.C, while 1.5.D contains telegraphic summaries of some other situations in which ample cones have been studied. Section 1.5.E is concerned with results of Campana and Peternell describing the local structure of the nef cone. We conclude in 1.5.F with a brief summary (without proofs) of Mori's cone theorem. We warn the reader that the present section assumes a somewhat broader background than has been required up to now.

1.5.A Ruled Surfaces

As a first example, we work out the nef and effective cones for ruled surfaces. At one point we draw on some facts concerning semistability that are discussed and established later in Section 6.4. The reader may consult [280, Chapter V, §2] or [114, Chapter 1.9] for a somewhat different perspective.

Let E be a smooth projective curve of genus g, let U be a vector bundle on E of rank two, and set $X = \mathbf{P}(U)$ with

$$\pi : X = \mathbf{P}(U) \longrightarrow E$$

the bundle projection. For ease of computation we assume that U has even degree. After twisting by a suitable divisor, we can then suppose without loss of generality that $\deg U = 0$.

Recall that $N^1(X)_{\mathbf{R}}$ is generated by the two classes

$$\xi \;=\; c_1\big(\mathcal{O}_{\mathbf{P}(U)}(1)\big) \;\text{ and }\; f \;=\; [F],$$

where F is a fibre of π. The intersection form on X is determined by the relations

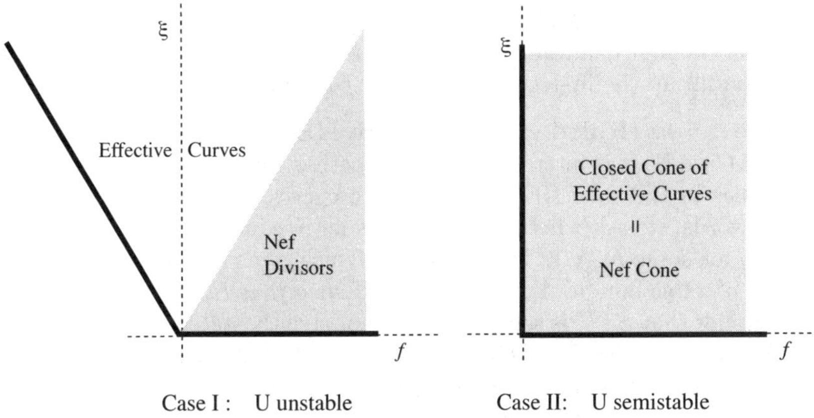

Figure 1.5. Néron–Severi group of a ruled surface

$$(\xi^2) = \deg U = 0 \quad , \quad (\xi \cdot f) = 1 \quad , \quad (f^2) = 0.$$

In particular, $((af + b\xi)^2) = 2ab$. If we represent the class $(af + b\xi)$ by the point (a, b) in the f-ξ plane, it follows that the nef cone $\mathrm{Nef}(X)$ must lie within the first quadrant $a, b \geq 0$. Moreover, the fibre F is evidently nef (e.g. by Example 1.4.6). Therefore the non-negative "f-axis" forms one of the two boundaries of $\mathrm{Nef}(X)$. Equivalently, f lies on the boundary of $\overline{\mathrm{NE}}(X)$.

The second ray bounding $\mathrm{Nef}(X)$ depends on the geometry of U. Specifically, there are two possibilities:

Case I: U is unstable. By definition, a rank-two bundle U of degree 0 is unstable if it has a line bundle quotient A of negative degree $a = \deg(A) < 0$. Assuming such a quotient exists,

$$C \;=\; \mathbf{P}(A) \;\subset\; \mathbf{P}(U) \;=\; X$$

is an effective curve in the class $af + \xi$. One has $(C^2) = 2a < 0$, and it follows from Example 1.4.33 that the ray spanned by $[C]$ bounds $\overline{\mathrm{NE}}(X)$. Therefore $\mathrm{Nef}(X)$ is bounded by the dual ray generated by $(-af + \xi)$. The situation is illustrated in Figure 1.5.

Case II: U is semistable. By definition, a bundle of degree 0 is semistable if it does not admit any quotients of negative degree. It is a basic fact that if U is semistable then so too are all the symmetric powers $S^m U$ of U (Corollary 6.4.14). In the present situation this implies that if A is a line bundle of degree a such that $H^0(E, S^m U \otimes A) \neq 0$, then $a \geq 0$. Now suppose that $C \subset X$ is an effective curve. Then C arises as a section of $\mathcal{O}_{\mathbf{P}(U)}(m) \otimes \pi^* A$ for some integer $m \geq 0$ and some line bundle A on E. On the other hand,

$$H^0\big(\mathbf{P}(U), \mathcal{O}_{\mathbf{P}(U)}(m) \otimes \pi^* A\big) \;=\; H^0\big(E, S^m U \otimes A\big),$$

so by what we have just said $a = \deg A \geq 0$. In other words, the class $(af + m\xi)$ of C lies in the first quadrant. So in this case $\mathrm{Nef}(X) = \overline{\mathrm{NE}}(X)$ and the cones in question fill up the first quadrant of the f-ξ plane.

Example 1.5.1. (Ruled surfaces where NE(X) is not closed). In the setting of Case II, it is interesting to ask whether the "positive ξ-axis" $\mathbf{R}_+ \cdot \xi$ actually lies in the cone $\mathrm{NE}(X)$ of effective curves, or merely in its closure. In other words, we ask whether there exists an irreducible curve $C \subset X$ with $[C] = m\xi$ for some $m \geq 1$. The presence of such a curve is equivalent to the existence of a line bundle A of degree 0 on E such that $H^0\big(E, S^mU \otimes A\big) \neq 0$, which implies that S^mU is semistable but not strictly stable. Using a theorem of Narasimhan and Seshadri [474] describing stable bundles in terms of unitary representations of the fundamental group $\pi_1(E)$, Hartshorne checks in [276, I.10.5] that if E has genus $g(E) \geq 2$ then there exist bundles U of degree 0 on E having the property that

$$H^0\big(E, S^mU \otimes A\big) = 0 \quad \text{for all } \; m \geq 1 \tag{1.20}$$

whenever $\deg A \leq 0$: in fact this holds for a "sufficiently general" semistable bundle U. Thus there is no effective curve C on the resulting surface $X = \mathbf{P}(U)$ with class $[C] = m\xi$, and therefore the positive ξ-axis does not itself lie in the cone of effective curves. This example is due to Mumford. □

Example 1.5.2. (Non-ample bundle that is positive on all curves). Mumford observed that the phenomenon just described also yields an example of a surface X carrying a line bundle L such that $\int_C c_1(L) > 0$ for every irreducible curve $C \subseteq X$ on X, but where L fails to be ample. In fact, let U be a bundle satisfying the condition in (1.20) and take $X = \mathbf{P}(U)$ and $L = \mathcal{O}_{\mathbf{P}(U)}(1)$. This shows that it is not enough to check intersections with curves in Nakai's criterion. By the same token it gives an example in which the linear functional ϕ_ξ determined by intersection with ξ is positive on the cone of curves $\mathrm{NE}(X)$ for a non-ample bundle ξ, explaining why one passes to the closed cone $\overline{\mathrm{NE}}(X)$ in Theorem 1.4.29. □

Example 1.5.3. The line bundle $L = \mathcal{O}_{\mathbf{P}(U)}(1)$ constructed in Example 1.5.1 is nef but not ample. It is instructive to see explicitly how the condition in Corollary 1.4.11 fails for L (as of course it must). In fact, a theorem of Segre, Nagata, and Ghione (Examples 7.2.13, 7.2.14) implies that for every m there is a line bundle A_m on E of degree $\leq g$ such that $H^0\big(E, S^mU \otimes A_m\big) \neq 0$. As above, this gives rise to a curve $C_m \subset X$ with $\int_{C_m} c_1(L) = \deg A_m$ bounded. For the reference ample class H one can take for instance $H = \xi + f$. Then one sees that the intersection numbers $\big(C_m \cdot H\big)$ go to infinity with m, and so

$$\lim_{m \to \infty} \frac{\big(C_m \cdot c_1(L)\big)}{\big(C_m \cdot H\big)} = 0. \quad \square$$

1.5.B Products of Curves

Our next examples involve products of curves, and we start by establishing notation. Denote by E a smooth irreducible complex projective curve of genus $g = g(E)$. We set

$$X = E \times E,$$

with projections $\mathrm{pr}_1, \mathrm{pr}_2 : X \longrightarrow E$. Fixing a point $P \in E$, consider in $N^1(X)_{\mathbf{R}}$ the three classes

$$f_1 = [\,\{P\} \times E\,] \ , \ f_2 = [\,E \times \{P\}\,] \ , \ \delta = [\,\Delta\,],$$

where $\Delta \subset E \times E$ is the diagonal (Figure 1.6). Provided that $g(E) \geq 1$ these

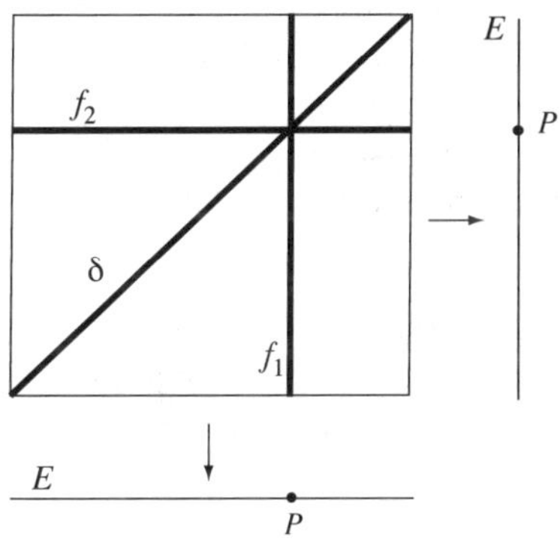

Figure 1.6. Cartesian product of curve with itself

classes are independent, and if E has general moduli then it is known that they span $N_1(X)_{\mathbf{R}}$. Intersections among them are governed by the formulae

$$\begin{aligned}(\delta \cdot f_1) \ = \ (\delta \cdot f_2) \ = \ (f_1 \cdot f_2) \ &= \ 1,\\ ((f_1)^2) \ = \ ((f_2)^2) \ &= \ 0,\\ (\delta^2) \ &= \ 2 - 2g.\end{aligned}$$

Elliptic curves. Assume that $g(E) = 1$. Then $X = E \times E$ is an abelian surface, and one has:

Lemma 1.5.4. *Any effective curve on X is nef, and consequently*

$$\overline{\mathrm{NE}}(X) \ = \ \mathrm{Nef}(X).$$

A class $\alpha \in N^1(X)_{\mathbf{R}}$ is nef if and only if

$$(\alpha^2) \geq 0 \ , \ (\alpha \cdot h) \geq 0,$$

for some ample class h. In particular, if

$$\alpha = x \cdot f_1 + y \cdot f_2 + z \cdot \delta,$$

then α is nef if and only if

$$\begin{aligned} xy + xz + yz &\geq 0 \\ x + y + z &\geq 0. \end{aligned} \qquad (*)$$

If we identify $x \cdot f_1 + y \cdot f_2 + z \cdot \delta$ in the natural way with the point $(x, y, z) \in \mathbf{R}^3$, then the equations (*) define a circular cone $\mathcal{K} \subset \mathbf{R}^3$. When $\rho(X) = 3$ — which as we have noted is the case for a sufficiently general elliptic curve E — it is precisely the nef cone, i.e. $\mathcal{K} = \text{Nef}(X)$. In general, $\mathcal{K}$ is the intersection of $\text{Nef}(X)$ with a linear subspace of $N^1(X)_{\mathbf{R}}$. In either event, the proposition shows that $\text{Nef}(X)$ is not polyhedral. (See also Example 1.5.6 and Proposition 1.5.17.) Kollár analyzes this example in more detail in [363, Chapter II, Exercise 4.16].

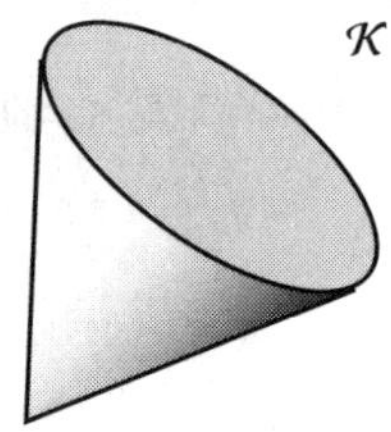

Proof of Lemma 1.5.4. The first statement is a special case of Example 1.4.7. A standard and elementary argument with Riemann–Roch (cf. [280, V.1.8]) shows that if D is an integral divisor on X such that $(D^2) > 0$ and $(D \cdot H) > 0$ for some ample H, then for $m \gg 0$, mD is linearly equivalent to an effective divisor. The second statement follows, and one deduces (*) by taking $h = f_1 + f_2 + \delta$. □

Remark 1.5.5. (Irrational polyhedral cones). One can use this example to construct a projective variety V with $\rho(V) = 2$ for which $\text{Nef}(V) \subseteq \mathbf{R}^2$ is an irrational polyhedron: see Example 5.4.17. □

Remark 1.5.6. (Arbitrary abelian surfaces). An analogous statement holds on an arbitrary abelian surface X. Specifically, $\overline{\text{NE}}(X) = \text{Nef}(X)$ and $\alpha \in N^1(X)_{\mathbf{R}}$ is nef if and only if $(\alpha^2) \geq 0$ and $(\alpha \cdot h) \geq 0$ for some ample class h. Moreover, if $\rho(X) = r$ then in suitable linear coordinates $x_1, \ldots, x_r$ on $N^1(X)_{\mathbf{R}}$, $\text{Nef}(X)$ is the cone given by

$$x_1^2 - x_2^2 - \ldots - x_r^2 \geq 0 \ , \ x_1 \geq 0.$$

(Note that in any event $r \leq 4$.) Abelian varieties of arbitrary dimension are discussed below. □

Example 1.5.7. (An example of Kollár). If D is an ample divisor on a variety Y, then by definition there is a positive integer $m(D)$ such that $\mathcal{O}_Y(mD)$ is very ample when $m \geq m(D)$. We reproduce from [152, Example 3.7] Kollár's example of a surface Y on which the integer $m(D)$ cannot be bounded independently of D. (By contrast, if Y is a smooth curve of genus g then one can take $m(D) = 2g + 1$.) Keeping notation as above, start with the product $X = E \times E$ of an elliptic curve with itself, and for each integer $n \geq 2$, form the divisor

$$A_n = n \cdot F_1 + (n^2 - n + 1) \cdot F_2 - (n-1) \cdot \Delta,$$

F_1, F_2 being fibres of the two projections $\mathrm{pr}_1, \mathrm{pr}_2 : X \longrightarrow E$. One has $\big(A_n \cdot A_n\big) = 2$ and $\big(A_n \cdot (F_1 + F_2)\big) = n^2 - 2n + 3 > 0$. It follows from 1.5.4 that A_n is ample.

Now set $R = F_1 + F_2$, let $B \in |2R|$ be a smooth divisor, and take for Y the double cover $f : Y \longrightarrow X$ of X branched along B (see Proposition 4.1.6 for the construction of such coverings). Let $D_n = f^*A_n$. Then D_n is ample, and we claim that the natural inclusion

$$H^0\big(X, \mathcal{O}_X(nA_n)\big) \longrightarrow H^0\big(Y, \mathcal{O}_Y(nD_n)\big) \tag{*}$$

is an isomorphism. It follows that $n \cdot D_n$ cannot be very ample, and hence that $m\big(D_n\big) > n$. For the claim, observe that $f_*\mathcal{O}_Y = \mathcal{O}_X \oplus \mathcal{O}_X(-R)$, and therefore

$$f_*\big(\mathcal{O}_Y(nD_n)\big) \;=\; \mathcal{O}_X(nA_n) \oplus \mathcal{O}_X(nA_n - R).$$

So to verify that (*) is bijective, it suffices to prove that

$$H^0\big(X, \mathcal{O}_X(nA_n - R)\big) \;=\; 0.$$

But this follows from the computation that $\big(\,(n \cdot A_n - R)^2\,\big) < 0$. □

Curves of higher genus. Suppose now that $g = g(E) \geq 2$. In this case the ample cone of $X = E \times E$ is already quite subtle, and not fully understood in general. Here we will give a few computations emphasizing the interplay between amplitude on X and the classical geometry of E. Some related results appear in Section 5.3.A.

Following [602], it is convenient to make a change of variables and replace δ by the class

$$\delta' = \delta - (f_1 + f_2).$$

This brings the intersection product into the simpler form

$$\big(\delta' \cdot f_1\big) = \big(\delta' \cdot f_2\big) = 0 \;\;,\;\; \big(\delta' \cdot \delta'\big) = -2g.$$

Given a positive real number $t > 0$, we focus particularly on understanding when the class

$$e_t \;=\; t(f_1 + f_2) - \delta'$$

is nef.[19] Since $(f_1 + f_2)$ is ample, e_t becomes ample for $t \gg 0$. For the purpose of the present discussion set

[19] This amounts to determining part of the ample cone of the symmetric product S^2E: see Example 1.5.14.

$$t(E) \;=\; \inf\big\{ t > 0 \mid e_t \text{ is nef } \big\}. \tag{1.21}$$

Since $\big(e_t \cdot e_t\big) = 2t^2 - 2g$, we find that in any event $t(E) \geq \sqrt{g}$.

An interesting result of Kouvidakis [370] shows that for many classes of curves, the invariant $t(E)$ reflects in a quite precise way the existence of special divisors on E. As a matter of terminology, we say that a branched covering

$$\pi : E \longrightarrow \mathbf{P}^1$$

is *simple* if π has only simple ramification (locally given by $z \mapsto z^2$) and if no two ramification points in E lie over the same point of $\mathbf{P}^1$. Writing $B \subset \mathbf{P}^1$ for the branch locus of π, this condition guarantees that $\pi_1(\mathbf{P}^1 - B)$ acts via monodromy as the full symmetric group on a general fibre $\pi^{-1}(y)$ (cf. [54, Lemma 1.3]).

Theorem 1.5.8. (Theorem of Kouvidakis). *Assume that E admits a simple branched covering*

$$\pi : E \longrightarrow \mathbf{P}^1$$

of degree $d \leq [\sqrt{g}\,] + 1$. Then $t(E) = \frac{g}{d-1}$.

Corollary 1.5.9. *If E is a very general curve of genus g, then*

$$\sqrt{g} \;\leq\; t(E) \;\leq\; \frac{g}{[\sqrt{g}\,]}.$$

The assumption on E means that the conclusion holds for all curves parameterized by the complement of a countable union of proper subvarieties of the moduli space M_g.

Remark 1.5.10. It is natural to conjecture that $t(E) = \sqrt{g}$ for very general E of sufficiently large genus. When g is a perfect square, this follows from the corollary. Ciliberto and Kouvidakis [92] have shown that the statement is implied by a conjecture of Nagata (Remark 5.1.14) provided that $g \geq 10$. When E is a general curve of genus 3 it follows from computations of Kouvidakis [370] and Bauer and Szemberg [41] that $t(E) = \frac{9}{5}$. □

Remark 1.5.11. If E is a very general curve of large genus, Kollár asks whether the diagonal $\Delta \subseteq E \times E$ is the only irreducible curve of negative self-intersection. □

Proof of Corollary 1.5.9. By the Riemann existence theorem, there exists a curve E_0 of genus g admitting a simple covering $\pi : E_0 \longrightarrow \mathbf{P}^1$ of degree $d = [\sqrt{g}] + 1$. Theorem 1.5.8 shows that

$$t(E_0) = \frac{g}{[\sqrt{g}\,]}.$$

The corollary then follows by letting E_0 vary in a complete family of curves of genus g, and applying Proposition 1.4.14 to the corresponding family of products. □

For the theorem the essential observation is the following

Lemma 1.5.12. *Assume that there exists a reduced irreducible curve*

$$C_0 \subset E \times E$$

with $[C_0] = e_s = s(f_1 + f_2) - \delta'$ *for some* $s \leq \sqrt{g}$. *Then* e_t *is nef if and only if* $t \geq \frac{g}{s}$, *and consequently* $t(E) = \frac{g}{s}$.

Proof. This could be deduced from Example 1.4.33 (ii), but it is simplest to argue directly. Suppose then that $t \geq \frac{g}{s}$. One has $\left(e_t \cdot C_0\right) = 2st - 2g \geq 0$, so it remains to show that if $C_1 \subset E \times E$ is any irreducible curve distinct from C_0, then $\left(e_t \cdot C_1\right) \geq 0$. The intersection product on $E \times E$ being non-degenerate, we can write

$$[C_1] \;=\; x_1 f_1 + x_2 f_2 - y\delta' + \alpha$$

where $\alpha \in N^1(X)_{\mathbf{R}}$ is a class orthogonal to f_1, f_2, and δ'. By intersecting with f_2 and f_1, we find that $x_1, x_2 \geq 0$. Moreover $\left(C_1 \cdot C_0\right) \geq 0$ since C_0 and C_1 meet properly, which yields

$$s(x_1 + x_2) - 2gy \geq 0. \tag{*}$$

But $\left(C_1 \cdot e_t\right) = t(x_1 + x_2) - 2gy$, and the two inequalities in the hypothesis of the Lemma imply that $t \geq s$. Therefore (*) shows that $\left(C_1 \cdot e_t\right) \geq 0$, as required. □

Proof of Theorem 1.5.8. Thanks to the lemma, it suffices to produce an irreducible reduced curve $C = C_f \subset E \times E$ having class $(d-1)(f_1 + f_2) - \delta'$, and these exist very naturally. Specifically, consider in $E \times E$ the fibre product

$$E \times E \;\supset\; E \times_{\mathbf{P}^1} E \;=\; \big\{(x, y) \mid \pi(x) = \pi(y)\big\}.$$

This contains the diagonal Δ_E as a component, and we take C to be the residual divisor: set theoretically, C is the closure of the set of all pairs $(x, y) \in E \times E$ with $x \neq y$ such that $\pi(x) = \pi(y)$. Thus

$$\begin{aligned} [C] &= d(f_1 + f_2) - \delta \\ &= (d-1)(f_1 + f_2) - \delta'. \end{aligned}$$

Moreover C is clearly reduced (being generically so), and it is irreducible thanks to the fact that the monodromy of f is the full symmetric group. □

Example 1.5.13. (Characterization of hyperelliptic curves). It follows from Theorem 1.5.8 that if E is hyperelliptic, then $t(E) = g$. In fact, this characterizes hyperelliptic curves: if E is non-hyperelliptic, then

$$t(E) \;\leq\; g - 1.$$

This strengthens and optimizes a result of Taraffa [564], who showed that

$$t(E) \leq 2\sqrt{g^2 - g} \approx 2g$$

for any curve E of genus g. (The main point is to check that any non-hyperelliptic curve E carries a simple covering $\pi : E \longrightarrow \mathbf{P}^1$ of degree g (cf. [54, Lemma 1.4]). Then, as in the proof of (1.5.8), let $C = C_\pi$ be the curve residual to Δ_E in $E \times_{\mathbf{P}^1} E \subset E \times E$. The class of C is given by $[C] = (g-1)(f_1+f_2) - \delta'$, and C is irreducible since f is simple. But $\big(C^2\big) \geq 0$, and consequently C is nef.) □

Example 1.5.14. (Symmetric products). Following [370] let $Y = S^2E$ be the second symmetric product of E, so that there is a natural double cover $p : X = E \times E \longrightarrow S^2E = Y$. This gives rise to an inclusion $p^* : N^1(Y)_{\mathbf{R}} \longrightarrow N^1(X)_{\mathbf{R}}$ realizing the Néron–Severi space of Y as a subspace in that of X. The classes

$$f = f_1 + f_2 \ \text{ and } \ \delta'$$

lie in $N^1(Y)_{\mathbf{R}}$, and when E has general moduli they span it. The invariant $t(E)$ determines one of the rays bounding the intersection

$$\mathrm{Nef}(Y) \cap \big(\mathbf{R}\cdot\delta' + \mathbf{R}\cdot f\big)$$

of Nef(Y) with the subspace spanned by δ' and f. The other bounding ray is generated by $gf + \delta'$. In other words, $(sf + \delta') \in N^1(Y)_{\mathbf{R}}$ is nef iff $s \geq g$. (The class $gf + \delta'$ in question is the pullback of the theta divisor Θ_E of the Jacobian of E under the Abel–Jacobi map $u : S^2E \longrightarrow \mathrm{Jac}(E)$. Hence $gf + \delta'$ is nef, and since f is ample so too is $(sf + \delta')$ when $s \geq g$. On the other hand, $\delta = [\Delta]$ is effective and

$$\begin{aligned} \big((sf + \delta')\cdot(\delta)\big) &= \big((sf + \delta')\cdot(\delta' + f)\big) \\ &= 2s - 2g. \end{aligned}$$

Consequently $(sf + \delta')$ is not nef when $s < g$.) □

Remark 1.5.15. (Higher-dimensional symmetric product). Let E be a general curve of even genus $g = 2k$. Using deep results of Voisin from [598], Pacienza [490] works out the nef cone of the symmetric product $Y = S^kE$. Since $\rho(Y) = 2$, the cone in question is determined as above by two slopes: in the case at hand, Pacienza shows that they are rational. □

Example 1.5.16. (Vojta's divisor). Fix real numbers $r, s > 0$ and let

$$a_1 = a_1(r) = \sqrt{(g+s)r} \ \ , \ \ a_2 = a_2(r) = \sqrt{\frac{(g+s)}{r}}.$$

Put $v_r = a_1 f_1 + a_2 f_2 + \delta'$, so that $\big(v_r \cdot v_r\big) = 2s$. If

$$r > \frac{(g+s)(g-1)}{s},$$

then v_r is nef. (See [602, Proposition 1.5].) This divisor — or more precisely the height function it determines — plays an important role in Vojta's proof of the Mordell conjecture (i.e. Faltings' theorem). See [382] for a nice overview. □

1.5.C Abelian Varieties

In this subsection, following [78], we describe the ample and nef cones of an abelian variety of arbitrary dimension.

Let X be an abelian variety of dimension n, and let H be a fixed ample divisor on X. The essential point is the following, which generalizes Lemma 1.5.4.

Proposition 1.5.17. *An arbitrary* **R**-*divisor* D *on* X *is ample if and only if*

$$\left(D^k \cdot H^{n-k}\right) > 0 \tag{1.22}$$

for all $0 \leq k \leq n$*, and* D *is nef if and only if* $\left(D^k \cdot H^{n-k}\right) \geq 0$ *for all* k.

Sketch of Proof. If D is ample or nef, then the stated inequalities follow from Examples 1.2.5 and 1.4.16. Conversely, we show that if D is an integral divisor satisfying (1.22), then D is ample: since there are only finitely many inequalities involved, the corresponding statement for **R**-divisors follows. To this end, consider the polynomial $P(t) = P_D(t) \in \mathbf{Z}[t]$ defined in the usual way by the expression

$$P_D(t) = \Big((D + t \cdot H)^n \Big).$$

The given inequalities imply that $P(t) > 0$ for every real number $t \geq 0$. So it is enough to show that if D is *not* ample, then $P_D(t)$ has a non-negative real root $t_0 \geq 0$. But this follows from the theory of the index of a line bundle on an abelian variety ([447, pp. 154–155]). In brief, write $X = V/\Lambda$ as the quotient of an n-dimensional complex vector space modulo a lattice, and let h_D, h_H be the Hermitian forms on V determined by D and H. To say that D fails to be ample means that h_D is not positive definite. On the other hand, $h_D + t \cdot h_H$ is positive definite for $t \gg 0$. Hence there exists $t_0 \geq 0$ such that $h_D + t_0 \cdot h_H$ is indefinite. But then $P_D(t_0) = 0$. (See [383, p. 79] for a more detailed account.) □

Corollary 1.5.18. *If* $\delta \in N^1(X)_{\mathbf{R}}$ *is a nef class that is not ample, then* $\left(\delta^n\right) = 0$.

Proof. If δ is nef but not ample, then the proposition implies that

$$\left(\delta^k \cdot h^{n-k}\right) = 0 \text{ for some } k \in [1, n],$$

h being a fixed ample class. But the Hodge-type inequalities in Corollary 1.6.3 show that

$$\left(\delta^k \cdot h^{n-k}\right)^n \geq \left(\delta^n\right)^k \cdot \left(h^n\right)^{n-k}.$$

Since both terms on the right are non-negative, the assertion follows. □

Remark 1.5.19. (Examples of hypersurfaces of large degree bounding the nef cone). Corollary 1.5.18 gives rise to explicit examples in which the nef cone is locally bounded by polynomial hypersurfaces of large degree: see [78, Example 2.6]. General results of Campana–Peternell about this nef boundary appear in Section 1.5.E. □

Remark 1.5.20. Bauer [37] shows that the nef cone $\mathrm{Nef}(X)$ of an abelian variety is rational polyhedral if and only if X is isogeneous to a product of abelian varieties of mutually distinct isogeny types, each having Picard number one. □

1.5.D Other Varieties

We survey here a few other examples of varieties whose ample cones have been studied in the literature. Our synopses are very brief: relevant definitions and details can be found in the cited references.

Blow-ups of $\mathbf{P}^2$. Let X be the blowing up of the projective plane at ten or more very general points. Denote by $e_i \in N^1(X)$ the classes of the exceptional divisors, and let ℓ be the pullback to X of the hyperplane class on $\mathbf{P}^2$. We may fix $0 < \varepsilon \ll 1$ such that $h =_{\mathrm{def}} \ell - \varepsilon \cdot \sum e_i$ is an ample class.

It is known that one can find (-1)-curves of arbitrarily high degree on X (see [280, Exercise V.4.15]). In other words, there exists a sequence $C_i \subseteq X$ of smooth rational curves with

$$\left(C_i \cdot C_i\right) = -1 \quad \text{and} \quad \left(C_i \cdot h\right) \to \infty \quad \text{with} \quad i.$$

By 1.4.32, each $[C_i]$ generates an extremal ray in $\overline{\mathrm{NE}}(X)$. On the other hand, let K_X denote as usual a canonical divisor on X. Then $\left(C_i \cdot K_X\right) = -1$ does not grow with i. This means that the rays $\mathbf{R}_+ \cdot [C_i]$ generated by the C_i cluster in $N_1(X)_{\mathbf{R}}$ towards the plane $K_X^{\perp}$ defined by the vanishing of K_X. The situation — which is an illustrative instance of Mori's cone theorem (Theorem 1.5.33) — is illustrated schematically in Figure 1.7. It is conjectured that $\overline{\mathrm{NE}}(X)$ is circular on the region $(K_X)_{>0}$ — i.e. that $\overline{\mathrm{NE}}(X) \cap (K_X)_{>0}$ consists of classes of non-negative self-intersection — but this is not known. (The cone of curves of X is governed by a conjecture of Hirschowitz [289]. See [426] for an account of some recent work on this question using classical methods, and Remarks 5.1.14 and 5.1.23 for a related conjecture of Nagata and an analogue in symplectic geometry.)

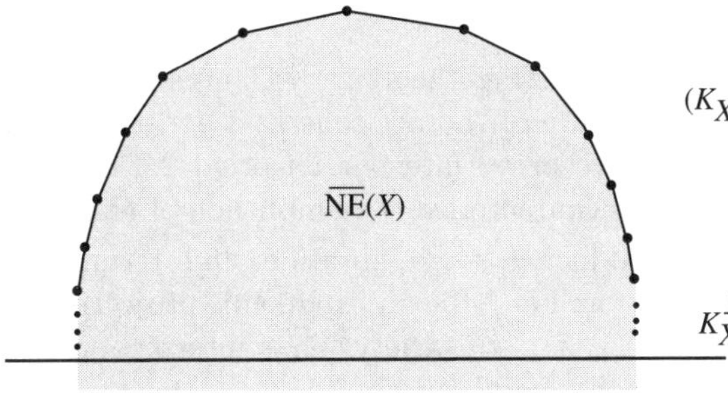

Figure 1.7. Cone of curves on blow-up of $\mathbf{P}^2$.

K3 surfaces. Kovács [371] has obtained very precise information about the cone of curves of a $K3$ surface X. He shows for example that if $\rho(X) \geq 3$, then either X does not contain any curves of negative self-intersection, or else $\overline{\mathrm{NE}}(X)$ is spanned by the classes of smooth rational curves on X. He deduces as a corollary that either $\overline{\mathrm{NE}}(X)$ is circular, or else has no circular part at all.

Holomorphic symplectic fourfolds. Hassett and Tschinkel [283] have studied the ample cone on holomorphic symplectic varieties of dimension four. They formulate a conjecture giving a Hodge-theoretic description of this cone (at least in important cases), and they present some evidence for this conjecture. Huybrechts [298] [299] establishes some related results on the Kähler cone of hyper-Kähler manifolds. Divisors on hyper-Kähler manifolds are also studied by Boucksom in [67].

$\overline{\mathbf{M}}_{\mathbf{g,n}}$. A very basic question, going back to Mumford, is to describe the ample cone of the Deligne–Mumford compactification $\overline{M}_{g,n}$ of the moduli space parameterizing n-pointed curves of genus g. The case of $g = 0$ is already very interesting and subtle: the variety $\overline{M}_{0,n}$ has a rich combinatorial structure, and has been the focus of considerable attention (cf. [328], [310]). Fulton conjectured that the closed cone of curves $\overline{\mathrm{NE}}(\overline{M}_{0,n})$ is generated by certain natural one-dimensional boundary strata in $\overline{M}_{0,n}$. This has been verified for $n \leq 6$ by Farkas and Gibney [183] following earlier work of Farber and Keel. Gibney, Keel, and Morrison [222] show that the truth of Fulton's conjecture for all n would in fact lead to a description of the ample cone of $\overline{M}_{g,n}$ in all genera. Fulton conjectured analogously that the closed cone $\overline{\mathrm{Eff}}(\overline{M}_{0,n})$ of pseudoeffective divisors (Definition 2.2.25) should be generated by boundary divisors, but counterexamples were given by Keel and by Vermeire [587]. This pseudoeffective cone has been studied for small n by Hassett and Tschinkel [284]. □

1.5.E Local Structure of the Nef Cone

We present here an interesting theorem of Campana and Peternell [78] describing the local structure of the nef cone at a general point. It depends on their result — which we prove later as Theorem 2.3.18 — that the Nakai–Moishezon inequalities characterize the amplitude of real divisor classes.

We start by introducing some notation and terminology intended to streamline the discussion. Let X be an irreducible projective variety or scheme of dimension n and $V \subseteq X$ a subvariety. Then intersection with V determines a real homogeneous polynomial on $N^1(X)_{\mathbf{R}}$:

$$\varphi_V : N^1(X)_{\mathbf{R}} \longrightarrow \mathbf{R} \ , \ \ \varphi_V(\xi) \ = \ \int_V \xi^{\dim V}.$$

Thus $\deg \varphi_V = \dim V$, and φ_V is given by an integer polynomial with respect to the natural integral structure on $N^1(X)_{\mathbf{R}}$. By Kleiman's theorem (Theorem 1.4.9), all the φ_V are non-negative on $\mathrm{Nef}(X)$.

Definition 1.5.21. (Null cone). The *null cone* $\mathcal{N}_V \subseteq N^1(X)_{\mathbf{R}}$ determined by V is the zero-locus of φ_V:

$$\mathcal{N}_V \ = \ \big\{ \xi \in N^1(X)_{\mathbf{R}} \mid \varphi_V(\xi) = 0 \big\}. \quad \square$$

Note that φ_V and $\mathcal{N}_V$ depend only on the homology class of V. Consequently as V varies over all subvarieties of X, only countably many distinct functions and cones occur.

Example 1.5.22. When X is a smooth surface, $\mathcal{N}_X$ is the familiar quadratic cone of classes of self-intersection zero. □

Example 1.5.23. If X is the blow-up of $\mathbf{P}^n$ at k points, then in suitable coordinates $\mathcal{N}_X$ is the Fermat-type hypersurface defined by the equation

$$x^n \ = \ y_1^n + \ldots + y_k^n. \quad \square$$

Example 1.5.24. (Singularities of the null cone). Let $f : X \longrightarrow Y$ be a surjective morphism of irreducible projective varieties or schemes with $\dim X = n$ and $\dim Y = d$. Then φ_X vanishes with multiplicity $n - d$ along the image of $f^* : N^1(Y)_{\mathbf{R}} \longrightarrow N^1(X)_{\mathbf{R}}$. (Given $\xi = f^*\eta$, consider $\varphi_X(\xi + h)$ for any ample $h \in \mathbf{N}^1(X)_{\mathbf{R}}$.) This result is due to Wiśniewski [610]: see that paper or [363, Exercise III.1.9] for some applications. □

Remark 1.5.25. (Calabi–Yau threefolds). Wilson [608], [609] uses a deep and careful analysis of the null cone $\mathcal{N}_X$ — especially its arithmetic properties — to prove some interesting results about the geometry of a Calabi–Yau threefold X. For example, he proves that if $\rho(X) > 19$, then X is a resolution of singularities of a "Calabi–Yau model" (i.e. a Calabi–Yau threefold that may have mild singularities) of smaller Picard number. Wilson also shows that the ample cones of Calabi–Yau threefolds are invariant under deformations if and only if none of the manifolds in question contains a smooth elliptic ruled surface. □

Definition 1.5.26. (Nef boundary). The *nef boundary* $\mathcal{B}_X \subseteq N^1(X)_{\mathbf{R}}$ of X is the boundary of the nef cone:

$$\mathcal{B}_X = \partial \operatorname{Nef}(X).$$

It is topologized as a subset of the Euclidean space $N^1(X)_{\mathbf{R}}$. □

Example 1.5.27. We may restate Corollary 1.5.18 as asserting that if X is an abelian variety then $\mathcal{B}_X \subseteq \mathcal{N}_X$, i.e. its nef boundary lies on the null cone of X. □

The results of Campana and Peternell are summarized in the next two statements:

Theorem 1.5.28. *Given any point $\xi \in \mathcal{B}_X$ on the nef boundary, there is a subvariety $V \subseteq X$ such that $\varphi_V(\xi) = 0$.*

In other words, any point in the nef boundary actually lies on one of the null cones $\mathcal{N}_V \subseteq N^1(X)_{\mathbf{R}}$. The possibility that $V = X$ is of course not excluded.

Because there are only countably many such cones, at most points the nef boundary $\mathcal{B}_X$ must then look locally like one of them:

Theorem 1.5.29. *There is an open dense set $\operatorname{CP}(X) \subseteq \mathcal{B}_X$ with the property that for every point $\xi \in \operatorname{CP}(X)$, there is an open neighborhood $U = U(\xi)$ of ξ in $N^1(X)_{\mathbf{R}}$, together with a subvariety $V \subseteq X$ of X (depending on ξ), such that*

$$\mathcal{B}_X \cap U(\xi) = \mathcal{N}_V \cap U(\xi).$$

In other words, $\mathcal{B}_X$ is cut out in $U(\xi)$ by the polynomial φ_V.

The conclusion of the theorem is illustrated schematically in Figure 1.8, which shows (with the convention of Remark 1.4.22) null cones bounding $\operatorname{Nef}(X)$. Note however that in general there may be infinitely many cones $\mathcal{N}_V$ required to fill out the nef boundary in Theorem 1.5.28, leading to clustering phenomena not shown in the picture.

Remark 1.5.30. The proof of 1.5.29 will show that we can take

$$d\varphi_V(\xi) \neq 0 \ \text{ for } \ \xi \in \operatorname{CP}(X).$$

It follows that $\mathcal{N}_V$ is smooth at ξ, and moreover that we can choose $U(\xi)$ so that $\varphi_V < 0$ on $U(\xi) - \operatorname{Nef}(X)$. In other words, $U(\xi) \cap \operatorname{Nef}(X)$ is defined on $U(\xi)$ by the inequality $\{\varphi_V \geq 0\}$. □

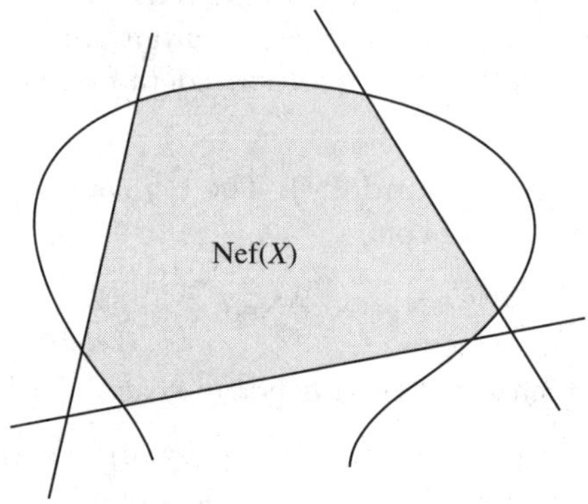

Figure 1.8. Null cones bounding Nef(X)

Example 1.5.31. If $\xi \in \mathrm{CP}(X)$, and if $\mathcal{N}_V$ is the corresponding null cone as in 1.5.30, then the dual hypersurface

$$\mathcal{N}_V{}^* \subseteq N^1(X)^*_{\mathbf{R}} = N_1(X)_{\mathbf{R}}$$

of hyperplanes tangent to $\mathcal{N}_V$ bounds $\overline{\mathrm{NE}}(X)$ in a neighborhood of the tangent plane to $\mathcal{N}_V$ at ξ. □

Remark 1.5.32. (Numerical characterization of the Kähler cone). Demailly and Paun [132] have recently established the striking result that an analogue of Theorem 1.5.28 remains true on any compact Kähler manifold. Given such a manifold X one considers the cone Kahler(X) of all Kähler classes in $H^{1,1}(X, \mathbf{R})$ (Definition 1.2.39). The main result of [132] is that Kahler(X) is a connected component of the set of all classes $\alpha \in H^{1,1}(X, \mathbf{R})$ such that

$$\int_V \alpha^{\dim V} > 0 \qquad (*)$$

for every irreducible analytic subvariety $V \subseteq X$ of positive dimension. It follows for instance that if X contains no proper subvarieties, then the Kähler cone of X is a connected component of the set of classes with positive self-intersection. When X is projective, Demailly and Paun prove moreover that Kahler(X) actually coincides with the set of classes satisfying (*). These results of course imply 1.5.28 (at least on smooth varieties), but in fact they are stronger since in general $N^1(X)_{\mathbf{R}}$ might span a proper subspace of $H^{1,1}(X, \mathbf{R})$. □

Proof of Theorem 1.5.28. This is a restatement of the Nakai criterion for **R**-divisors (Theorem 2.3.18). In fact, if $\xi \in \mathcal{B}_X$ then certainly $\varphi_V(\xi) \geq 0$ for all $V \subseteq X$. But if $\varphi_V(\xi) > 0$ for every V then the result in question implies that $\xi \in \mathrm{Amp}(X)$. □

Proof of Theorem 1.5.29. Given a subvariety $V \subseteq X$, let

$$O_V = \Big\{ \xi_1 \in \mathcal{B}_X \mid \text{some neighborhood of } \xi_1 \text{ in } \mathcal{B}_X \text{ lies in } \mathcal{N}_V \Big\},$$

and put

$$O = \bigcup_{V \subseteq X} O_V.$$

Clearly O is open, and we claim that it is dense in $\mathcal{B}_X$. For in the contrary case we would find a point $z \in \mathcal{B}_X$ having a compact neighborhood K in $\mathcal{B}_X$ such that $\big(K - \mathcal{N}_V\big)$ is dense in K for every V. Since there are only countably many distinct $\mathcal{N}_V$ as V varies over all subvarieties of X, Baire's theorem implies that

$$\bigcap_V \big(K - \mathcal{N}_V\big) = K - \bigcup_V \mathcal{N}_V$$

is dense, hence non-empty. But by Theorem 1.5.28, $K \subseteq \bigcup_V \mathcal{N}_V$, a contradiction.

Now let $P \subseteq \mathcal{B}_X$ denote the set of all points satisfying the conclusion of the theorem. Again P is open by construction, so the issue is to show that it is dense. To this end, fix any point $\xi_1 \in O$ and a subvariety $V \subseteq X$ of minimal dimension such that $\xi_1 \in O_V$. Choose also a very ample divisor H on X meeting V properly, and let $h \in N^1(X)_{\mathbf{R}}$ denote its numerical equivalence class. Then $\xi_1 \notin O_{(V \cap H)}$ thanks to the minimality of V. Therefore ξ_1 is a limit of points $\xi \in O_V$ such that

$$\int_V \big(\xi^{\dim V - 1} \cdot h\big) > 0. \tag{*}$$

We will show that if $\xi \in O_V$ is any point satisfying (*), then $\xi \in P$. This implies the required density of P, and will complete the proof.

So fix such a point ξ. We claim first that $d\varphi_V(\xi) \neq 0$, and hence that $\mathcal{N}_V$ is non-singular near ξ. In fact,

$$\big(\dim V\big) \cdot \int_V \big(\xi^{\dim V - 1} \cdot h\big) = \lim_{t \to 0} \frac{1}{t} \cdot \int_V \big(\xi + t \cdot h\big)^{\dim V}$$

is the directional derivative of φ_V at ξ in the direction h, which is non-vanishing by (*). We now argue that if $U(\xi)$ is a small convex neighborhood of ξ in $N^1(X)_{\mathbf{R}}$, then

$$\mathcal{B}_X \cap U(\xi) = \mathcal{N}_V \cap U(\xi). \tag{**}$$

This will show that $\xi \in P$, as required. For (**), the point roughly speaking is that $\mathcal{B}_X \cap U(\xi)$ — being a piece of the boundary of a closed convex set with non-empty interior — is a topological manifold, and since $\mathcal{B}_X \cap U(\xi) \subseteq \mathcal{N}_V \cap U(\xi)$ for sufficiently small $U(\xi)$, the sets in question must coincide. In more detail, let $L \subseteq N^1(X)_{\mathbf{R}}$ be the embedded affine tangent space to $\mathcal{N}_V$ at ξ, and let

$$\pi : N^1(X)_{\mathbf{R}} \longrightarrow L$$

be an affine linear projection. Thus π restricts to an isomorphism $\mathcal{N}_V \longrightarrow L$ in a neighborhood of ξ. On the other hand, since $\mathcal{B}_X$ is the boundary of a closed convex set, the image under π of a convex neighborhood of ξ in $\mathcal{B}_X$ is a convex neighborhood of $\pi(\xi)$ in L. But $\mathcal{B}_X \cap U(\xi) \subseteq \mathcal{N}_V$ by construction, so this implies that $\mathcal{B}_X \cap U(\xi)$ contains an open neighborhood of ξ in $\mathcal{N}_V$. After possibly shrinking $U(\xi)$, (**) follows. □

1.5.F The Cone Theorem

Let X be a smooth complex projective variety, and K_X a canonical divisor on X. In his seminal paper [438], Mori proved that the closed cone of curves $\overline{\mathrm{NE}}(X)$ has a surprisingly simple structure on the subset of $N_1(X)_{\mathbf{R}}$ of classes having negative intersection with K_X. He showed moreover that this has important structural implications for X. We briefly summarize here some of these results, and state some simple consequences, but we don't give proofs. Chapter 1 of the book [368] contains an excellent introduction to this circle of ideas, and full proofs appear in Chapters 1 and 3 of that book. See also [114] for an account aimed at novices, and [419] for a very detailed exposition in the spirit of [368].

As above, let X be a smooth projective variety. Given any divisor D on X write

$$\overline{\mathrm{NE}}(X)_{D\geq 0} \;=\; \overline{\mathrm{NE}}(X) \cap D_{\geq 0}$$

for the subset of $\overline{\mathrm{NE}}(X)$ lying in the non-negative half-space determined by D. Mori [438] first of all proved

Theorem 1.5.33. (Cone Theorem). *Assume that* $\dim X = n$ *and that* K_X *fails to be nef.*

(i). *There are countably many rational curves* $C_i \subseteq X$*, with*

$$0 \;\leq\; -\big(C_i \cdot K_X\big) \;\leq\; n+1,$$

that together with $\overline{\mathrm{NE}}(X)_{K_X \geq 0}$ *generate* $\overline{\mathrm{NE}}(X)$*, i.e.*

$$\overline{\mathrm{NE}}(X) \;=\; \overline{\mathrm{NE}}(X)_{K_X\geq 0} \;+\; \sum_i \mathbf{R}_+ \cdot [C_i].$$

(ii). *Fix an ample divisor* H*. Then given any* $\varepsilon > 0$*, there are only finitely many of these curves — say* $C_1, \ldots, C_t$ *— whose classes lie in the region* $(K_X + \varepsilon \cdot H)_{\leq 0}$*. Therefore*

$$\overline{\mathrm{NE}}(X) \;=\; \overline{\mathrm{NE}}(X)_{(K_X+\varepsilon H)\geq 0} \;+\; \sum_{i=1}^{t} \mathbf{R}_+ \cdot [C_i].$$

The theorem was illustrated in the example of $\mathbf{P}^2$ blown up at ten or more points: see Figure 1.7.

Example 1.5.34. (Fano varieties). Let X be a Fano variety, i.e. a smooth projective variety such that $-K_X$ is ample. Then $\overline{\mathrm{NE}}(X) \subseteq N_1(X)_{\mathbf{R}}$ is a finite rational polytope, spanned by the classes of rational curves. (Apply statement (ii) of the cone theorem.) □

Example 1.5.35. (Adjoint bundles). Let X be a smooth projective variety of dimension n, and H any ample integer divisor on X. Then $K_X + (n+1)H$ is nef, and $K_X + (n+2)H$ is ample. More generally, if D is any ample divisor such that

$$\big(D \cdot C\big) \geq n+1 \quad \text{(respectively } \big(D \cdot C\big) \geq n+2\text{)}$$

for every irreducible curve $C \subseteq X$, then $K_X + D$ is nef (respectively $K_X + D$ is ample). (This follows from the inequalities on the intersection numbers $\big(C_i \cdot K_X\big)$ appearing in statement (i) and the fact that the intersection numbers $\big(C_i \cdot H\big)$ are positive integers.) See Example 1.8.23 and Section 10.4.A for some remarks on the interest in statements of this sort. □

Mori proved Theorem 1.5.33 via his "bend and break" method. This in turn has found a multitude of other applications, for example to the circle of ideas involving rational connectedness. We refer to Kollár's book [363] for an extensive survey of some of these developments. An alternative cohomological approach to the cone theorem — which works also on mildly singular varieties, but does not recognize the curves C_i as rational — was developed by Kawamata and Shokurov following ideas of Reid ([317], [534], [517]). We again refer to [368, Chapter 3] for details.

The rational curves C_i appearing in the cone theorem generate extremal rays of $\overline{\mathrm{NE}}(X)$ in the sense of Definition 1.4.32. Mori showed that if $\dim X = 3$, and if $\mathbf{r}$ is an extremal ray in $\overline{\mathrm{NE}}(X)_{(K+\varepsilon H)\leq 0}$, then there is a mapping

$$\mathrm{cont}_{\mathbf{r}} : X \longrightarrow \overline{X}$$

that contracts every curve whose class lies in $\mathbf{r}$. This was extended to smooth projective varieties of all dimensions, as well as to varieties with mild singularities, by Kawamata and Shokurov, again following ideas of Reid. This contraction theorem is of great importance, because it opens the door to the possibility of constructing minimal models for varieties of arbitrary dimension. Specifically, given a projective variety X whose canonical bundle is not nef, these results guarantee that X carries an extremal curve which contracts under the corresponding morphism $\mathrm{cont}_{\mathbf{r}} : X \longrightarrow \overline{X}$. Then one would like to repeat the process starting on $\overline{X}$. Unfortunately, this naive idea runs into problems coming from the singularities of $\overline{X}$. However this *minimal model program* has led to many important developments in recent years. Besides the book [368] of Kollár and Mori, the reader can consult [326], [114], or [419] for an account of some of this work.

1.6 Inequalities

In recent years, generalizations and analogues of the classical Hodge index inequality have arisen in several contexts, starting with work of Khovanskii and Teissier (cf. [334], [566], [567], [420], [411], [126]). This section is devoted to a presentation of some of this material. We start with some inequalities of Hodge type among the intersection numbers of nef divisors on a complete variety, and then discuss briefly some related results of Teissier for the mixed multiplicities of $\mathfrak{m}$-primary ideals.

1.6.A Global Results

The basic global result is a generalization of the Hodge index theorem on surfaces. The statement was known (at least by experts) to follow from inequalities of Khovanskii, Matsusaka, and Teissier (Example 1.6.4); it achieved wide circulation in Demailly's paper [124]. The direct argument we give is due to Beltrametti and Sommese [50, Chapter 2.5], and independently to Fulton and Ein (see [205, p. 120]).

Theorem 1.6.1. (Generalized inequality of Hodge type). *Let X be an irreducible complete variety (or scheme) of dimension n, and let*

$$\delta_1, \ldots, \delta_n \in N^1(X)_{\mathbf{R}}$$

be nef classes. Then

$$\Big(\delta_1 \cdot \ldots \cdot \delta_n\Big)^n \geq \Big((\delta_1)^n\Big) \cdot \ldots \cdot \Big((\delta_n)^n\Big). \tag{1.23}$$

Proof. We can assume first of all that X is reduced since in general its cycle satisfies $[X] = a \cdot [X_{\text{red}}]$ for some $a > 0$. By Chow's lemma (Remark 1.4.3), we can suppose also that X is projective. In this case, it suffices to prove the theorem under the additional assumption that the δ_i are ample. In fact, if the stated inequality holds for $\delta_i \in \text{Amp}(X)$, then by continuity it holds also for $\delta_i \in \text{Nef}(X) = \overline{\text{Amp}(X)}$ (Theorem 1.4.23).

We now argue by induction on $n = \dim X$. If X is a smooth projective surface, then the stated inequality is a version of the Hodge index theorem (cf. [280], Exercise V.1.9). As in the previous paragraph, once one knows (1.23) for ample classes, it follows also for nef ones. This being said, if X is singular, one deduces (1.23) by passing to a resolution. We assume henceforth that $n \geq 3$, and that (1.23) is already known on all irreducible varieties of dimension $n-1$.

We next claim that given any ample classes

$$\beta_1, \ldots, \beta_{n-1}, h \in N^1(X)_{\mathbf{R}},$$

one has the inequality

$$\Big(\beta_1 \cdot \ldots \cdot \beta_{n-1} \cdot h\Big)^{n-1} \geq \Big((\beta_1)^{n-1} \cdot h\Big) \cdot \ldots \cdot \Big((\beta_{n-1})^{n-1} \cdot h\Big). \tag{1.24}$$

In fact, it suffices by continuity to prove this when h and the β_i are rational ample classes. So we may assume that they are represented by very ample divisors $B_1, \ldots, B_{n-1}$ and H on X, and the issue is to verify the inequality

$$\Big(B_1 \cdot \ldots \cdot B_{n-1} \cdot H\Big)^{n-1} \geq \Big((B_1)^{n-1} \cdot H\Big) \cdot \ldots \cdot \Big((B_{n-1})^{n-1} \cdot H\Big).$$

We suppose that H is an irreducible projective scheme of dimension $n-1$, and that each B_i meets H properly. Denoting by $\overline{B}_i$ the restriction of B_i to H, the inequality in question is equivalent term by term to the relation

$$\Big(\overline{B}_1 \cdot \ldots \cdot \overline{B}_{n-1}\Big)^{n-1} \geq \Big((\overline{B}_1)^{n-1}\Big) \cdot \ldots \cdot \Big((\overline{B}_{n-1})^{n-1}\Big)$$

of intersection numbers on H. But this follows by applying the induction hypothesis to H.

We now show that the desired inequality (1.23) follows formally from (1.24). So let

$$\delta_1, \ldots, \delta_n \in N^1(X)_{\mathbf{R}}$$

be n ample classes on X.[20] Fix some index $j \in [1, n]$ and apply (1.24) with $h = \delta_j$ and $\beta_1, \ldots, \beta_{n-1}$ the remaining δ_i. One finds

$$\Big(\delta_1 \cdot \ldots \cdot \delta_n\Big)^{n-1} \geq \prod_{i \neq j} \Big(\delta_i^{n-1} \cdot \delta_j\Big).$$

Taking the product over j yields

$$\Big(\delta_1 \cdot \ldots \cdot \delta_n\Big)^{n(n-1)} \geq \prod_j \prod_{i \neq j} \Big(\delta_i^{n-1} \cdot \delta_j\Big). \tag{1.25}$$

But now apply (1.24) with $h = \beta_1 = \ldots = \beta_{n-2} = \delta_i$ and $\beta_{n-1} = \delta_j$ to obtain

$$\Big(\delta_i^{n-1} \cdot \delta_j\Big)^{n-1} \geq \Big(\delta_i^n\Big)^{n-2} \Big(\delta_i \cdot \delta_j^{n-1}\Big).$$

Therefore

$$\begin{aligned} \prod_j \prod_{i \neq j} \Big(\delta_i^{n-1} \cdot \delta_j\Big)^{n-1} &\geq \prod_j \prod_{i \neq j} \Big(\delta_i^n\Big)^{n-2} \cdot \Big(\delta_i \cdot \delta_j^{n-1}\Big) \\ &= \left(\prod_i \Big(\delta_i^n\Big)^{(n-1)(n-2)}\right) \left(\prod_j \prod_{i \neq j} \Big(\delta_j \cdot \delta_i^{n-1}\Big)\right). \end{aligned}$$

[20] The amplitude of the δ_i guarantees that all of the intersection numbers appearing in the computations that follow are positive. Therefore we can cancel and take roots without further thought.

The second term on the right cancels against the left-hand side, and taking $(n-2)^{\text{nd}}$ roots one arrives at

$$\prod_j \prod_{i \neq j} \left(\delta_i^{n-1} \cdot \delta_j\right) \geq \prod_i \left(\delta_i^n\right)^{(n-1)}.$$

The inequality (1.23) follows by plugging this into (1.25) and taking $(n-1)^{\text{st}}$ roots. □

We record some variants and special cases. To begin with, one has the following:

Variant 1.6.2. Let X be an irreducible complete variety or scheme of dimension n, and fix an integer $0 \leq p \leq n$. Let

$$\alpha_1, \ldots, \alpha_p, \ \beta_1, \ldots, \beta_{n-p} \ \in \ N_1(X)_{\mathbf{R}}$$

be nef classes. Then

$$\left(\alpha_1 \cdot \ldots \cdot \alpha_p \cdot \beta_1 \cdot \ldots \cdot \beta_{n-p}\right)^p \geq \left(\alpha_1^p \cdot \beta_1 \cdot \ldots \cdot \beta_{n-p}\right) \cdot \ldots \cdot \left(\alpha_p^p \cdot \beta_1 \cdot \ldots \cdot \beta_{n-p}\right). \quad (1.26)$$

Indication of Proof. Following the argument leading to the inequality (1.24) in the proof of the theorem, one applies 1.6.1 to the complete intersection $B_1 \cap \ldots \cap B_{n-p}$ of suitable very ample divisors on X. □

The case of two classes is particularly useful:

Corollary 1.6.3. (Inequalities for two classes). *Let X be an irreducible complete variety or scheme of dimension n, and let $\alpha, \ \beta \in N_1(X)_{\mathbf{R}}$ be nef classes on X. Then the following inequalities are satisfied:*

(i). *For any integers $0 \leq q \leq p \leq n$,*

$$\left(\alpha^q \cdot \beta^{n-q}\right)^p \geq \left(\alpha^p \cdot \beta^{n-p}\right)^q \cdot \left(\beta^n\right)^{p-q}. \quad (1.27)$$

(ii). *For any $0 \leq i \leq n$,*

$$\left(\alpha^i \cdot \beta^{n-i}\right)^n \geq \left(\alpha^n\right)^i \cdot \left(\beta^n\right)^{n-i}. \quad (1.28)$$

(iii).

$$\left(\left(\alpha + \beta\right)^n\right)^{1/n} \geq \left(\left(\alpha^n\right)\right)^{1/n} + \left(\left(\beta^n\right)\right)^{1/n}. \quad (1.29)$$

Proof. For (i), take $\alpha_1 = \ldots = \alpha_q = \alpha$ and $\alpha_{q+1} = \ldots = \alpha_p = \beta_1 = \ldots = \beta_{n-p} = \beta$ in Variant 1.6.2. Statement (ii) is the special case $q = i$ and $p = n$ of (i). For (iii) expand out $(\alpha+\beta)^n$, apply (ii), and take n^{th} roots. □

Example 1.6.4. (Inequalities of Khovanskii and Teissier). In the situation of 1.6.3, put $s_i = (\alpha^i \cdot \beta^{n-i})$. Then for all $1 \le i \le n-1$,

$$s_i^2 \ \ge \ s_{i-1} \cdot s_{i+1}.$$

In other words, the function $i \mapsto \log s_i$ is concave. (Apply 1.6.2 with $p = 2$, $\alpha_1 = \alpha$, $\alpha_2 = \beta$, $\beta_1 = \ldots = \beta_{i-1} = \alpha$, and $\beta_i = \ldots = \beta_{n-p} = \beta$.) See [363, VI.2.15.8] for some applications due to Matsusaka. The papers [253] and [489] of Gromov and Okounkov have some interesting ideas on concavity statements of this sort. Okounkov also establishes an inequality [489, (3.6)] closely related to (1.29). □

Remark 1.6.5. (Positive characteristics). The material in this section remains valid for varieties defined over an algebraically closed field of arbitrary characteristic. □

1.6.B Mixed Multiplicities

We now sketch some local analogues originating with Teissier. Let X be a variety or scheme of pure dimension n, and let $x \in X$ be a (closed) point with maximal ideal $\mathfrak{m} \subseteq \mathcal{O}_X$. We suppose given $k \le n$ ideal sheaves

$$\mathfrak{a}_1, \ldots, \mathfrak{a}_k \ \subseteq \ \mathcal{O}_X \quad \text{with} \quad \mathrm{Supp}(\mathcal{O}_X/\mathfrak{a}_i) = \{x\},$$

i.e. we assume that the $\mathfrak{a}_i$ are $\mathfrak{m}$-primary. Fix also non-negative integers $d_1, \ldots, d_k$ with $\sum d_i = n$. Then one can define the *mixed multiplicity*

$$e\big(\mathfrak{a}_1^{[d_1]}\,;\, \mathfrak{a}_2^{[d_2]}\,;\, \ldots\,;\, \mathfrak{a}_k^{[d_k]}\big) \in \mathbf{N}$$

of the $\mathfrak{a}_i$, for instance by the property that the lengths

$$\mathrm{length}_{\mathcal{O}_X}\big(\mathcal{O}_X/\mathfrak{a}_1^{t_1} \cdot \ldots \cdot \mathfrak{a}_k^{t_k}\big)$$

are given for $t_i \gg 0$ by a polynomial of the form

$$\sum_{d_1+\ldots+d_k=n} \frac{n!}{d_1! \cdot \ldots \cdot d_k!}\, e\big(\mathfrak{a}_1^{[d_1]}\,;\, \mathfrak{a}_2^{[d_2]}\,;\, \ldots\,;\, \mathfrak{a}_k^{[d_k]}\big) \cdot t_1^{d_1} \cdot \ldots \cdot t_k^{d_k} \\ + \Big(\text{ lower degree terms } \Big).$$

So for example $e(\mathfrak{a}^{[n]}) = e(\mathfrak{a})$ is the classic Samuel multiplicity of $\mathfrak{a}$ (viewed as an ideal in the local ring $\mathcal{O} = \mathcal{O}_{X,x}$). When X is affine, and one takes d_i "general" elements $g_{i,1}, \ldots, g_{i,d_i} \in \mathfrak{a}_i$,[21] then $e\big(\mathfrak{a}_1^{[d_1]}\,;\, \mathfrak{a}_2^{[d_2]}\,;\, \ldots\,;\, \mathfrak{a}_k^{[d_k]}\big)$ computes the intersection multiplicity at x of the corresponding divisors. More geometrically, let

[21] For instance, one can take general $\mathbf{C}$-linear combinations of a system of generators of the ideal in question: see Definition 9.2.27.

$$\mu : X' \longrightarrow X$$

be a proper birational map that dominates the blow-up of each $\mathfrak{a}_i$, so that $\mathfrak{a}_i \cdot \mathcal{O}_{X'} = \mathcal{O}_{X'}(-f_i)$ for some effective Cartier divisor f_i on X' contracting to x. Then

$$e\big(\mathfrak{a}_1^{[d_1]} \, ; \, \mathfrak{a}_2^{[d_2]} \, ; \, \ldots \, ; \, \mathfrak{a}_k^{[d_k]}\big) \;=\; (-1) \cdot \Big((-f_1)^{d_1} \cdot \ldots \cdot (-f_k)^{d_k} \Big).$$

Note that by allowing repetitions, it is enough to study the multiplicities $e(\mathfrak{a}_1 \, ; \, \ldots \, ; \, \mathfrak{a}_n)$ associated to n (possibly non-distinct) ideals $\mathfrak{a}_1, \ldots, \mathfrak{a}_n \subseteq \mathcal{O}_X$. We refer to [565], [566], [567], [514], [519], [349], and [350] for fuller accounts and further developments.

Example 1.6.6. (Samuel multiplicity of a product). Given $\mathfrak{m}$-primary ideals $\mathfrak{a}, \mathfrak{b} \subseteq \mathcal{O}_X$ as above, the Samuel multiplicity $e(\mathfrak{a}\mathfrak{b})$ of their product has the expression

$$e\big(\mathfrak{a}\mathfrak{b}\big) \;=\; \sum_{i=0}^{n} \binom{n}{i} \, e\big(\mathfrak{a}^{[i]} \, ; \, \mathfrak{b}^{[n-i]}\big)$$

in terms of the mixed multiplicities of $\mathfrak{a}$ and $\mathfrak{b}$. (This follows immediately from the definition.) □

Teissier [565], [566] and Rees–Sharp [514] proved some inequalities among these mixed multiplicities that one can view as local analogues of the global statements appearing in the previous subsection.

Theorem 1.6.7. (Inequalities for mixed multiplicities). *Fix $x \in X$ as above.*

(i). *For any $\mathfrak{m}$-primary ideals $\mathfrak{a}_1, \ldots, \mathfrak{a}_n \subseteq \mathcal{O}_X$, one has*

$$e(\mathfrak{a}_1 \, ; \, \ldots \, ; \, \mathfrak{a}_n)^n \;\le\; e(\mathfrak{a}_1) \cdot \ldots \cdot e(\mathfrak{a}_n).$$

(ii). *Given $\mathfrak{m}$-primary ideals $\mathfrak{a}, \mathfrak{b} \subseteq \mathcal{O}_X$, and any integers $0 \le q \le p \le n$,*

$$e\big(\mathfrak{a}^{[q]} \, ; \, \mathfrak{b}^{[n-q]}\big)^p \;\le\; e\big(\mathfrak{a}^{[p]} \, ; \, \mathfrak{b}^{[n-p]}\big)^q \cdot e\big(\mathfrak{b}\big)^{p-q}.$$

(iii). *With $\mathfrak{a}, \mathfrak{b} \subseteq \mathcal{O}_X$ as in* (ii), *and $0 \le i \le n$,*

$$e\big(\mathfrak{a}^{[i]} \, ; \, \mathfrak{b}^{[n-i]}\big)^n \;\le\; e\big(\mathfrak{a}\big)^i \cdot e\big(\mathfrak{b}\big)^{n-i}.$$

(iv). *In the situation of* (iii), *set $m_i = e\big(\mathfrak{a}^{[i]} \, ; \, \mathfrak{b}^{[n-i]}\big)$. Then for $1 \le i \le n-1$,*

$$m_i^2 \;\le\; m_{i-1} \cdot m_{i+1}.$$

Remark 1.6.8. (Reversal of inequalities). Note that compared to the global statements, the direction of the inequalities here is reversed. This may be explained by noting that the global results finally come down to the fact that the intersection form has signature $(1,-1)$ on a two-dimensional space of ample classes on a surface. One can view the local inequalities, on the other hand, as ultimately springing from the negativity of the intersection form on the space spanned by the exceptional curves in a birational map of surfaces. Of course, the classical Hodge index theorem is at work in both cases. □

Example 1.6.9. (An inequality of Teissier). Given $\mathfrak{m}$-primary ideals $\mathfrak{a}, \mathfrak{b} \subseteq \mathcal{O}_X$ as in the theorem, the Samuel multiplicities of $\mathfrak{a}$, $\mathfrak{b}$, and $\mathfrak{ab}$ satisfy the inequality

$$e\big(\mathfrak{ab}\big)^{1/n} \ \le \ e\big(\mathfrak{a}\big)^{1/n} + e\big(\mathfrak{b}\big)^{1/n}. \quad \square$$

Example 1.6.10. Teissier gives the following nice example in [566, Example A]. Consider polynomials $f_1, \ldots, f_n \in \mathbf{C}[z_1, \ldots, z_n]$ defining a finite map $f : \mathbf{C}^n \longrightarrow \mathbf{C}^n$, and let $g_1, \ldots, g_{n-1}$ be general elements in the ideal generated by the f_i. Assume that $f_1, \ldots, f_n$ all vanish at the origin, so that $f(0) = 0$. The g_i cut out a curve $\Gamma \subseteq \mathbf{C}^n$ which is typically singular at 0. But the multiplicity of Γ at 0 is bounded by the degree of f:

$$\operatorname{mult}_0(\Gamma) \ \le \ \big(\deg f\big)^{1-\frac{1}{n}}.$$

(Take $\mathfrak{a} = (z_1, \ldots, z_n)$, $\mathfrak{b} = (f_1, \ldots, f_n)$, and $i = 1$ in (iii).) One can replace the polynomials in question by convergent power series. □

Remark 1.6.11. It will follow from the proof that the inequalities in the theorem hold for $\mathfrak{m}$-primary ideals $\mathfrak{a}_1, \ldots, \mathfrak{a}_n \subseteq \mathcal{O}$ in any Noetherian n-dimensional local ring $(\mathcal{O}, \mathfrak{m})$ having infinite residue field. □

Sketch of Proof of Theorem 1.6.7. We focus on (i). When $n = 2$ the assertion is that $e(\mathfrak{a};\mathfrak{b})^2 \le e(\mathfrak{a}) \cdot e(\mathfrak{b})$: this was established by Teissier [565] for normal surfaces and by Rees and Sharp [514], Theorem 2.2, for $\mathfrak{m}$-primary ideals in any local Noetherian ring $(\mathcal{O}, \mathfrak{m})$ of dimension two.[22] As in the cited papers, we proceed by induction on $n = \dim X$. Specifically, Teissier ([565], p. 306) shows[23] that if $f \in \mathfrak{a}_n$ is sufficiently general, and if one denotes by

$$Y \ = \ \operatorname{div}(f) \ \subseteq \ X$$

the hypersurface cut out by f, and by $\overline{\mathfrak{a}}_i \subseteq \mathcal{O}_Y$ the ideals determined by the $\mathfrak{a}_i$, then

$$e_X\big(\mathfrak{a}_1\,;\,\ldots\,;\,\mathfrak{a}_n\big) \ = \ e_Y\big(\overline{\mathfrak{a}}_1\,;\,\ldots\,;\,\overline{\mathfrak{a}}_{n-1}\big), \qquad (*)$$

[22] In the geometric setting, the idea roughly speaking is to resolve the singularities of the blow-up of $\mathfrak{ab}$, and to use the fact that the intersection form on the exceptional fibre is negative.

[23] Teissier proves this assuming only that the residue field of $(\mathcal{O}, \mathfrak{m})$ is infinite.

where we are using subscripts to indicate the scheme on which the multiplicities are computed. By induction, one finds that

$$\begin{aligned} e_X\big(\mathfrak{a}_1\,;\,\dots\,;\,\mathfrak{a}_n\big)^{n-1} &= e_Y\big(\overline{\mathfrak{a}}_1\,;\,\dots\,;\,\overline{\mathfrak{a}}_{n-1}\big)^{n-1} \\ &\leq e_Y\big(\overline{\mathfrak{a}}_1\big)\cdot\ldots\cdot e_Y\big(\overline{\mathfrak{a}}_{n-1}\big) \\ &= e_X\big(\mathfrak{a}_1^{[n-1]}\,;\,\mathfrak{a}_n\big)\cdot\ldots\cdot e_X\big(\mathfrak{a}_{n-1}^{[n-1]}\,;\,\mathfrak{a}_n\big), \end{aligned}$$

where in the last line we have used (*) again to express the multiplicity of $\overline{\mathfrak{a}}_i$ on Y as a mixed multiplicity on X. The resulting inequality

$$e\big(\mathfrak{a}_1\,;\,\dots\,;\,\mathfrak{a}_n\big)^{n-1} \leq e\big(\mathfrak{a}_1^{[n-1]}\,;\,\mathfrak{a}_n\big)\cdot\ldots\cdot e\big(\mathfrak{a}_{n-1}^{[n-1]}\,;\,\mathfrak{a}_n\big)$$

of mixed multiplicities on X is the analogue of equation (1.24) in the proof of Theorem 1.6.1, and as in that argument the inequality appearing in (i) now follows formally. □

1.7 Amplitude for a Mapping

In this section we outline the basic facts concerning amplitude relative to a mapping. We follow the conventions of Grothendieck in [255], which differ slightly from those adopted by Hartshorne in [280].

By way of preparation, consider a proper mapping $f : X \longrightarrow T$ of schemes, and a coherent sheaf $\mathcal{F}$ on X. Then $f_*\mathcal{F}$ is a coherent sheaf on T, and so one can form the T-scheme

$$\mathbf{P}(\mathcal{F}) =_{\text{def}} \operatorname{Proj}_{\mathcal{O}_T}\big(\operatorname{Sym}(f_*\mathcal{F})\big) \longrightarrow T$$

whose fibre over a given point $t \in T$ is the projective space of one-dimensional quotients of the fibre $f_*(\mathcal{F}) \otimes \mathbf{C}(t)$. This is the analogue in the relative setting of the projective space of sections of a sheaf on a fixed complete variety. Moreover, there is a natural mapping $f^*f_*\mathcal{F} \longrightarrow \mathcal{F}$ whose surjectivity is the analogue of the global generation of $\mathcal{F}$ in the absolute situation.

These remarks motivate

Definition 1.7.1. (Amplitude for a map). Let $f : X \longrightarrow T$ be a proper mapping of algebraic varieties or schemes, and let L be a line bundle on X.

(i). L is *very ample relative to* f, or f-*very ample*, if the canonical map

$$\rho : f^*f_*L \longrightarrow L$$

is surjective and defines an embedding

$$\begin{array}{ccc} X & \xhookrightarrow{\quad j \quad} & \mathbf{P}(f_*L) \\ & {\scriptstyle f}\searrow \quad \swarrow & \\ & T & \end{array}$$

of schemes over T.

(ii). L is *ample relative to* f, or *f-ample*, if $L^{\otimes m}$ is f-very ample for some $m > 0$.

A Cartier divisor D on X is very ample for f if the corresponding line bundle is so, and f-amplitude for Cartier $\mathbf{Q}$-divisors is defined by clearing denominators. □

Example 1.7.2. If E is a vector bundle on a scheme T, then the Serre line bundle $\mathcal{O}_{\mathbf{P}(E)}(1)$ on $X = \mathbf{P}(E)$ is ample for the natural mapping $\pi : \mathbf{P}(E) \longrightarrow T$. □

We start by recording several useful observations.

Remark 1.7.3. (Amplitude is local on the base). Observe that both properties in 1.7.1 are local on T. In other words, either condition holds for $f : X \longrightarrow T$ if and only it holds for each of the restrictions $f_i : X_i = f^{-1}(U_i) \longrightarrow U_i$ of f to the inverse images of the members of an open covering $\{U_i\}$ of T. □

Remark 1.7.4. (Equivalent condition for f-very ample). The condition in Definition 1.7.1 (i) is equivalent to the existence of a coherent sheaf $\mathcal{F}$ on T, plus an embedding $i : X \hookrightarrow \mathbf{P}(\mathcal{F})$ over T, such that $L = \mathcal{O}_{\mathbf{P}(\mathcal{F})}(1) \,|\, X$. In fact, such an embedding gives rise to a surjection $\tau : f^*\mathcal{F} \longrightarrow L$, which determines a homomorphism $\sigma : \mathcal{F} \longrightarrow f_*L$ together with a factorization

$$\begin{array}{ccc} f^*\mathcal{F} & \xrightarrow{\quad \tau \quad} & L \\ & {\scriptstyle f^*\sigma}\searrow \quad \nearrow{\scriptstyle \rho} & \\ & f^*f_*L & \end{array}$$

of τ. It follows in the first place that ρ is surjective. Moreover, the given embedding i is realized as the composition of the morphism $j : X \longrightarrow \mathbf{P}(f_*L)$ arising from ρ with a linear projection

$$\Big(\mathbf{P}(f_*L) - \mathbf{P}(\operatorname{coker} \sigma)\Big) \longrightarrow \mathbf{P}(\mathcal{F})$$

of T-schemes. This shows that j is an embedding. (See [255, II.4.4.4] for details.)

It follows in particular that if T is *affine* — so that f_*L is globally generated — then L is very ample for f if and only if there is an embedding $j : X \hookrightarrow \mathbf{P}^N \times T$ such that $L = j^*\mathcal{O}_{\mathbf{P}^N \times T}(1)$. This is taken as the definition of very ample relative to a mapping in [280]. □

Remark 1.7.5. The condition in 1.7.1 (ii) does not appear as the definition of f-amplitude in [255], but it is equivalent to that definition by virtue of [255, II.4.6.11]. □

An analogue of Theorem 1.2.6 holds in the present setting:

Theorem 1.7.6. *Let $f : X \longrightarrow T$ be a proper morphism of schemes, and L a line bundle on X. Then the following are equivalent:*

(i). *L is ample for f.*

(ii). *Given any coherent sheaf $\mathcal{F}$ on X, there exists a positive integer $m_1 = m_1(\mathcal{F})$ such that*

$$R^i f_*\big(\mathcal{F} \otimes L^{\otimes m}\big) = 0 \qquad \textit{for all } \; i > 0 \; , \; m \geq m_1(\mathcal{F}).$$

(iii). *Given any coherent sheaf $\mathcal{F}$ on X, there is a positive integer $m_2 = m_2(\mathcal{F})$ such that the canonical mapping*

$$f^* f_*\big(\mathcal{F} \otimes L^{\otimes m}\big) \longrightarrow \mathcal{F} \otimes L^{\otimes m}$$

is surjective whenever $m \geq m_2$.

(iv). *There is a positive integer $m_3 > 0$ such that $L^{\otimes m}$ is f-very ample for every $m \geq m_3$.*

References for Proof. As in the proof of Theorem 1.2.6, for (i) $\Longrightarrow$ (ii) one reduces first to the case in which L is very ample for f. The assertion being local on T, one can suppose that T is affine, and then [280, III.5.2] applies. Note that if T is affine, then the condition in (iii) is equivalent to asking that $\mathcal{F} \otimes L^{\otimes m}$ be globally generated. This being said, (ii) $\Leftrightarrow$ (iii) follows (at least when f is projective) from [280, III.5.3], and under the same hypothesis (iii) $\Rightarrow$ (iv) is a consequence of [280, II.7.6]. For an arbitrary proper mapping f, and the remaining implications, consult [255, II.4.6.8, II.4.6.11] and [256, III.2.6]. □

Example 1.7.7. Suppose given a diagram

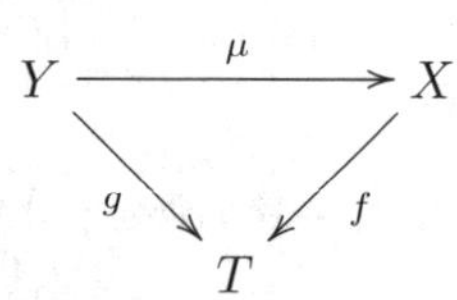

of schemes over T, with μ finite. If L is an f-ample line bundle on X, then $\mu^* L$ is ample relative to g. □

The following useful result allows one in practice to reduce to the absolute setting:

Theorem 1.7.8. (Fibre-wise amplitude). *Let $f : X \longrightarrow T$ be a proper morphism of schemes, let L be a line bundle on X, and for $t \in T$ set*

$$X_t = f^{-1}(t) \quad , \quad L_t = L \mid X_t.$$

Then L is ample for f if and only if L_t is ample on X_t for every $t \in T$.

Proof. If L is ample for f, then the previous Example 1.7.7 shows that each of the restrictions L_t is ample on X_t. Conversely, suppose that L_t is ample for every $t \in T$. As noted in Remark 1.2.18, the proof of Theorem 1.2.17 shows that every point $t \in T$ has a neighborhood U with the property that $L|f^{-1}(U) = \phi^*\big(\mathcal{O}_{\mathbf{P}\times T}(1)\big)$ for a finite mapping $X_U = f^{-1}(U) \longrightarrow \mathbf{P}^r \times U$ of schemes over U. Recalling that f-amplitude is local on T, it follows from Example 1.7.7 that L is indeed ample with respect to f. □

Corollary 1.7.9. (Nakai's criterion for a mapping). *In the setting of the previous theorem, a $\mathbf{Q}$-divisor D on X is ample with respect to f if and only if $\big(D^{\dim V} \cdot V\big) > 0$ for every irreducible subvariety $V \subset X$ of positive dimension that maps to a point in T.* □

The next result summarizes the connection between relative and global amplitude.

Proposition 1.7.10. *Consider a morphism*

$$f : X \longrightarrow T$$

of projective schemes. Let L be a line bundle on X, and let A be an ample line bundle on T. Then L is f-ample if and only if $L \otimes f^*\big(A^{\otimes m}\big)$ is an ample line bundle on X for all $m \gg 0$.

Proof. Assume that L is f-ample. Replacing L by a high power, we can suppose that $f^*f_*L \longrightarrow L$ is surjective. Since A is ample, $f_*(L) \otimes A^{\otimes p}$ is globally generated if p is sufficiently large. Therefore its pullback $f^*f_*(L) \otimes f^*A^{\otimes p}$ is likewise generated by its global sections, and choosing generators gives rise to a morphism

$$\phi : X \longrightarrow \mathbf{P} \times T$$

of schemes over T with the property that $L \otimes f^*(A^{\otimes p}) = \phi^*\mathrm{pr}_1^*\mathcal{O}_{\mathbf{P}}(1)$. Moreover, ϕ is finite since it is evidently so on each fibre of f. Therefore

$$L \otimes f^*(A^{\otimes p+1}) = \phi^*\big(\mathrm{pr}_1^*\mathcal{O}_{\mathbf{P}}(1) \otimes \mathrm{pr}_2^*A\big)$$

is the pullback of an ample line bundle under a finite map, and consequently is ample. The converse follows from 1.7.8. □

Finally, we say a few words about nefness for a mapping. Here one takes the analogue of 1.7.8 as the definition:

Definition 1.7.11. (Nefness relative to a mapping). Given a proper morphism $f : X \longrightarrow T$ as above, a line bundle L on X is *nef relative to* f if the restriction $L_t = L \,|\, X_t$ of L to each fibre is nef, or equivalently if $\big(c_1(L) \cdot C\big) \geq 0$ for every curve $C \subseteq X$ mapping to a point under T. □

Note that the analogue of Proposition 1.7.10 need not be true.

Example 1.7.12. The following example is taken from [368, Example 1.46]. Let $X = E \times E$ be the product of an elliptic curve with itself and $f : X \longrightarrow E$ the first projection. Let $D = E \times \{\text{point}\}$ and denote by $\Delta \subseteq E \times E$ the diagonal. Then $D - \Delta$ is f-nef. However, for any divisor A on E the divisor

$$D - \Delta + f^*(A)$$

has self-intersection -2, and hence cannot be nef. □

Remark 1.7.13. (Fujita vanishing for a mapping). In his paper [331], Keeler extends Fujita's vanishing theorem 1.4.35 to the relative setting. □

Remark 1.7.14. (Other ground fields). Everything in this section goes through to varieties defined over an algebraically closed field of arbitrary characteristic. □

1.8 Castelnuovo–Mumford Regularity

The Cartan–Serre–Grothendieck theorems imply that all the cohomological subtleties that may be associated to a coherent sheaf $\mathcal{F}$ on a projective space $\mathbf{P}$ disappear after twisting by a sufficiently high multiple of the hyperplane line bundle. Specifically, for $m \gg 0$:

- the higher cohomology groups of $\mathcal{F}(m)$ vanish;
- $\mathcal{F}(m)$ is generated by its global sections;
- the maps $H^0(\mathbf{P}, \mathcal{F}(m)) \otimes H^0(\mathbf{P}, \mathcal{O}_{\mathbf{P}}(k)) \longrightarrow H^0(\mathbf{P}, \mathcal{F}(m+k))$ are surjective for every $k > 0$.

Castelnuovo–Mumford regularity gives a quantitative measure of how much one has to twist in order that these properties take effect. It then governs the algebraic complexity of a coherent sheaf, and for this reason has been the focus of considerable recent activity. As we shall see, regularity is also well adapted to arguments involving vanishing theorems.

In the first subsection, we give the definition and basic properties, and describe some variants. The complexity-theoretic meaning is indicated in the second, and in the third we survey without proof several results giving bounds on regularity. Section 1.8.D gives a brief overview — also without proof — of a circle of ideas surrounding syzygies of algebraic varieties.

1.8.A Definitions, Formal Properties, and Variants

Fix a complex vector space V of dimension $r+1$, and denote by $\mathbf{P} = \mathbf{P}(V)$ the corresponding r-dimensional projective space. We start with the definition of m-regularity of a coherent sheaf on $\mathbf{P}$.

Definition 1.8.1. (Castelnuovo–Mumford regularity of a coherent sheaf). Let $\mathcal{F}$ be a coherent sheaf on the projective space $\mathbf{P}$, and let m be an integer. One says that $\mathcal{F}$ is *m-regular in the sense of Castelnuovo–Mumford* if

$$H^i\big(\mathbf{P}, \mathcal{F}(m-i)\big) \;=\; 0 \quad \text{for all } i > 0.$$

As long as the context is clear, we usually speak simply of an m-regular sheaf.

Example 1.8.2. (i). The line bundle $\mathcal{O}_{\mathbf{P}}(a)$ is $(-a)$-regular.

(ii). The ideal sheaf $\mathcal{I}_L \subseteq \mathcal{O}_{\mathbf{P}}$ of a linear subspace $L \subseteq \mathbf{P}$ is 1-regular.

(iii). If $X \subseteq \mathbf{P}$ is a hypersurface of degree d, then its structure sheaf $\mathcal{O}_X$ — viewed via extension by zero as a coherent sheaf on $\mathbf{P}$ — is $(d-1)$-regular. □

While the formal definition may seem rather non intuitive, a result of Mumford gives a first indication of the fact that Castelnuovo–Mumford regularity measures the point at which cohomological complexities vanish.

Theorem 1.8.3. (Mumford's theorem, I). *Let $\mathcal{F}$ be an m-regular sheaf on $\mathbf{P}$. Then for every $k \geq 0$:*

(i). *$\mathcal{F}(m+k)$ is generated by its global sections.*

(ii). *The natural maps*

$$H^0(\mathbf{P}, \mathcal{F}(m)) \otimes H^0(\mathbf{P}, \mathcal{O}_{\mathbf{P}}(k)) \longrightarrow H^0(\mathbf{P}, \mathcal{F}(m+k))$$

are surjective.

(iii). *$\mathcal{F}$ is $(m+k)$-regular.*

Proof. Since $\mathcal{F}(m+\ell)$ is in any event globally generated for $\ell \gg 0$ (Theorem 1.2.6), the surjectivities in (ii) imply that $\mathcal{F}(m)$ itself must already be generated by its global sections. The same is then true of $\mathcal{F}(m+k)$ whenever $k \geq 0$. Hence we need only prove (ii) and (iii), and thanks to (iii) it suffices to treat the case $k = 1$.

For this we consider the canonical Koszul complex of bundles on $\mathbf{P} = \mathbf{P}(V)$ (see Appendix B.2). Denote by $V_{\mathbf{P}} = V \otimes_{\mathbf{C}} \mathcal{O}_{\mathbf{P}}$ the trivial vector bundle on $\mathbf{P}$ with fibre V. Starting with the surjective bundle map

$$V_{\mathbf{P}}(-1) = V \otimes \mathcal{O}_{\mathbf{P}}(-1) \longrightarrow \mathcal{O}_{\mathbf{P}},$$

form the exact sequence

$$0 \longrightarrow \Lambda^{r+1} V_{\mathbf{P}}(-r-1) \longrightarrow \ldots \longrightarrow \Lambda^2 V_{\mathbf{P}}(-2) \longrightarrow V_{\mathbf{P}}(-1) \longrightarrow \mathcal{O}_{\mathbf{P}} \longrightarrow 0. \qquad (K_\bullet)$$

Twisting through by $\mathcal{F}(m+1)$ yields an exact complex

$$\ldots \longrightarrow \Lambda^3 V_{\mathbf{P}} \otimes \mathcal{F}(m-2) \longrightarrow \Lambda^2 V_{\mathbf{P}} \otimes \mathcal{F}(m-1) \longrightarrow V_{\mathbf{P}} \otimes \mathcal{F}(m) \longrightarrow \mathcal{F}(m+1) \longrightarrow 0. \qquad (*)$$

The m-regularity of $\mathcal{F}$ implies that $H^i\big(\mathbf{P}, \Lambda^{i+1} V_{\mathbf{P}} \otimes \mathcal{F}(m-i)\big) = 0$, and it follows by chasing through (*) that the map

$$H^0\big(\mathbf{P}, \mathcal{F}(m)\big) \otimes H^0\big(\mathbf{P}, \mathcal{O}_{\mathbf{P}}(1)\big) = H^0\big(\mathbf{P}, V_{\mathbf{P}} \otimes \mathcal{F}(m)\big) \to H^0\big(\mathbf{P}, \mathcal{F}(m+1)\big)$$

is surjective. This proves (ii). For (iii) one twists $(K_\bullet)$ by $\mathcal{F}(m)$ and argues similarly. □

Theorem 1.8.3 is equivalent to a variant involving globally generated ample line bundles on any projective variety. Since we frequently use it in this form, we give the relevant definition and statement explicitly:

Definition 1.8.4. (Regularity with respect to a globally generated ample line bundle). Let X be a projective variety and B an ample line bundle on X that is generated by its global sections. A coherent sheaf $\mathcal{F}$ on X is *m-regular with respect to B* if

$$H^i\big(X, \mathcal{F} \otimes B^{\otimes(m-i)}\big) = 0 \quad \text{for } i > 0. \quad \square$$

Then 1.8.3 becomes:

Theorem 1.8.5. (Mumford's theorem, II). *Let $\mathcal{F}$ be an m-regular sheaf on X with respect to B. Then for every $k \geq 0$:*

(i). *$\mathcal{F} \otimes B^{\otimes(m+k)}$ is generated by its global sections.*

(ii). *The natural maps*

$$H^0\big(X, \mathcal{F} \otimes B^{\otimes m}\big) \otimes H^0\big(X, B^{\otimes k}\big) \longrightarrow H^0\big(X, \mathcal{F} \otimes B^{\otimes(m+k)}\big)$$

are surjective.

(iii). *$\mathcal{F}$ is $(m+k)$-regular with respect to B.*

Proof. One repeats the proof of Theorem 1.8.3 with $\mathcal{O}_{\mathbf{P}}(1)$ replaced by B and V replaced by $H^0\big(X, B\big)$. Alternatively, one can apply 1.8.3 to the direct image of $\mathcal{F}$ under the finite map $X \longrightarrow \mathbf{P}$ determined by B. □

For most purposes, dealing with sheaves on projective space involves little loss in generality. This is therefore the context in which we shall work — unless stated otherwise — for the remainder of this section.

A number of additional concrete examples are worked out in the next subsection. Here we continue by presenting some formal properties of regularity.

Example 1.8.6. (Extensions). An extension of two m-regular sheaves on the projective space $\mathbf{P}$ is itself m-regular. □

Example 1.8.7. Suppose that a coherent sheaf $\mathcal{F}$ on $\mathbf{P}$ is resolved by a long exact sequence

$$\ldots \longrightarrow \mathcal{F}_2 \longrightarrow \mathcal{F}_1 \longrightarrow \mathcal{F}_0 \longrightarrow \mathcal{F} \longrightarrow 0$$

of coherent sheaves on $\mathbf{P}$. If $\mathcal{F}_i$ is $(m+i)$-regular for every $i \geq 0$, then $\mathcal{F}$ is m-regular. Moreover, in this case the mapping $H^0\big(\mathbf{P}, \mathcal{F}_0(m)\big) \longrightarrow H^0\big(\mathbf{P}, \mathcal{F}(m)\big)$ is surjective. □

A partial converse of Example 1.8.7 provides a useful characterization of m-regularity.

Proposition 1.8.8. (Linear resolutions). *Let $\mathcal{F}$ be a coherent sheaf on the projective space $\mathbf{P}$. Then $\mathcal{F}$ is m-regular if and only if $\mathcal{F}$ is resolved by a long exact sequence*

$$\ldots \longrightarrow \oplus \mathcal{O}_{\mathbf{P}}(-m-2) \longrightarrow \oplus \mathcal{O}_{\mathbf{P}}(-m-1) \longrightarrow \oplus \mathcal{O}_{\mathbf{P}}(-m) \longrightarrow \mathcal{F} \longrightarrow 0 \tag{1.30}$$

whose terms are direct sums of the indicated line bundles.

Proof. Given a long exact sequence (1.30), the m-regularity of $\mathcal{F}$ follows from Example 1.8.7 (or equivalently by reading off the vanishings $H^i\big(\mathbf{P}, \mathcal{F}(m-i)\big) = 0$ for $i > 0$). Conversely, supposing that $\mathcal{F}$ is m-regular, we construct the resolution (1.30) step by step. To this end, recall first from Proposition 1.8.3 (i) that $\mathcal{F}(m)$ is globally generated. Setting $W = H^0\big(\mathbf{P}, \mathcal{F}(m)\big)$, one therefore has a surjective sheaf homomorphism $e : W_{\mathbf{P}}\mathcal{O}_{\mathbf{P}}(-m) \longrightarrow \mathcal{F}$. Let $\mathcal{F}_1 = \ker e$:

$$0 \longrightarrow \mathcal{F}_1 \longrightarrow W_{\mathbf{P}} \otimes \mathcal{O}_{\mathbf{P}}(-m) \longrightarrow \mathcal{F} \longrightarrow 0. \tag{*}$$

We will show momentarily that $\mathcal{F}_1$ is $(m+1)$-regular. Granting this, set $W' = H^0\big(\mathbf{P}, \mathcal{F}_1(m+1)\big)$ and map the corresponding trivial bundle $W'_{\mathbf{P}}$ to $\mathcal{F}_1(m+1)$ to construct an exact sequence

$$W'_{\mathbf{P}} \otimes \mathcal{O}_{\mathbf{P}}(-m-1) \longrightarrow W_{\mathbf{P}} \otimes \mathcal{O}_{\mathbf{P}}(-m) \longrightarrow \mathcal{F} \longrightarrow 0.$$

Then continue until one arrives finally at (1.30). As for the $(m+1)$-regularity of $\mathcal{F}_1$, it is almost immediate. In fact, by construction the homomorphism $H^0\big(\mathbf{P}, W_{\mathbf{P}}\big) \longrightarrow H^0\big(\mathbf{P}, \mathcal{F}(m)\big)$ determined by (*) is surjective, and therefore $H^1\big(\mathbf{P}, \mathcal{F}_1(m)\big) = 0$. On the other hand, it follows from (*) and the m-regularity of $\mathcal{F}$ that

$$\begin{aligned} H^{i+1}\left(\mathbf{P}, \mathcal{F}_1\big((m+1)-(i+1)\big)\right) &= H^{i+1}\left(\mathbf{P}, \mathcal{F}_1(m-i)\right) \\ &= H^{i}\left(\mathbf{P}, \mathcal{F}(m-i)\right) \\ &= 0 \end{aligned}$$

for $i > 0$. Therefore $\mathcal{F}_1$ is $(m+1)$-regular, as required. □

We next use Proposition 1.8.8 to show that at least for vector bundles, regularity has pleasant tensorial properties.

Proposition 1.8.9. (Regularity of tensor products). *Let $\mathcal{F}$ be a coherent sheaf on* $\mathbf{P}$*, and let E be a locally free sheaf on* $\mathbf{P}$*. If $\mathcal{F}$ is m-regular and E is ℓ-regular, then $E \otimes \mathcal{F}$ is $(\ell + m)$-regular.*

Corollary 1.8.10. (Wedge and symmetric products). *If E is an m-regular locally free sheaf, then the p-fold tensor power T^pE is (pm)-regular. In particular, Λ^pE and S^pE are likewise (pm)-regular.* □

Proof of Proposition 1.8.9. Starting with the resolution (1.30) of $\mathcal{F}$ appearing in Proposition 1.8.8, tensor through by E to obtain a complex

$$\ldots \longrightarrow \oplus E(-m-2) \longrightarrow \oplus E(-m-1) \longrightarrow \oplus E(-m) \longrightarrow E \otimes \mathcal{F} \longrightarrow 0.$$

Since tensoring by a locally free sheaf preserves exactness, this sequence is in fact exact. But $E(-m-i)$ is $(\ell + m + i)$-regular thanks to the ℓ-regularity of E, and so the $(\ell + m)$-regularity of $E \otimes \mathcal{F}$ follows from Example 1.8.7. □

Remark 1.8.11. This proof shows that it suffices for Proposition 1.8.9 to assume that at every point of $\mathbf{P}$ either E or $\mathcal{F}$ is locally free. □

Example 1.8.12. Chardin informs us that he has found examples for which Proposition 1.8.9 fails when the sheaves in question are not locally free along a set of dimension ≥ 2. □

Example 1.8.13. In the situation of Proposition 1.8.9, the natural map

$$H^0\big(\mathbf{P}, \mathcal{F}(m)\big) \otimes H^0\big(\mathbf{P}, E(\ell)\big) \longrightarrow H^0\big(\mathbf{P}, \mathcal{F} \otimes E(m + \ell)\big)$$

is surjective. □

Remark 1.8.14. (Regularity in positive characteristics). Everything we have said so far except Corollary 1.8.10 goes through without change for varieties defined over an algebraically closed field of arbitrary characteristic. (In positive characteristics, the symmetric and alternating products appearing in 1.8.10 may no longer be direct summands of the tensor product.) □

Example 1.8.15. (Green's theorem). Let $W \subseteq H^0\big(\mathbf{P}, \mathcal{O}_{\mathbf{P}}(d)\big)$ be a subspace of codimension c giving a free linear series. Then the map

$$s_k : W \otimes H^0\big(\mathbf{P}, \mathcal{O}_{\mathbf{P}}(k)\big) \longrightarrow H^0\big(\mathbf{P}, \mathcal{O}_{\mathbf{P}}(d + k)\big)$$

determined by multiplication of polynomials is surjective for $k \geq c$. In other words, any homogeneous polynomial of degree $\geq d+c$ lies in the ideal spanned by W. (Let M_d be the vector bundle on $\mathbf{P} = \mathbf{P}(V)$ arising as the kernel of the evaluation map on forms of degree d:

$$0 \longrightarrow M_d \longrightarrow S^dV \otimes \mathcal{O}_{\mathbf{P}} \longrightarrow \mathcal{O}_{\mathbf{P}}(d) \longrightarrow 0.$$

Then M_d is 1-regular, and so $\Lambda^k M_d$ is k-regular thanks to 1.8.10. On the other hand, W determines an analogous bundle M_W,

$$0 \longrightarrow M_W \longrightarrow W \otimes \mathcal{O}_{\mathbf{P}} \longrightarrow \mathcal{O}_{\mathbf{P}}(d) \longrightarrow 0,$$

and the two bundles in question sit in a sequence $0 \longrightarrow M_W \longrightarrow M_d \longrightarrow \mathcal{O}_{\mathbf{P}}^c \longrightarrow 0$. One then uses the Eagon–Northcott complex (EN_1) from Appendix B to show that M_W is $(c+1)$-regular, which implies that s_k is surjective as soon as $k \geq c$.) This result and its generalizations have interesting applications to infinitesimal computations in Hodge theory: see [236], [237], or [83, Chapter 7]. □

Remark 1.8.16. (Warning on generalized regularity). Owing to the fact that a globally generated ample line bundle B on a projective variety X need not itself be (-1)-regular, one should not expect Propositions 1.8.8 and 1.8.9 or Corollary 1.8.10 to extend to the setting of Definition 1.8.4. For example, $\mathcal{F} = \mathcal{O}_X$ has a (trivial) linear resolution, but it is not in general 0-regular. However, Arapura [14, Corollary 3.2] observes that if $R = \max\{1, \mathrm{reg}_B(\mathcal{O}_X)\}$, then $\mathcal{F} \otimes B^{\otimes m}$ has an "R-linear" resolution. □

It is often useful to know the best possible regularity of a sheaf:

Definition 1.8.17. (Regularity of a sheaf). The *Castelnuovo–Mumford regularity* $\mathrm{reg}(\mathcal{F})$ of a coherent sheaf $\mathcal{F}$ on $\mathbf{P}$ is the least integer m for which $\mathcal{F}$ is m-regular (or $-\infty$ if $\mathcal{F}$ is supported on a finite set, and hence m-regular for all $m \ll 0$). □

We conclude this subsection by discussing some variants.

Example 1.8.18. (Regularity with respect to a vector bundle). Let X be an irreducible projective variety of dimension n, and let U be a vector bundle on X having the property that for every point $x \in X$, there is a section of U whose zero locus is a finite set containing x. (Example: $U = B \oplus \cdots \oplus B$ (n times), where B is a globally generated ample line bundle on X.) If $\mathcal{F}$ is a coherent sheaf on X such that

$$H^i\big(X, \Lambda^i U^* \otimes \mathcal{F}\big) \ = \ 0 \quad \text{for } \ i > 0,$$

then $\mathcal{F}$ is globally generated. (Let $s \in \Gamma(X, U)$ be a section vanishing on a finite subscheme $Z \subset X$. Form the Koszul complex $K_\bullet$ determined by the resulting map $U^* \longrightarrow \mathcal{O}_X$ and suppose for the moment that $K_\bullet$ is exact (which will be the case e.g. if X is smooth and $\mathrm{rank}\, U = \dim X$) and that $\mathcal{F}$ is locally free. Tensoring through by $\mathcal{F}$, the given vanishings imply that the restriction map

$$H^0\big(X, \mathcal{F}\big) \longrightarrow H^0\big(X, \mathcal{F} \otimes \mathcal{O}_Z\big)$$

is surjective, and hence that $\mathcal{F}$ is globally generated at every point of Z. Thanks to Example B.1.3 in Appendix B, the same argument works even if $K_\bullet$ is not exact or $\mathcal{F}$ is not locally free, since in any event $K_\bullet \otimes \mathcal{F}$ is exact off the finite set Z.) □

Example 1.8.19. (Global generation on Grassmannians). Let $\mathbf{G} = \mathbf{G}(k, m)$ be the Grassmannian of k-dimensional quotients of an m-dimensional vector space, and denote by Q the tautological rank k quotient bundle on $\mathbf{G}$. Suppose that $\mathcal{F}$ is a coherent sheaf on $\mathbf{G}$ satisfying the condition that for every $i > 0$,

$$H^i\big(\mathbf{G}, \Lambda^{i_1}Q^* \otimes \cdots \otimes \Lambda^{i_{m-k}}Q^* \otimes \mathcal{F}\big) = 0 \quad \text{whenever} \quad i_1 + \cdots + i_{m-k} = i.$$

Then $\mathcal{F}$ is globally generated. (Given any point $x \in \mathbf{G}$, the $(m-k)$-fold direct sum $V = Q \oplus \cdots \oplus Q$ has a section vanishing precisely at x.) This result is due to M. Kim [335], who used it to study branched coverings of Grassmannians. (See Sections 6.3.D and 7.1.C.) □

Remark 1.8.20. (Regularity on Grassmannians). A general theory of regularity on Grassmannians has been developed and studied by Chipalkatti [90]. □

Remark 1.8.21. (Regularity on abelian varieties). Pareschi and Popa have introduced some very interesting notions of regularity on an abelian variety X with respect to an arbitrary ample line bundle L on X [496], [497]. For example, let $\mathcal{F}$ be a coherent sheaf on X, and assume that $\mathcal{F}$ satisfies the vanishing

$$H^i\big(X, \mathcal{F} \otimes L^{-1} \otimes P\big) \ = \ 0 \quad \text{for all } \ i > 0 \ \text{ and } \ P \in \operatorname{Pic}^0(X).$$

It is established in [496] that then $\mathcal{F}$ is globally generated. This generalizes a useful lemma of Kempf. Pareschi and Popa establish similar statements under weaker hypotheses, and use them to prove several striking results about the geometry of abelian varieties and their subvarieties, as well as the equations defining their projective embeddings. □

Just as regularity can be valuable for proving that a sheaf is globally generated, it is also useful for establishing that certain line bundles are very ample:

Example 1.8.22. (Criterion for very ample bundles). Let X be an irreducible projective variety of dimension n, and B an ample line bundle that is generated by its global sections. One has then the following

> **Proposition.** Let N be a line bundle on X that is 0-regular with respect to B in the sense of Definition 1.8.4. Then $N \otimes B$ is very ample.

(As $N \otimes B$ is free, it is enough to show that the sheaf $N \otimes B \otimes \mathfrak{m}_x$ is globally generated for any point $x \in X$, $\mathfrak{m}_x$ being the maximal ideal of x. As in Example 1.8.18 this in turn will follow if we show that given any $x \in X$, there is a finite scheme $Z = Z(x)$ containing x such that $N \otimes B \otimes \mathcal{I}_Z$ is 0-regular with respect to B, for this implies the 0-regularity of $N \otimes B \otimes \mathfrak{m}_x$. Having fixed x, take Z to be the zero-scheme cut out by n general sections of B that vanish at x. As in Example 1.8.18, form the resulting Koszul complex and twist by $B \otimes N$: one reads off the required 0-regularity of $B \otimes N \otimes \mathcal{I}_Z$ from the vanishings giving the 0-regularity of N.) □

The next example shows that Kodaira-type vanishing theorems are well suited to regularity statements.

Example 1.8.23. (Adjoint bundles). Let X be a smooth complex projective variety of dimension n and let D be an ample divisor on X. As usual denote by K_X a canonical divisor on X, and let P be an arbitrary nef divisor on X. There has been a great deal of interest recently in adjoint-type bundles of the form

$$L_k = \mathcal{O}_X(K_X + kD + P).$$

One can view these as the analogues of line bundles of large degree on curves (see Section 10.4.A). Under various hypotheses on D, divisors of this shape have also been extensively studied by Sommese and his school (see [50]), as well as by Fujita (see [197]).

Assume now that D is *free* (and ample). Then the divisor L_k above is free when $k \geq n+1$ and very ample when $k \geq n+2$. (The Kodaira vanishing theorem 4.2.1 asserts that if A is any ample divisor on X, then $H^i\big(X, \mathcal{O}_X(K_X + A)\big) = 0$ for $i > 0$. One uses this as input to Theorem 1.8.5 and Example 1.8.22.) These results for L_k are elementary special cases of a celebrated conjecture of Fujita, which asserts that the same statements should hold — at least when $P = 0$ — assuming only that D is ample. Fujita's conjecture is discussed in more detail in Section 10.4. See also Theorem 1.8.60. □

We conclude by outlining a relative notion of regularity.

Example 1.8.24. (Regularity with respect to a mapping). Let $f : X \longrightarrow Y$ be a proper surjective mapping of varieties (or schemes). We suppose given a line bundle A on X satisfying:

(a). A is ample for f;
(b). The canonical mapping $f^* f_* A \longrightarrow A$ is surjective.

For example, condition (b) holds if A is globally generated. If Y is normal and X is the normalized blowing-up of a sheaf of ideals $\mathfrak{a} \subseteq \mathcal{O}_Y$, with exceptional divisor $E \subseteq \mathcal{O}_X$, then (a) and (b) hold with $A = \mathcal{O}_X(-E)$. (For (b), use that $\mathfrak{a} \subseteq f_* \mathcal{O}_X(-E)$.)

Given a coherent sheaf $\mathcal{F}$ on X, we define $\mathcal{F}$ to be m-regular with respect to A and f if

$$R^i f_* \big(\mathcal{F} \otimes A^{\otimes(m-i)}\big) = 0 \quad \text{for } i > 0.$$

Then the natural analogue of Theorem 1.8.3 holds in this setting. Specifically, assume that $\mathcal{F}$ is m-regular with respect to A and f. Then for every $k \geq 0$:

(i). The homomorphism

$$f^*f_*\big(\mathcal{F}\otimes A^{\otimes(m+k)}\big) \longrightarrow \mathcal{F}\otimes A^{\otimes(m+k)}$$

is surjective;

(ii). The map

$$f_*\big(\mathcal{F}\otimes A^{\otimes m}\big)\otimes f_*\big(A^{\otimes k}\big) \longrightarrow f_*\big(\mathcal{F}\otimes A^{\otimes(m+k)}\big)$$

is surjective;

(iii). $\mathcal{F}$ is $(m+k)$-regular with respect to A and f.

(The statement being local on Y, one can assume that Y is affine. Then there exists a finite-dimensional vector space $V\subseteq H^0\big(Y,f_*A\big)$ of sections that generate f_*A, giving rise to a surjective bundle map $V_Y = V\otimes\mathcal{O}_Y \longrightarrow f_*A$. Pulling back and composing with $f^*f_*A\longrightarrow A$, one arrives at a surjective map $V_X \longrightarrow A$ of bundles on X, and now one argues as in the proof of Theorem 1.8.3. Namely, form the corresponding Koszul complex and twist by $\mathcal{F}\otimes A^{\otimes m}$:

$$\ldots\longrightarrow \Lambda^2V_X\otimes\mathcal{F}\otimes A^{\otimes(m-1)} \longrightarrow V_X\otimes\mathcal{F}\otimes A^{\otimes m}\longrightarrow \mathcal{F}\otimes A^{\otimes(m+1)}\longrightarrow 0.$$

Chasing through the complex and using the hypothesis of m-regularity, one finds that the map $V_Y\otimes f_*\big(\mathcal{F}\otimes A^{\otimes m}\big)\longrightarrow f_*\big(\mathcal{F}\otimes A^{\otimes(m+1)}\big)$ is surjective. But this factors through

$$f_*A\otimes f_*\big(\mathcal{F}\otimes A^{\otimes m}\big)\longrightarrow f_*\big(\mathcal{F}\otimes A^{\otimes(m+1)}\big),$$

and (ii) follows. The proof of (iii) is similar, while for (i) one uses (ii) plus the fact that $f^*f_*\big(\mathcal{F}\otimes A^{\otimes(m+k)}\big)\longrightarrow\mathcal{F}\otimes A^{\otimes(m+k)}$ is surjective for $k\gg 0$ by virtue of the f-amplitude of A.) □

Example 1.8.25. (Regularity on a projective bundle). Let X be a variety or scheme, and E a vector bundle on X, with projectivization $\pi : \mathbf{P}(E)\longrightarrow X$. A coherent sheaf $\mathcal{F}$ on $\mathbf{P}(E)$ is *m-regular with respect to π* if

$$R^i\pi_*\big(\mathcal{F}\otimes\mathcal{O}_{\mathbf{P}(E)}(m-i)\big) \;=\; 0$$

for $i>0$. If this condition holds, then:

(i). $\pi^*\pi_*\big(\mathcal{F}\otimes\mathcal{O}_{\mathbf{P}(E)}(m)\big)$ surjects onto $\mathcal{F}\otimes\mathcal{O}_{\mathbf{P}(E)}(m)$;

(ii). The mapping

$$\pi_*\big(\mathcal{F}\otimes\mathcal{O}_{\mathbf{P}(E)}(m)\big)\otimes\pi_*\mathcal{O}_{\mathbf{P}(E)}(k)\longrightarrow\pi_*\big(\mathcal{F}\otimes\mathcal{O}_{\mathbf{P}(E)}(m+k)\big)$$

is surjective for $k\geq 0$; and

(iii). $\mathcal{F}$ is $(m+1)$-regular for π.

(This follows from the previous example with $A=\mathcal{O}_{\mathbf{P}(E)}(1)$.) □

1.8.B Regularity and Complexity

We now present some results suggesting that the regularity of a sheaf governs the complexity[24] of the algebraic objects associated with it.

Continuing to work on the projective space $\mathbf{P} = \mathbf{P}(V)$, denote by $S = \mathrm{Sym}(V)$ the homogeneous coordinate ring of $\mathbf{P}$, so that S is a polynomial ring in $r+1$ variables. Fix a coherent sheaf $\mathcal{F}$ on $\mathbf{P}$, and let

$$F \;=\; \bigoplus_k H^0\big(\mathbf{P}, \mathcal{F}(k)\big)$$

be the corresponding graded S-module. We assume for simplicity that

$$H^0\big(\mathbf{P}, \mathcal{F}(k)\big) \;=\; 0 \quad \text{for all} \;\; k \ll 0,$$

so that F is finitely generated.[25]

Like any finitely generated S-module, F admits a minimal graded free resolution $E_\bullet$:

$$0 \longrightarrow E_{r+1} \longrightarrow \ldots \longrightarrow E_1 \longrightarrow E_0 \longrightarrow F \longrightarrow 0. \tag{1.31}$$

Here each E_p is a free graded S-module,

$$E_p \;=\; \bigoplus_i S(-a_{p,i}),$$

and minimality means that the maps of $E_\bullet$ are given by matrices of homogeneous polynomials containing no non-zero constants as entries. The integers $a_{p,i} \in \mathbf{Z}$ specifying the degrees of the generators of E_p are uniquely determined by F and hence $\mathcal{F}$. Set

$$a_p \;=\; a_p(\mathcal{F}) \;=\; \max_i\{a_{p,i}\},$$

so that a_p is the largest degree of a generator of the pth module of syzygies of F. Algorithms for computing with homogeneous polynomials typically proceed degree by degree, so the integers $a_0, \ldots, a_{r+1}$ in effect serve as a measure of the algebraic complexity of F or the underlying sheaf $\mathcal{F}$.

From the present point of view, the basic meaning of regularity is that it is equivalent to an upper bound on all of the a_p.

Theorem 1.8.26. (Regularity and syzygies). *The sheaf $\mathcal{F}$ is m-regular if and only if each of the integers $a_p = a_p(\mathcal{F})$ satisfies the inequality*

$$a_p \;\leq\; p + m. \tag{1.32}$$

[24] We stress that we are using the term "complexity" in a non-technical sense.

[25] This is equivalent to the assumption that none of the associated primes of $\mathcal{F}$ have zero-dimensional support. If this condition is not satisfied, then one should work instead with a truncation $F_{\geq k_0} = \oplus_{k \geq k_0} H^0\big(\mathbf{P}, \mathcal{F}(k)\big)$ of F.

So knowing the regularity of a coherent sheaf is the same thing as having bounds on the degrees of generators of all the modules of syzygies of the corresponding module.

Example 1.8.27. (Complete intersections). Let $X \subseteq \mathbf{P}$ be the complete intersection of hypersurfaces of degrees $d_1, \ldots, d_e$, and let $\mathcal{I}_X$ be the ideal sheaf of X. Then the Koszul resolution of the homogeneous ideal I_X of X shows that

$$\operatorname{reg}(\mathcal{I}_X) \;=\; (d_1 + \cdots + d_e - e + 1).$$

In this example, the regularity is governed by the highest module of syzygies of I_X rather than by the degrees of its generators. □

Indication of Proof of Theorem 1.8.26. The argument parallels the proof of Proposition 1.8.8. If F admits a resolution satisfying the degree bound (1.32), then sheafifying yields a resolution of $\mathcal{F}$ to which Example 1.8.7 applies. Assume conversely that $\mathcal{F}$ is m-regular. Then it follows from statement (ii) of Mumford's theorem that all the generators of F occur in degrees $\leq m$. Choosing generators then gives rise to an exact sequence of S-modules

$$0 \longrightarrow F_1 \longrightarrow \oplus S(-a_{0,i}) \longrightarrow F \longrightarrow 0$$

with all $a_{0,i} \leq m$. The sheafification of F_1 is $(m+1)$-regular, and hence all generators of F_1 occur in degrees $\leq (m+1)$. Continuing the process step by step leads to the required resolution. □

A particularly interesting case occurs when $\mathcal{F}$ is the ideal sheaf of a subvariety (or subscheme) of projective space:

Definition 1.8.28. (Regularity of a projective subvariety). We say that a subvariety (or subscheme) $X \subseteq \mathbf{P}$ is m-regular if its ideal sheaf $\mathcal{I}_X$ is. The *regularity* of X is the regularity $\operatorname{reg}(\mathcal{I}_X)$ of its ideal. □

Thus if X is m-regular, then its saturated homogeneous ideal

$$I_X \;=\; \oplus H^0\big(\mathbf{P}, \mathcal{I}_X(k)\big)$$

is generated by forms of degrees $\leq m$, and the pth syzygies among these generators appear in degrees $\leq m+p$. In the next subsection we will discuss the problem of bounding the regularity of X in terms of geometric data. Section 1.8.D centers around some more subtle invariants associated to syzygies.

We conclude this subsection with some examples. The first shows that in checking the regularity of a subvariety, some of the vanishings in the definition are automatic:

Example 1.8.29. If $X \subseteq \mathbf{P}$ has dimension n, then (for $m > 0$) X is m-regular if and only if $H^i\big(\mathbf{P}, \mathcal{I}_X(m-i)\big) = 0$ for $1 \leq i \leq n+1$. □

Example 1.8.30. (Regularity of finite sets). Suppose that $X \subseteq \mathbf{P}$ is a finite subset consisting of d distinct (reduced) points. Then X is d-regular, and if the points of X are collinear then X is not $(d-1)$-regular. The analogous statement holds if X is a finite scheme of length d. □

Example 1.8.31. (Regularity of some monomial curves). Suppose that $C \subset \mathbf{P}^r$ is a smooth rational curve, embedded by a possibly incomplete linear series. Then (for $m > 0$) C is m-regular if and only if hypersurfaces of degree $m-1$ cut out a complete linear series on C, i.e. the map

$$H^0(\mathbf{P}, \mathcal{O}_{\mathbf{P}}(m-1)) \longrightarrow H^0(C, \mathcal{O}_C(m-1))$$

is surjective. If C is the image of the embedding

$$\mathbf{P}^1 \hookrightarrow \mathbf{P}^3 \quad , \quad [s,t] \mapsto [s^d, s^{d-1}t, st^{d-1}, t^d],$$

then C is $(d-1)$-regular but not $(d-2)$-regular. □

Example 1.8.32. (Regularity of a disjoint union). Let $X, Y \subset \mathbf{P}$ be disjoint subvarieties or subschemes. If X is m-regular and Y is ℓ-regular, then $X \cup Y$ is $(m+\ell)$-regular. (Use Remark 1.8.11.) Sidman proves some related results for homogeneous ideals in [535]. □

Remark 1.8.33. (Algebraic pathology). Castelnuovo–Mumford regularity does not behave very well with respect to natural algebraic operations, but examples are hard to come by. For example, given a homogeneous ideal $I \subseteq S$, Ravi [513] raised the question whether $\text{reg}(\sqrt{I}) \leq \text{reg}(I)$.[26] However counter-examples were only recently given, by Chardin and D'Cruz [86]. This paper also gives an example in which the regularity of an ideal increases after removing some positive-dimensional embedded components. In general, the absence of systematic techniques for constructing examples is one of the biggest lacunae in the current state of the theory. □

Remark 1.8.34. (Work of Bayer and Stillman). Bayer and Stillman [43] establish a more precise connection between regularity and complexity. Namely, suppose that $I \subseteq S$ is the saturated homogeneous ideal of a scheme $X \subseteq \mathbf{P}$. Most computational algorithms in algebraic geometry are based on choosing coordinates on $\mathbf{P}$, and working with Gröbner bases. Bayer and Stillman show that the regularity $\text{reg}(X)$ of X is equal to the largest degree of a generator of the initial ideal $\text{in}(I)$ of I with respect to the reverse-lex order on generic coordinates. In other words, the regularity of X is already detected as soon as one computes Gröbner bases. In the same paper [43], Bayer and Stillman give a computationally efficient method of calculating the regularity of an ideal. □

[26] Concerning the meaning of regularity for an arbitrary homogeneous ideal, see the comments following Example 1.8.38.

1.8.C Regularity Bounds

The results of the previous section point to the interest in finding upper bounds on the regularity of a projective scheme $X \subseteq \mathbf{P}$ in terms of geometric data. Here we survey without proof some of the main results in this direction. While the picture is not yet complete, a rather fascinating dichotomy emerges, as emphasized in the influential survey [42] of Bayer and Mumford. On the one hand, for arbitrary schemes X — for which essentially best-possible bounds are known — the regularity can grow doubly exponentially as a function of the input parameters. On the other hand, the situation for "nice" varieties is very different: in particular, the regularity of non-singular varieties is known or expected to grow linearly in terms of geometric invariants. (The situation for reduced but possibly singular varieties remains somewhat unclear.)

Gotzmann's bound. The earliest results bounded regularity in terms of Hilbert polynomials. Given a projective scheme $X \subseteq \mathbf{P}$, with ideal sheaf $\mathcal{I} = \mathcal{I}_X \subseteq \mathcal{O}_{\mathbf{P}}$, write

$$Q(k) \;=\; \chi\big(X, \mathcal{O}_X(k)\big)$$

for the Hilbert polynomial of X. In the course of his construction of Grothendieck's Hilbert schemes, Mumford [445] bounded the regularity of X in terms of Q.[27] Although one could render the statements effective, Mumford's arguments were not intended to give sharp estimates. Brodmann has extended Mumford's boundedness theorem in various ways (see [72, Chapters 16 and 17] and the references therein).

Using a different approach, Gotzmann [228] subsequently found the optimal statement in this direction:

Theorem 1.8.35. (Gotzmann's regularity theorem). *There are unique integers*

$$a_1 \,\geq\, a_2 \,\geq\, \ldots \,\geq\, a_s \,\geq\, 0$$

such that $Q(k)$ can be expressed in the form

$$Q(k) \;=\; \binom{k+a_1}{a_1} + \binom{k+a_2-1}{a_2} + \ldots + \binom{k+a_s-(s-1)}{a_s},$$

and then $\mathcal{I}$ is s-regular. □

We refer to [238], [240, §3], and [74] for discussion, proofs, and generalizations.

Example 1.8.36. (One-dimensional schemes). Suppose that $X \subseteq \mathbf{P}^r$ is a one-dimensional scheme of degree d and arithmetic genus $p = 1 - \chi\big(X, \mathcal{O}_X\big)$, so that

$$Q(k) \;=\; dk + (1-p).$$

[27] In fact, Mumford introduced Definition 1.8.1 and proved Theorem 1.8.3 in order to establish the boundedness of the family of all projective subschemes having given Hilbert polynomial.

Then the Gotzmann representation is obtained by taking

$$s = \binom{d}{2} + (1-p),$$

$$a_1 = \ldots = a_d = 1 \ , \quad a_{d+1} = \ldots = a_s = 0.$$

In particular, X is $\left(\binom{d}{2} + 1 - p\right)$-regular. One can contrast this statement with the Castelnuovo-type bound from [259] for reduced irreducible curves: if $X \subseteq \mathbf{P}^r$ is a non-degenerate curve of degree d, then X is $(d+2-r)$-regular (see 1.8.46). □

Bounds from defining equations. As Bayer remarks, in an actual computation — where a scheme is described by explicit equations — the degrees of generators of an ideal will be known or bounded from the given input data. So it is very natural to look for regularity bounds in terms of these degrees.

We start with a definition:

Definition 1.8.37. (Generating degree of an ideal). The *generating degree* $d(\mathcal{I})$ of an ideal sheaf $\mathcal{I} \subseteq \mathcal{O}_{\mathbf{P}}$ is the least integer d such that $\mathcal{I}(d)$ is generated by its global sections. Similarly, if

$$J \subseteq S = \mathbf{C}[T_0, \ldots, T_r]$$

is a homogeneous ideal, the *generating degree* $d(J)$ of J is the largest degree of a minimal generator of J. □

Note that if $I \subseteq S$ is the saturated homogeneous ideal determined by an ideal sheaf $\mathcal{I} \subseteq \mathcal{O}_{\mathbf{P}}$, then $d(I) \geq d(\mathcal{I})$.

Example 1.8.38. (Generating degree of a smooth variety). Let $X \subseteq \mathbf{P}^r$ be a smooth variety of dimension n, and write $\mathcal{I}_X$ for the ideal sheaf of X in $\mathbf{P}^r$. Then

$$d(\mathcal{I}_X) \leq \deg(X),$$

i.e. X is cut out scheme-theoretically by hypersurfaces of degree $\leq \deg(X)$. (Let $\Lambda \subseteq \mathbf{P}^r$ be a linear space of dimension $r-n-1$ disjoint from X. Then the cone $C_\Lambda(X)$ over X centered on Λ is a hypersurface of degree $d = \deg(X)$ passing through X. As Λ varies, these hypersurfaces generate $\mathcal{I}_X(d)$.) This result is due to Mumford: see [448] for details. □

The first bounds are most naturally stated for homogeneous ideals. One can develop the general theory in this context (cf. [42] or [164]), but for present purposes it is simplest to recall that a homogeneous ideal $I \subseteq S$ is m-regular if I is saturated in degrees $\geq m$ and if the corresponding ideal sheaf $\mathcal{I}$ is m-regular. Bayer and Mumford [42, Proposition 3.8] give a very elementary proof of an essentially doubly exponential bound:

Theorem 1.8.39. (Bound for arbitrary ideals). *If $I \subseteq \mathbf{C}[T_0, \ldots, T_r]$ is any homogeneous ideal, then*

$$\operatorname{reg}(I) \ \leq \ \big(2d(I)\big)^{r!}. \quad \square$$

They also observe [42, Theorem 3.7] that work of Giusti and Galligo leads to the stronger bound $\operatorname{reg}(I) \leq \big(2d(I)\big)^{2^{r-1}}$.

Quite surprisingly, the shape of these statements cannot be avoided: there are examples of ideals whose regularity actually grows doubly exponentially in the generating degree. Indeed, Bayer and Stillman [44] show that a construction used by Mayr and Meyer leads to the following remarkable fact:

> Set $r = 10n$ and fix an integer $d \geq 3$. Then there exists an ideal $I \subseteq \mathbf{C}[T_0, \ldots, T_r]$ with $d(I) = d + 2$ and
>
> $$\operatorname{reg}(I) \ \geq \ (d)^{2^{n-1}}.$$

Note that by introducing a new variable and working on $\mathbf{P}^{r+1}$ one can then get ideal sheaves whose regularity satisfies the analogous bounds. We remark that the example of Mayr–Meyer–Bayer–Stillman is rather combinatorial in nature: it would be interesting to have also more geometric constructions.

By contrast, for the ideals of *non-singular* varieties the picture is completely different. Specifically, a result of Bertram, Ein, and the author from [55] leads to a linear inequality in the generating degree:

Theorem 1.8.40. (Linear bound for smooth ideals). *Let $X \subset \mathbf{P}^r$ be a smooth irreducible complex projective variety of dimension n and codimension $e = r - n$, and set $d = d(\mathcal{I})$. Then*

$$H^i\big(\mathbf{P}^r, \mathcal{I}_X(k)\big) \ = \ 0 \quad \text{for } i \geq 1 \ \text{ and } \ k \geq e \cdot d - r.$$

In particular, X is $(ed - e + 1)$-regular.

The proof — which is a very quick application of vanishing theorems — appears in Section 4.3.B below. A more algebraic approach has recently been given by Chardin and Ulrich [87].

Example 1.8.41. The bound in the theorem is achieved by the complete intersection of e hypersurfaces of degree d, so the statement is the best possible. In fact, these are the only borderline cases (Example 1.8.43). □

Example 1.8.42. (Criterion for projective normality). As in Theorem 1.8.40, suppose that $X \subseteq \mathbf{P}^r$ is a smooth subvariety of dimension n and codimension e that is cut out by hypersurfaces of degree d. If $ed \leq r + 1$ then X is projectively normal, and if $ed \leq r$ then X is projectively Cohen–Macaulay. These inequalities apply for instance to n-folds $X \subseteq \mathbf{P}^{2n+1}$ or $X \subseteq \mathbf{P}^{2n}$ that are cut out by quadrics. □

Example 1.8.43. (Borderline cases of regularity bound). In the situation of Theorem 1.8.40, X fails to be $(ed-e)$-regular if and only if it is the transversal complete intersection of e hypersurfaces of degree d. So we may say that complete intersections have the worst regularity among all smooth varieties cut out by hypersurfaces of a given degree. (The vanishings in Theorem 1.8.40 show that if X fails to be $(ed-e)$-regular, then necessarily

$$H^{n+1}\big(\mathbf{P}^r, \mathcal{I}_X(ed-e-n-1)\big) \;=\; H^n\big(X, \mathcal{O}_X(ed-r-1)\big) \neq 0,$$

or equivalently $H^0\big(X, \omega_X \otimes \mathcal{O}_X(r+1-ed)\big) \neq 0$. Assuming for simplicity that X has dimension ≥ 1, choose e general hypersurfaces $D_1, \ldots, D_e$ of degree d passing through X, so that

$$D_1 \cap \cdots \cap D_e \;=\; X \cup X'.$$

But $X \cap X'$ is an effective divisor on X representing the line bundle

$$\mathcal{O}_X(de) \otimes \det(N^*_{X/\mathbf{P}}) \;=\; \omega_X^{-1}(ed-r-1).$$

It follows that $X \cap X' = \varnothing$, and since $X \cup X'$ is in any event connected (e.g. by Theorem 3.3.3), we conclude that $X = D_1 \cap \cdots \cap D_e$. See [55] for details.) □

Remark 1.8.44. (Generators of different degrees). The result established in [55] takes into account generators of different degrees. Specifically, suppose that $X \subset \mathbf{P}^r$ is a non-singular variety of codimension e cut out scheme-theoretically by hypersurfaces of degrees $d_1 \geq d_2 \geq \cdots \geq d_m$. Then

$$H^i\big(\mathbf{P}^r, \mathcal{I}_X^a(k)\big) = 0 \quad \text{for } \; i > 0 \;\text{ and }\; k \,\geq\, ad_1 + d_2 + \cdots + d_e - r,$$

and in particular X is $(d_1 + \cdots + d_e - e + 1)$-regular. Again this regularity statement is sharp (exactly) for complete intersections. □

Remark 1.8.45. (Regularity of singular subvarieties). It would be very interesting to know to what extent these results remain valid for reduced (but possibly singular) varieties $X \subseteq \mathbf{P}$. One can use multiplier ideals to construct sheaves $\mathcal{J} \subseteq \mathcal{I}_X$ that have the expected regularity, but $\mathcal{J}$ may differ from $\mathcal{I}_X$ along the singular locus of X. See Example 10.1.5 for further discussion. □

Castelnuovo-type bounds. Consider a subvariety $X \subseteq \mathbf{P}^r$ with ideal sheaf $\mathcal{I}_X$. For $i \geq 2$ (and $k \geq -r$) there is an isomorphism

$$H^i\big(\mathbf{P}, \mathcal{I}_X(k)\big) \;=\; H^{i-1}\big(X, \mathcal{O}_X(k)\big).$$

Therefore these groups depend only on the line bundle on X defining the embedding, and in practice they can often be handled relatively easily. Thus the essential point for regularity bounds is usually to control the groups $H^1\big(\mathbf{P}, \mathcal{I}_X(k)\big)$, which measure the failure of hypersurfaces of degree k to cut out a complete linear series on X. This leads to a connection with some classical results of Castelnuovo.

Specifically, let $C \subseteq \mathbf{P}^r$ be a smooth irreducible curve of degree d that doesn't lie in any hyperplanes. Castelnuovo proved that hypersurfaces of degrees $\geq d-2$ cut out a complete linear series on C. By the argument leading up to his celebrated bound on the genus of a space curve, this implies that C is $(d-1)$-regular. The optimal statement along these lines was established by Gruson, Peskine, and the author in [259]:

Theorem 1.8.46. (Regularity for curves). *Let $C \subseteq \mathbf{P}^r$ be an irreducible (but possibly singular) reduced curve of degree d. Assume that C is non-degenerate, i.e. that it doesn't lie in any hyperplanes. Then C is $(d+2-r)$-regular.*

Example 1.8.31 exhibits some borderline examples in $\mathbf{P}^3$. The paper [259] also treats the case of possibly reducible curves.

The natural extrapolation of 1.8.46 to smooth varieties of higher dimension occurred to several mathematicians at the time of [259].

Conjecture 1.8.47. (Castelnuovo-type regularity conjecture). Consider a smooth non-degenerate subvariety $X \subseteq \mathbf{P}^r$ of dimension n and degree d. Then X is $(d+n+1-r)$-regular.

This has been established for surfaces by Pinkham and the author in [505] and [391], and for threefolds by Ran in [511]. We refer to [375] for a nice survey and some extensions. Eisenbud and Goto conjecture in [165] that the bound should hold for any reduced and irreducible non-degenerate variety. For some evidence in this direction see the papers [499], [137] of Peeva–Sturmfels and Derksen–Sidman.

Example 1.8.48. (Mumford's bound). Suppose that $X \subseteq \mathbf{P}^r$ is a smooth subvariety of degree d. Then X is $\big((n+1)(d-1)+1\big)$-regular. (By taking a generic projection, one reduces to the case $r = 2n+1$ and $e = n+1$. Then use the fact that X is scheme-theoretically cut out by hypersurfaces of degree d (Example 1.8.38) and apply 1.8.40.) □

Asymptotic regularity of powers of an ideal. It is a basic principle in commutative algebra that the powers of an ideal often exhibit better behavior than the ideal itself. Recently it has become clear that this holds in particular for Castelnuovo–Mumford regularity.

Specifically, the regularity of powers of an ideal is studied in several papers ([55], [560], [220], [84], [100], [354], [99], [535]). Asymptotically the picture becomes very clean. Most notably, Cutkosky, Herzog, and Trung [100], and independently Kodiyalam [354] prove the appealing result:

Theorem 1.8.49. *Let $I \subset \mathbf{C}[T_0, \dots, T_r]$ be an arbitrary homogeneous ideal. Then*

$$\lim_{k\to\infty} \frac{\operatorname{reg}(I^k)}{k} \;=\; \lim_{k\to\infty} \frac{d(I^k)}{k} \;\leq\; d(I),$$

where as above, $d(J)$ is the generating degree of a homogeneous ideal J. □

These authors also show that the limit in question is an integer. The essential point is to exploit the finite generation of certain Rees rings. Analogous results for ideal sheaves — deduced from Fujita's vanishing theorem — were given in [99], and are reproduced in Section 5.4.

1.8.D Syzygies of Algebraic Varieties

There are a number of natural settings — particularly involving embeddings defined by complete linear series — in which the regularity of a variety carries only coarse geometric information (cf. Example 1.8.52). In his pioneering paper [235], Green observed that in such cases it is interesting to study more delicate algebraic invariants, involving syzygies. This circle of ideas has generated considerable work in recent years, and we present here a quick overview — entirely without proof — of some of the highlights. Eisenbud's forthcoming book [163] contains a detailed introduction from a more algebraic perspective.

Let X be an irreducible projective variety, and L a very ample line bundle on X defining an embedding

$$\phi_L : X \hookrightarrow \mathbf{P} \;=\; \mathbf{P}H^0\big(X, L\big).$$

Consider the graded ring $R_L \;=\; R(X,L) \;=\; \oplus H^0\big(X, L^{\otimes m}\big)$ determined by L (Definition 2.1.17), and write $S \;=\; \operatorname{Sym} H^0\big(X, L\big)$ for the homogeneous coordinate ring of $\mathbf{P}$. Viewed as an S-module, R_L admits as in (1.31) a minimal graded free resolution $E_\bullet$:

$$\begin{array}{ccccccccc} \cdots \longrightarrow & \oplus S(-a_{2,j}) & \longrightarrow & \oplus S(-a_{1,j}) & \longrightarrow & \begin{matrix} S \\ \oplus \\ \oplus S(-a_{0,j}) \end{matrix} & \longrightarrow & R_L & \longrightarrow 0. \\ & \| & & \| & & \| & & & \\ & E_2 & & E_1 & & E_0 & & & \end{array}$$

Here the first summand in E_0 corresponds to the degree zero generator of R_L given by the constant function 1, and $a_{0,j} \geq 2$ for every j thanks to the fact that ϕ_L defines a linearly normal embedding. Similarly, a moment's thought reveals that $a_{i,j} \geq i+1$ for all $i \geq 1$ and every j. Observe that ϕ_L defines a projectively normal embedding of X if and only if $E_0 = S$, i.e. the summands $\oplus S(-a_{0,j})$ are not actually present. In this case, the remainder of $E_\bullet$ determines a resolution of the homogeneous ideal $I_{X/\mathbf{P}}$ of X in $\mathbf{P}$.

We ask when the first p terms in $E_\bullet$ are as simple as possible:

Definition 1.8.50. (Property ($\mathbf{N_p}$)). The embedding line bundle L satisfies *Property* (N_p) if $E_0 = S$, and

$$a_{i,j} = i+1 \quad \text{for all } j$$

whenever $1 \leq i \leq p$. A divisor D satisfies (N_p) if (N_p) holds for the corresponding line bundle $\mathcal{O}_X(D)$. □

Thus, very concretely:

(N_0) holds for L $\iff$ For every $m \geq 0$, the natural maps

$$S^m H^0(X,L) \longrightarrow H^0(X, L^{\otimes m})$$

are surjective;

(N_1) holds for L $\iff$ (N_0) is satisfied, and the homogeneous ideal $I = I_{X/\mathbf{P}}$ of X in $\mathbf{P}$ is generated by quadrics;

(N_2) holds for L $\iff$ (N_1) is satisfied, and the first module of syzygies among quadratic generators $Q_\alpha \in I$ is spanned by relations of the form

$$\sum L_\alpha Q_\alpha = 0,$$

where the L_α are linear forms;

and so on. Properties (N_0) and (N_1) were studied by Mumford [448], who called them "normal generation" and "normal presentation" respectively. The present terminology was introduced in [241].

Example 1.8.51. (Rational and elliptic normal curves). A rational normal cubic $C \subseteq \mathbf{P}^3$ satisfies (N_2). An elliptic normal curve $E \subseteq \mathbf{P}^3$ of degree 4 satisfies (N_1) but not (N_2). (C is defined by the maximal minors of a 2×3 matrix of linear forms, and so admits an Eagon–Northcott resolution (EN_0) from Appendix B. Similarly, E is the complete intersection of two quadrics.) □

Example 1.8.52. (Regularity of curves of large degree). Let X be a smooth curve of genus $g \geq 1$, and let L be a line bundle of degree $d \geq 2g+1$. Then L defines an embedding $X \subseteq \mathbf{P}^{d-g}$ in which X is 3-regular but not 2-regular. This uniform behavior of Castelnuovo–Mumford regularity contrasts with a number of interesting results and questions relating the syzygies to the geometry of X and L: see 1.8.53, 1.8.54, and 1.8.58. □

Curves. Consider to begin with a non-singular projective curve X of genus g, and a line bundle L on X. A classical result of Castelnuovo, Mattuck [421] and Mumford [448] states that if $\deg L \geq 2g+1$ then L is normally generated, while Fujita [191] and Saint-Donat [521] showed that if $\deg L \geq 2g+2$ then

L is normally presented. Green [235] proved that these are special cases of a general result for higher syzygies:

Theorem 1.8.53. (Syzygies of curves of large degree). *Assume that* $\deg L \geq 2g + 1 + p$. *Then* (N_p) *holds for* L.

The crucial point for the proof is to interpret syzygies via Koszul cohomology. A quick formulation using vector bundles appears in [392].

Remark 1.8.54. (Borderline cases). The inequality in 1.8.53 is best possible. In fact, it is established in [243] that if L is a line bundle of degree $2g + p$, then (N_p) fails for L if and only if either X is hyperelliptic or X has a $(p + 2)$-secant p-plane in the embedding defined by L. □

The most interesting statements concern canonical curves, where $L = \mathcal{O}_X(K_K)$ is the canonical bundle. Here there are again two classical results: Noether's theorem that K_X defines a projectively normal embedding if X is non-hyperelliptic, and Petri's theorem that the homogeneous ideal of a canonical curve $X \subseteq \mathbf{P}^{g-1}$ is generated by quadrics unless X is trigonal or a smooth plane quintic (cf. [15, Chapter III, §2, §3]). Green realized that this should generalize to higher syzygies via the Clifford index:

Definition 1.8.55. (Clifford index). Let A be a line bundle on a smooth curve X. The *Clifford index* of A is

$$\mathrm{Cliff}(A) \;=\; \deg(A) - 2r(A),$$

where as usual $r(A) = h^0(X, A) - 1$. The Clifford index of X itself is

$$\mathrm{Cliff}(X) \;=\; \min\left\{\mathrm{Cliff}(A) \mid h^0(X, A) \geq 2\ ,\ h^1(X, A) \geq 2.\right\}. \quad \square$$

Thus Clifford's theorem states that $\mathrm{Cliff}(X) \geq 0$, with equality if and only if X is hyperelliptic. Similarly, $\mathrm{Cliff}(X) = 1$ if and only if X is trigonal or a smooth plane quintic. If X is a general curve of genus g, then $\mathrm{Cliff}(X) = \left[\frac{g-1}{2}\right]$.

The natural extension of the theorems of Noether and Petri is contained in a celebrated conjecture of Green:

Conjecture 1.8.56. (Green's conjecture on canonical curves). The Clifford index $\mathrm{Cliff}(X)$ is equal to the least integer p for which Property (N_p) fails for the canonical divisor K_X.

One direction is elementary: it was established by Green and the author in [235, Appendix] that if $\mathrm{Cliff}(X) = e$, then (N_e) fails. What seems very difficult is to start with a syzygy and produce a line bundle. The first non-classical case $p = 2$ was treated by Schreyer [527] and Voisin [596].

The most significant progress to date on Green's conjecture is due to Voisin [598], [601], who proves that it holds for general curves:

Theorem 1.8.57. (Voisin's theorem on canonical curves). *If* X *is a general curve of genus* g, *then Green's conjecture holds for* X.

At least in the case of even genus, the idea is to study curves lying on a suitably generic $K3$ surface: it was established in [390] that such curves are Brill–Noether general, so this is a natural place to look. By a number of deep calculations, Voisin is able to establish a vanishing on the surface that implies 1.8.57. Cases of 1.8.57 had been obtained previously by Teixiodor.

Remark 1.8.58. (Further conjectures for curves). Green and the author proposed in [241] a more general conjecture. Specifically, suppose that L is a very ample line bundle with

$$\deg L \;\geq\; 2g + 1 + p - 2 \cdot h^1\big(X, L\big) - \text{Cliff}(X).$$

Then (N_p) should hold for L unless ϕ_L embeds X with a $(p{+}2)$-secant p-plane. The case $p = 0$ is treated in [241]. In another direction, one can consider the full resolution of a curve X of large degree. Conjecture 3.7 of [241] asserts that its grading depends (in a precise way) only on the gonality of X. This has recently been established for general curves of large gonality by Aprodu and Voisin [12] using the ideas of Voisin from [598]. □

Abelian varieties. Embeddings of abelian varieties were also considered classically from an algebraic perspective. Let X be an abelian variety of dimension g, and let L be an ample line bundle on X. It is a classical theorem of Lefschetz that $L^{\otimes m}$ is very ample provided that $m \geq 3$, and Mumford, Koizumi [355] and Sekiguchi [529] proved that $L^{\otimes m}$ is normally generated in this case. Mumford [448] and Kempf [333] proved that if $m \geq 4$, then X is cut out by quadrics under the embedding defined by $L^{\otimes m}$. The author remarked that these statements admit a natural extrapolation to higher syzygies, and the resulting conjecture was established by Pareschi [495]:

Theorem 1.8.59. (Pareschi's theorem). *Property (N_p) holds for $L^{\otimes m}$ as soon as $m \geq p + 3$.*

Pareschi's theorem has been systematized and extended by Pareschi and Popa through their work on regularity for abelian varieties ([496], [497]).

Varieties of arbitrary dimension. Inspired by Fujita's conjectures (Section 10.4.A), Mukai observed that one can rephrase Green's Theorem 1.8.53 as asserting that if A is an ample divisor on a curve C then $D = K_C + (p+3)A$ satisfies (N_p). Given an ample divisor A on a smooth projective variety X of dimension n, he remarked that it is then natural to wonder whether $D = K_X + (n+p+2)A$ satisfies (N_p). At the moment this seems completely out of reach: even Fujita's conjecture that the divisor in question is very ample when $p = 0$ remains very much open as of this writing.

However, the situation becomes much simpler if one works with very ample instead of merely ample divisors. Specifically, Ein and the author [153] used vanishing theorems for vector bundles to establish:

Theorem 1.8.60. (Syzygies of "hyperadjoint" divisors). *Let X be a smooth projective variety of dimension n, let B be a very ample divisor on X, and let P be any nef divisor. Then*

$$D = K_X + (n+1+p)B + P$$

satisfies Property (N_p).

When $p = 0$ the assertion is that $K_X + (n+1)B$ defines a projectively normal embedding of X: a simple proof appears in Example 4.3.19. (Note that if $X = \mathbf{P}^n$, $p = 0$, and B is a hyperplane divisor, then D is trivial, so the statement needs to be properly interpreted. However in all other cases the divisor in question is actually very ample.) Syzygies of surfaces have been studied by Gallego and Purnaprajna [218], [217].

Notes

The material in Sections 1.1–1.4 is for the most part classical, although the contemporary outlook puts greater emphasis on nef bundles and $\mathbf{Q}$-divisors than earlier perspectives. Chapter 1 of Hartshorne's notes [276] remains an excellent source for the basic theory of ample and nef divisors. We have also drawn on [363, Chapter VI] and [368, Chapter 1.5], as well as Debarre's presentation [114]. Theorem 1.4.40 (at least for locally free sheaves on smooth varieties) appears in [126].

The essential facts about Castelnuovo–Mumford regularity are present or implicit in [445] and [448]. Examples 1.8.18, 1.8.22, and 1.8.24 are generalized folklore, while Proposition 1.8.9 was noted by the author some years ago in response to a question from Ein. Special cases of Mumford's Theorem 1.8.5 have been rediscovered repeatedly in the literature in connection with vanishing theorems.

2

Linear Series

This chapter presents some of the basic facts and examples concerning linear series on a projective variety X. The theme is to use the theory developed in the previous chapter to study the complete linear series $|mD|$ associated to a divisor D on X that may not be ample or nef.

After giving the basic definitions, we focus in Section 2.1 on the asymptotic behavior of $|mD|$ as $m \to \infty$. The heart of the chapter appears in the second section, where we study big divisors. These are the birational analogues of ample line bundles, and they play a central role in many parts of this book. Section 2.3 is devoted to examples and complements: in particular, we give several concrete examples of the sort of challenging (i.e. interesting!) behavior that can occur for big linear series. Finally we discuss in Section 2.4 a formalism for extending some of these ideas to the setting of possibly incomplete linear series, as well as an algebraic construction that yields local models of several global phenomena associated with linear series.

We recall that according to Convention 1.1.2, the term "divisor" without further adjectives always means a Cartier divisor.

2.1 Asymptotic Theory

Let L be a line bundle on a projective variety X. We investigate in this section the asymptotic behavior of the linear series $|L^{\otimes m}|$ as $m \to \infty$. The basic fact, due to Iitaka and Ueno, is that up to birational equivalence they define an essentially unique rational mapping $\phi : X \dashrightarrow Y$, and L is close to trivial (in a suitable sense) along the fibres of ϕ.

The first subsection is devoted to definitions and examples. In the second we construct the fibration ϕ in the important case in which some power of L is globally generated. The general Iitaka fibration appears in the third subsection.

Throughout this section X denotes unless otherwise stated an irreducible complex projective variety.

2.1.A Basic Definitions

This subsection is devoted to some basic definitions and examples concerning linear series.

We start with the semigroup associated to a line bundle or divisor.

Definition 2.1.1. (Semigroup and exponent of a line bundle). Let L be a line bundle on the irreducible projective variety X. The *semigroup* of L consists of those non-negative powers of L that have a non-zero section:

$$\mathbf{N}(L) \;=\; \mathbf{N}(X,L) \;=\; \big\{m \geq 0 \mid H^0\big(X, L^{\otimes m}\big) \neq 0\big\}.$$

(In particular, $\mathbf{N}(L) = (0)$ if $H^0\big(X, L^{\otimes m}\big) = 0$ for all $m > 0$.) Assuming $\mathbf{N}(L) \neq (0)$, all sufficiently large elements of $\mathbf{N}(X,L)$ are multiples of a largest single natural number $e = e(L) \geq 1$, which we may call the *exponent* of L, and all sufficiently large multiples of $e(L)$ appear in $\mathbf{N}(X,L)$.[1] The semigroup $\mathbf{N}(X,D)$ and exponent $e = e(D)$ of a Cartier divisor D are defined analogously, or equivalently by passing to $L = \mathcal{O}_X(D)$. □

Example 2.1.2. Let T be a projective variety of dimension $d \geq 1$ that carries a non-trivial torsion line bundle η, having order e in $\mathrm{Pic}(T)$. Let Y be any projective variety of dimension k, and B a very ample line bundle on Y. Set

$$X \;=\; Y \times T \quad , \quad L \;=\; \mathrm{pr}_1^*(B) \otimes \mathrm{pr}_2^*(\eta).$$

Then $e(L) = e$ and $\mathbf{N}(L) = \mathbf{N}e$. Note that in this example $L^{\otimes m}$ is globally generated if $m \in \mathbf{N}(L)$, whereas by definition $H^0\big(X, L^{\otimes m}\big) = 0$ otherwise. □

Given $m \in \mathbf{N}(X,L)$, consider the rational mapping

$$\phi_m = \phi_{|L^{\otimes m}|} \,:\, X \dashrightarrow \mathbf{P}H^0\big(X, L^{\otimes m}\big)$$

associated to the complete linear series $|L^{\otimes m}|$. We denote by

$$Y_m \;=\; \phi_m(X) \;\subseteq\; \mathbf{P}H^0\big(X, L^{\otimes m}\big)$$

the closure of its image, i.e. the image of the closure of the graph of ϕ_m. One of our objects will be to understand the asymptotic birational behavior of these mappings as $m \to \infty$.

Definition 2.1.3. (Iitaka dimension). Assume that X is normal. Then the *Iitaka dimension* of L is defined to be

[1] The exponent e is the g.c.d. of all the elements of $\mathbf{N}(L)$.

$$\kappa(L) = \kappa(X,L) = \max_{m\in\mathbf{N}(L)} \big\{ \dim \phi_m(X) \big\},$$

provided that $\mathbf{N}(L) \neq (0)$. If $H^0\big(X, L^{\otimes m}\big) = 0$ for all $m > 0$, one puts $\kappa(X,L) = -\infty$. If X is non-normal, pass to its normalization $\nu : X' \longrightarrow X$ and set

$$\kappa(X,L) = \kappa(X', \nu^* L).$$

Finally, for a Cartier divisor D one takes $\kappa(X,D) = \kappa(X, \mathcal{O}_X(D))$.[2] □

Thus either $\kappa(X,L) = -\infty$, or else

$$0 \leq \kappa(X,L) \leq \dim X.$$

Example 2.1.4. For the pair (X,L) considered in Example 2.1.2, $\kappa(X,L) = \dim Y = k$. □

Example 2.1.5. (Kodaira dimension). Let X be a smooth projective variety, and K_X a canonical divisor on X. Then $\kappa(X) = \kappa(X, K_X)$ is the *Kodaira dimension* of X: it is the most basic birational invariant of a variety. The Kodaira dimension of a singular variety is defined to be the Kodaira dimension of any smooth model. □

Example 2.1.6. (Kodaira dimension of singular varieties). If X is smooth, then $\mathcal{O}_X(K_X) = \omega_X$ is the dualizing line bundle on X. However when X is singular, it may happen that the dualizing sheaf ω_X exists as a line bundle on X, but $\kappa(X, \omega_X) > \kappa(X)$. (This occurs for instance when $X \subseteq \mathbf{P}^3$ is the cone over a smooth plane curve of large degree.) The impact of singularities on Kodaira dimension and other birational invariants plays an important role in the minimal model program. Reid's survey [516] gives a nice introduction to this circle of ideas. □

Example 2.1.7. (Iitaka dimensions of restrictions). The Iitaka dimension of a line bundle can vary unpredictably under restriction to subvarieties. For example let $X = \mathrm{Bl}_P(\mathbf{P}^2)$ be the blowing up of $\mathbf{P}^2$ at a point, and denote by E and H respectively the exceptional divisor and the pullback of a hyperplane divisor. Then $\mathcal{O}_X(H)$ and $\mathcal{O}_X(H+E)$ have maximal Iitaka dimension, but their restrictions to E have Iitaka dimensions 0 and $-\infty$ respectively. So in these cases the Iitaka dimension drops upon restriction. On the other hand, let $X = \mathbf{P}^1 \times \mathbf{P}^1$ and take $L = \mathrm{pr}_1^*\mathcal{O}_{\mathbf{P}^1}(-1) \otimes \mathrm{pr}_2^*\mathcal{O}_{\mathbf{P}^1}(1)$, so that $\kappa(X,L) = -\infty$. If $Y = \{\text{point}\} \times \mathbf{P}^1$ then $\kappa(Y, L \,|\, Y) = 1$, so that here Iitaka dimension has increased upon restriction. □

Example 2.1.8. (Non-normal varieties). Linear series on a non-normal variety may behave quite differently than their pullbacks to the normalization $\nu : X' \longrightarrow X$. As a simple example, let X be a nodal plane cubic curve, and let

[2] The Iitaka dimension of a line bundle L or divisor D is sometimes also called the L- or D-dimension of X.

$$L \in \operatorname{Pic}^0(X) \cong \mathbf{G}_m$$

be a non-torsion line bundle of degree zero. Then $H^0\big(X, L^{\otimes m}\big) = 0$ for all $m > 0$. On the other hand,

$$X' = \mathbf{P}^1 \quad , \quad L' = \nu^* L = \mathcal{O}_{\mathbf{P}^1},$$

and so $h^0\big(X', L'^{\otimes m}\big) = 1$ for every $m > 0$. By taking products as in Example 2.1.2, one can construct analogous examples with $H^0\big(X, L^{\otimes m}\big) = 0$ for all $m > 0$, and $\kappa(X', L')$ arbitrarily large. So if one did not pass to the normalization in Definition 2.1.3 one would not obtain a birationally invariant theory. □

Example 2.1.9. (Iitaka dimension and deformations). The Iitaka dimension of a line bundle L is not invariant under deformations of L. For example, if $L \in \operatorname{Pic}^0(X)$, then $\kappa(X, L) = 0$ if L is trivial or torsion, but $\kappa(X, L) = -\infty$ otherwise. Taking products as in (2.1.2) yields similar examples in which κ can be arbitrarily large. Other examples (on a variable family of varieties) appear in 2.2.13. On the other hand, we will see in the next section (Corollary 2.2.8) that bundles L with maximal Iitaka dimension $\kappa(X, L) = \dim X$ are recognized from their numerical equivalence classes, so that the Iitaka dimension is constant under deformations of L (on a fixed variety) in this case. □

Example 2.1.10. (Powers achieving the Iitaka dimension). Assume that X is normal. If $\kappa(X, L) = \kappa$, then $\dim \phi_m(X) = \kappa$ for all sufficiently large $m \in \mathbf{N}(X, L)$. (Replacing L by $L^{\otimes e}$ one can suppose that L has exponent $e(L) = 1$, so there exists a positive integer p_0 such that $H^0\big(X, L^{\otimes p}\big) \neq 0$ for all $p \geq p_0$. Fix some k such that $\dim \phi_k(X) = \kappa$. Multiplying by a non-zero section in $H^0\big(X, L^{\otimes p}\big)$ determines for every $p \geq p_0$ an embedding

$$H^0\big(X, L^{\otimes k}\big) \subseteq H^0\big(X, L^{\otimes k+p}\big). \tag{*}$$

This in turn gives rise to a factorization $\phi_k = \nu_p \circ \phi_{k+p}$, where

$$\nu_p : \mathbf{P}H^0\big(X, L^{\otimes k+p}\big) \dashrightarrow \mathbf{P}H^0\big(X, L^{\otimes k}\big)$$

is the rational mapping arising from the linear projection associated with (*) (Example 1.1.12). Therefore $\dim Y_{k+p} \geq \dim Y_k = \kappa$, and the reverse inequality holds by definition.) □

We will frequently need to deal with morphisms $f : X \longrightarrow Y$ having the property that $H^0\big(Y, L\big) = H^0\big(X, f^*L\big)$ for every line bundle L on Y. The next definition expresses a natural condition for this.

Definition 2.1.11. (Algebraic fibre space). An *algebraic fibre space* is a surjective projective mapping $f : X \longrightarrow Y$ of reduced and irreducible varieties such that $f_*\mathcal{O}_X = \mathcal{O}_Y$.[3] □

[3] Mori [439, p. 272] requires in addition that X and Y be non-singular and projective, but the present definition is more convenient for our purposes.

Note that given any projective surjective morphism $g : V \longrightarrow W$ of irreducible varieties, the Stein factorization [280, III.11.5] expresses g as a composition

$$V \xrightarrow{a} W' \xrightarrow{b} W,$$

where a is an algebraic fibre space and b is finite. So g itself is a fibre space if and only if the finite part of its Stein factorization is trivial. In particular, all the fibres of a fibre space f as above are connected. Conversely, if Y is normal then any projective surjective morphism $f : X \longrightarrow Y$ with connected fibres is a fibre space (see [280, proof of Corollary 11.4]). By the same token, if $f : X \longrightarrow Y$ is an algebraic fibre space with X normal, and if $\mu : X' \longrightarrow X$ is a birational mapping, then the composition $f \circ \mu : X' \longrightarrow Y$ is again a fibre space.

Example 2.1.12. (Fibre spaces and function fields). Let $f : X \longrightarrow Y$ be a projective surjective morphism of normal varieties, and

$$\mathbf{C}(Y) \subseteq \mathbf{C}(X)$$

the corresponding finitely generated extension of function fields. Then f is a fibre space if and only if $\mathbf{C}(Y)$ is algebraically closed in $\mathbf{C}(X)$. This allows one to attach a meaning to algebraic fibre spaces in the birational category: one would ask that a dominant rational mapping $X \dashrightarrow Y$ of normal varieties induce an algebraically closed extension of function fields. (Suppose first that $\mathbf{C}(Y)$ is not algebraically closed in $\mathbf{C}(X)$. Considered as a rational mapping, f then factors as a composition

$$X \overset{u}{\dashrightarrow} Y' \overset{v}{\dashrightarrow} Y,$$

where Y' is generically finite of degree ≥ 2 over Y. Replacing Y' and X by suitable birational modifications — which, by the normality of X, doesn't affect the fibre space hypothesis — one can suppose that u and v are in fact morphisms. But then f cannot be a fibre space, or else all the fibres of v would be connected. Conversely, if $\mathbf{C}(Y)$ is algebraically closed in $\mathbf{C}(X)$, then the finite part of the Stein factorization of f must be trivial (Y being normal), whence f is a fibre space.) □

Lemma 2.1.13. (Pullbacks via a fibre space). *Let $f : X \longrightarrow Y$ be an algebraic fibre space, and let L be a line bundle on Y. Then*

$$H^0\big(X, f^*L^{\otimes m}\big) \;=\; H^0\big(Y, L^{\otimes m}\big) \quad \textit{for all} \;\; m \geq 0.$$

In particular, $\kappa(Y, L) = \kappa(X, f^*L)$.

Proof. Quite generally, $H^0\big(X, f^*L^{\otimes m}\big) \;=\; H^0\big(Y, f_*(f^*L^{\otimes m})\big)$. But

$$f_*\big(f^*L^{\otimes m}\big) \;=\; f_*\mathcal{O}_X \otimes L^{\otimes m}$$

thanks to the projection formula, and $f_*\mathcal{O}_X = \mathcal{O}_Y$ since f is a fibre space. □

Example 2.1.14. (Injectivity of Picard groups). Let X and Y be irreducible projective varieties, and $f : X \longrightarrow Y$ an algebraic fibre space. Then the induced homomorphism

$$f^* : \operatorname{Pic}(Y) \longrightarrow \operatorname{Pic}(X)$$

is injective. (In fact, suppose that B is a line bundle on Y such that $f^*B \cong \mathcal{O}_X$. Then $H^0\big(Y, B\big) = H^0\big(X, f^*B\big) \neq 0$ thanks to 2.1.13, and similarly $H^0\big(Y, B^*\big) = H^0\big(X, f^*B^*\big) \neq 0$. Therefore $B = \mathcal{O}_Y$.) □

Example 2.1.15. (Normality of fibre spaces). Let $f : X \longrightarrow Y$ be an algebraic fibre space. If X is normal, then Y is also normal. (Let $\nu : Y' \longrightarrow Y$ be the normalization of Y. Then f factors through ν, and since f is a fibre space this forces ν to be an isomorphism.) □

Example 2.1.16. (Exceptional twists of birational pullbacks). Consider a proper birational mapping

$$\mu : X' \longrightarrow X$$

of irreducible varieties, with X' normal. Let L be a line bundle on X, and E a μ-exceptional effective divisor on X' (i.e. an effective Cartier divisor on X' with the property that each of the irreducible components of the corresponding Weil divisor maps under μ to a variety of codimension ≥ 2 in X). Then the natural inclusion

$$H^0\big(X', \mu^*L\big) \longrightarrow H^0\big(X', \mu^*L \otimes \mathcal{O}_{X'}(E)\big) \tag{*}$$

is an isomorphism. (One can assume that X is normal. Then the group on the left in (*) is identified with $H^0\big(X, L\big)$, while a section of the group on the right restricts to a section of L defined off the codimension ≥ 2 fundamental locus of μ. By normality, the latter section extends to all of X.) □

We close this subsection by introducing two further constructions. The first is an important algebra associated to a line bundle or divisor.

Definition 2.1.17. (Section ring associated to a line bundle). Given a line bundle L on a projective variety X, the *graded ring* or *section ring* associated to L is the graded $\mathbf{C}$-algebra

$$R(L) \;=\; R(X, L) \;=\; \bigoplus_{m \geq 0} H^0\big(X, L^{\otimes m}\big).$$

The graded ring $R(D) = R(X, D)$ associated to a Cartier divisor D is defined similarly. □

Example 2.1.18. If $X = \mathbf{P}^n$ and $L = \mathcal{O}_{\mathbf{P}^n}(1)$, then $R(L) = \mathbf{C}[T_0, \ldots, T_n]$ is the homogeneous coordinate ring of $\mathbf{P}^n$. □

The geometric properties of L are greatly influenced by whether or not $R(L)$ is finitely generated. We therefore propose:

Definition 2.1.19. (Finitely generated line bundles and divisors). A line bundle L on a projective variety X is *finitely generated* if its section ring $R(X, L)$ is a finitely generated $\mathbf{C}$-algebra. A divisor D is finitely generated if $\mathcal{O}_X(D)$ is so. □

Several examples in which finite generation fails appear in Section 2.3. Examples 2.1.30, 2.1.31 and Theorem 2.3.15 give some consequences and criteria for finite generation.

Passing to additive notation, we discuss next the stable base locus of a divisor. Suppose then that X is an irreducible projective variety and that D is a (Cartier) divisor on X with $\kappa(X, D) \geq 0$. Given $m \geq 1$ we consider the set-theoretic base locus $\mathrm{Bs}\big(|mD|\big)$ of $|mD|$, with the convention that $\mathrm{Bs}\big(|mD|\big) = X$ if $|mD| = \varnothing$.

Definition 2.1.20. (Stable base locus). The *stable base locus* of D is the algebraic set

$$\mathbf{B}\big(D\big) \ = \ \bigcap_{m \geq 1} \mathrm{Bs}\big(|mD|\big). \tag{2.1}$$

We emphasize that the stable base locus is defined only as a closed subset of X: we do not wish to view the right side of (2.1) as an intersection of schemes. (See Example 2.1.25.) □

The terminology stems from the fact that the base loci $\mathrm{Bs}\big(|mD|\big)$ actually stabilize to $\mathbf{B}\big(D\big)$ for sufficiently large and divisible m:

Proposition 2.1.21. *The stable base locus $\mathbf{B}\big(D\big)$ is the unique minimal element of the family of algebraic sets*

$$\big\{ \mathrm{Bs}\big(|mD|\big) \big\}_{m \geq 1}. \tag{2.2}$$

Moreover, there exists an integer m_0 such that

$$\mathbf{B}\big(D\big) \ = \ \mathrm{Bs}\big(|km_0D|\big) \quad \textit{for all} \ \ k \gg 0.$$

Proof. Given any natural numbers $m, \ell \geq 1$, there is a set-theoretic inclusion

$$\mathrm{Bs}\big(|\ell mD|\big) \ \subseteq \ \mathrm{Bs}\big(|mD|\big) \tag{*}$$

thanks to the reverse inclusion $\mathfrak{b}\big(|mD|\big)^{\ell} \subseteq \mathfrak{b}\big(|m\ell D|\big)$ on base ideals (Example 1.1.9). This implies right away that the family of closed sets appearing in (2.2) does indeed have a unique minimal element, which by definition must then coincide with $\mathbf{B}\big(D\big)$. In fact, the existence of at least one minimal element follows from the descending chain condition on Zariski-closed subsets of X. On the other hand, if $\mathrm{Bs}\big(|pD|\big)$ and $\mathrm{Bs}\big(|qD|\big)$ are each minimal, then thanks to (*) they both coincide with $\mathrm{Bs}\big(|pqD|\big)$. The second assertion of the proposition follows similarly from (*). □

Remark 2.1.22. (Realizing the stable base locus). It is not in general possible to take $m_0 = 1$ in the last statement of the proposition. To begin with, if $e(D) = e > 1$ then by definition $\mathrm{Bs}(|mD|) = X$ unless $e|m$. Therefore $\mathbf{B}(D) = \mathrm{Bs}(|mD|)$ only if $e|m$. However, even when $e(D) = 1$ it can happen that $\mathbf{B}(D) \neq \mathrm{Bs}(|mD|)$ for arbitrarily large values of m. For instance in Example 2.3.4 we construct a divisor D on a surface having the property that $|mD|$ is free if m is even, while $\mathrm{Bs}(|mD|)$ is a curve when m is odd. In this case $\mathbf{B}(D) = \varnothing$, so here again it does not happen that $\mathbf{B}(D) = \mathrm{Bs}(|mD|)$ for all sufficiently large m. (In fact, the divisor D in this example is even big and nef in the sense of Section 2.2.) □

Example 2.1.23. (Multiples). For any Cartier divisor D with $\kappa(X, D) \geq 0$ one has

$$\mathbf{B}(pD) \;=\; \mathbf{B}(D) \quad \text{for all } \; p \geq 1.$$

(This follows from the last statement of 2.1.21.) □

Remark 2.1.24. (Q-divisors). The previous example leads to a natural definition of $\mathbf{B}(D)$ for an arbitrary $\mathbf{Q}$-divisor D: choose p so that pD is integral, and put $\mathbf{B}(D) = \mathbf{B}(pD)$. □

Example 2.1.25. (Scheme structures). Recall that $\mathrm{Bs}(|mD|)$ carries a natural scheme structure determined by the base-ideal $\mathfrak{b}(|mD|) \subseteq \mathcal{O}_X$ (Definition 1.1.8). However, considered as schemes, these base loci may be distinct even when their underlying sets coincide. For a simple example, take X to be the blow-up of $\mathbf{P}^2$ at a point, denote by E and H respectively the exceptional divisor and the pullback of a line, and set $D = H + E$. Then

$$\mathfrak{b}(|mD|) \;=\; \mathcal{O}_X(-mE),$$

i.e. the scheme-theoretic base locus is the mth-order neighborhood mE of E. A more subtle example appears in 2.3.5: here the base scheme of $|mD|$ consists of a curve occurring "with multiplicity one" together with an embedded point that varies aperiodically with m. □

2.1.B Semiample Line Bundles

We now turn to the problem of understanding the asymptotic behavior of the mappings ϕ_k determined by $|L^{\otimes k}|$ for large $k \in \mathbf{N}(X, L)$. This subsection focuses on the important case that some power of L is free: here all the essential ideas occur in a particularly transparent setting. We also discuss finiteness properties of the section rings of such bundles.

We start with some definitions.

Definition 2.1.26. (Semiample bundles and divisors). A line bundle L on a complete scheme is *semiample* if $L^{\otimes m}$ is globally generated for some $m > 0$. A divisor D is *semiample* if the corresponding line bundle is so. □

Fixing a semiample line bundle L, for the purposes of this discussion we denote by $M(X,L) \subseteq \mathbf{N}(X,L)$ the sub-semigroup

$$M(X,L) \;=\; \big\{ m \in \mathbf{N} \mid L^{\otimes m} \text{ is free } \big\}.$$

We write $f = f(L)$ for the "exponent" of $M(X,L)$, i.e. the largest natural number such that every element of $M(X,L)$ is a multiple of f (so that in particular $L^{\otimes kf}$ is free for $k \gg 0$). Note that it can happen that $f(L) > 1$ even when $\kappa(X,L) = \dim X$ and $\mathbf{N}(X,L) = \mathbf{N}$: see Example 2.3.4 below.

Given $m \in M(X,L)$, write as above $Y_m = \phi_m(X)$ for the image of the morphism

$$\phi_m \;=\; \phi_{|L^{\otimes m}|} : X \longrightarrow Y_m \;\subseteq\; \mathbf{P} \;=\; \mathbf{P}H^0\big(X, L^{\otimes m}\big)$$

determined by the complete linear series $|\,L^{\otimes m}\,|$. By a slight abuse of notation, we view ϕ_m as a mapping from X to Y_m rather than to $\mathbf{P}$. Keeping this convention in mind, one has:

Theorem 2.1.27. (Semiample fibrations). *Let X be a normal projective variety, and let L be a semiample bundle on X. Then there is an algebraic fibre space*

$$\phi : X \longrightarrow Y$$

having the property that for any sufficiently large integer $m \in M(X,L)$,

$$Y_m \;=\; Y \;\; \textit{and} \;\; \phi_m \;=\; \phi.$$

Furthermore there is an ample line bundle A on Y such that $\phi^* A = L^{\otimes f}$, where $f = f(L)$ is the exponent of $M(X,L)$.

In other words, for $m \gg 0$ the mappings ϕ_m stabilize to define a fibre space structure on X (essentially characterized by the fact that $L^{\otimes f}$ is trivial on the fibres). Observe that Lemma 2.1.13 implies that $H^0\big(X, L^{\otimes kf}\big) = H^0\big(Y, A^{\otimes k}\big)$ for all $k \geq 0$, while it follows from Example 2.1.15 that Y is normal.

We start by making some observations that will be important in the proof of the theorem. So consider as above a normal projective variety X and a line bundle L on X such that $L^{\otimes m}$ is globally generated for some fixed integer $m > 0$. Then $S^k H^0\big(X, L^{\otimes m}\big)$ determines a free subseries of $|\,L^{\otimes km}\,|$, corresponding to the kth Veronese re-embedding of Y_m. This Veronese embedding of Y_m in turn arises as a projection of the morphism determined by the complete linear series $|\,L^{\otimes km}\,|$ (Example 1.1.12). In other words, ϕ_m factors as a composition $\phi_m = \nu_k \circ \phi_{km}$:

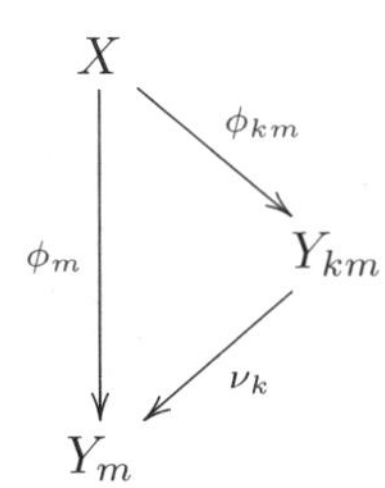

where $\nu_k : Y_{km} \longrightarrow Y_m$ is a finite mapping (again by 1.1.12). Note also that Y_m carries a very ample line bundle A_m — viz. the restriction of the hyperplane bundle on $\mathbf{P} = \mathbf{P}H^0\big(X, L^{\otimes m}\big)$ — such that $\phi_m^* A_m = L^{\otimes m}$ and $H^0\big(X, L^{\otimes m}\big) = H^0\big(Y_m, A_m\big)$.

The essential observation is now the following:

Lemma 2.1.28. *In the situation of Theorem 2.1.27, fix any $m \in M(X, L)$. Then for all sufficiently large integers $k \gg 0$, the composition*

$$X \xrightarrow{\phi_{km}} Y_{km} \xrightarrow{\nu_k} Y_m$$

gives the Stein factorization of ϕ_m, i.e. ϕ_{km} is an algebraic fibre space. In particular, Y_{km} and ϕ_{km} are independent of k for $k \gg 0$.

Proof. Let $X \xrightarrow{\psi} V \xrightarrow{\mu} Y_m$ be the Stein factorization of ϕ_m, so that ψ is a fibre space, V is normal, and μ is finite. Denote by A_m the ample bundle on Y_m that pulls back to $L^{\otimes m}$ on X. Since μ is finite, $B =_{\text{def}} \mu^* A_m$ is an ample line bundle on V. Therefore $B^{\otimes k}$ is very ample on V for all $k \gg 0$. On the other hand,

$$\psi^* B^{\otimes k} = L^{\otimes km} \ \text{ and } \ H^0\big(X, L^{\otimes km}\big) \ = \ H^0\big(V, B^{\otimes k}\big)$$

thanks to Lemma 2.1.13. But this means that V is the image of X under the morphism $\phi_{|kmL|} : X \longrightarrow \mathbf{P}H^0\big(X, L^{\otimes km}\big)$, i.e. that $Y_{km} = V$ and $\phi_{km} = \psi$ for $k \gg 0$. □

We now turn to the

Proof of Theorem 2.1.27. In order to lighten notation we assume (as we may by replacing L by $L^{\otimes f}$) that $f = f(L) = 1$, so that every sufficiently large power of L is free. Now fix relatively prime positive integers p and q such that ϕ_p and ϕ_q satisfy the conclusion of Lemma 2.1.28. In other words, assume that $Y_{kp} = Y_p$ and $\phi_{kp} = \phi_p$ for all $k \geq 1$, and similarly for Y_q and ϕ_q. Then in the first place $Y_p = Y_{pq} = Y_q$ and $\phi_p = \phi_{pq} = \phi_q$: write $\phi : X \longrightarrow Y$ for the resulting fibre space.

We next produce a line bundle A on Y that pulls back to $L = L^{\otimes f}$. In fact, Y carries very ample bundles A_p and A_q such that $\phi^* A_p = L^{\otimes p}$ and $\phi^* A_q = L^{\otimes q}$. But since p and q are relatively prime, we can write $1 = rp + sq$ for suitable $r, s \in \mathbf{Z}$, and it suffices to take

$$A = A_p^{\otimes r} \otimes A_q^{\otimes s}.$$

Now $\phi^* : \text{Pic}(Y) \longrightarrow \text{Pic}(X)$ is injective thanks to the fact that ϕ is a fibre space (Example 2.1.14). Therefore $A_p = A^{\otimes p}$ and $A_q = A^{\otimes q}$, and in particular A is ample.

It remains to show that $Y_m = Y$ and $\phi_m = \phi$ for all $m \gg 0$. To this end, fix positive integers $c, d \geq 1$. Then the product $S^c H^0(Y, A^{\otimes p}) \otimes S^d H^0(Y, A^{\otimes q})$ determines a free linear subseries of

$$H^0(Y, A^{\otimes (cp+dq)}) = H^0(X, L^{\otimes (cp+dq)}).$$

As in the discussion preceding Lemma 2.1.28, it follows that ϕ factors as the composition of ϕ_{cp+dq} with a finite map. Since ϕ is a fibre space, this implies that $\phi = \phi_{cp+dq}$. But any sufficiently large integer m is of the form $cp + qd$, and the theorem follows. □

Theorem 2.1.27 leads to some important finite generation properties of semiample bundles.

Example 2.1.29. (Surjectivity of multiplication maps). Let L be a line bundle on a normal projective variety X that is generated by its global sections. Then there exists an integer $m_0 = m_0(L)$ such that the mappings

$$H^0(X, L^{\otimes a}) \otimes H^0(X, L^{\otimes b}) \longrightarrow H^0(X, L^{\otimes (a+b)})$$

determined by multiplication are surjective whenever $a, b \geq m_0$. More generally, for any coherent sheaf $\mathcal{F}$ on X,

$$H^0(X, \mathcal{F} \otimes L^{\otimes a}) \otimes H^0(X, L^{\otimes b}) \longrightarrow H^0(X, \mathcal{F} \otimes L^{\otimes (a+b)})$$

is onto for $a, b \gg 0$. (Use the fibre space $\phi : X \longrightarrow Y$ associated to L to reduce to the case 1.2.22 of ample bundles. For the second statement, observe that $H^0(Y, \phi_* \mathcal{F} \otimes A^{\otimes a}) = H^0(X, \mathcal{F} \otimes \phi^*(A^{\otimes a}))$.) □

Example 2.1.30. (Finite generation of semiample bundles). The previous example implies a basic theorem of Zariski:

> **Theorem.** *Let L be a semiample line bundle on a normal projective variety X. Then L is finitely generated (Definition 2.1.19), i.e. the section ring $R(X, L)$ of L is finitely generated as a* **C***-algebra.*

(Supposing that $L^{\otimes k}$ is free, 2.1.29 implies that the Veronese subalgebra $R(X, L)^{(k)}$ is finitely generated. Applying 2.1.29 — taking $\mathcal{F}$ in turn to be each of the sheaves $L, L^{\otimes 2}, \ldots, L^{\otimes k-1}$ — yields the finite generation of $R(X, L)$ itself.) With a little more care one can prove that the theorem holds for a semiample bundle on any complete scheme: see for instance [22, Theorem 9.14]. There is also an analogous statement — likewise due to Zariski — for the multi-graded ring defined by several semiample bundles. □

Example 2.1.31. (Section rings of finitely generated line bundles). As a partial converse to the previous example, consider a finitely generated line bundle L (Definition 2.1.19) on a normal projective variety X. Then there exist a natural number $p > 0$, a projective birational map $\nu : X^+ \longrightarrow X$ with X^+ normal, and an effective divisor N^+ on X^+ such that

$$L^+ =_{\text{def}} \nu^*(L^{\otimes p}) \otimes \mathcal{O}_{X^+}(-N^+)$$

is globally generated, and

$$R(X,L)^{(p)} = R(X^+, L^+).$$

In other words, after passing to a Veronese subring, any finitely generated section ring becomes the section ring of a semiample bundle on a modification of the given variety. (Let $R = R(X,L)$ be the graded ring of L. Since R is finitely generated we may fix an integer $p \gg 0$ such that R_p generates the Veronese subring $R^{(p)}$ ([69, Chapter III, §1, Proposition 3]), i.e. so that

$$\mathrm{Sym}^m H^0\big(X, L^{\otimes p}\big) \longrightarrow H^0\big(X, L^{\otimes mp}\big)$$

is surjective for all $m \geq 0$. Writing $\mathfrak{b}_k \subseteq \mathcal{O}_X$ for the base-ideal of $|L^{\otimes k}|$, it follows that $\mathfrak{b}_{mp} = \mathfrak{b}_p^m$ for all $m \geq 0$. Let $\nu : X^+ \longrightarrow X$ be the normalization of the blow-up of X along $\mathfrak{b}_p$, with exceptional divisor N^+. Then $L^+ = \nu^*(L^{\otimes p}) \otimes \mathcal{O}_{X^+}(-N^+)$ is free. Moreover, $\nu_* \mathcal{O}_{X^+}(-mN^+) = \overline{\mathfrak{b}_p^m}$ is the integral closure of $\mathfrak{b}_p^m$ (Definition 9.6.2), and hence

$$\nu_*\big((L^+)^{\otimes m}\big) = L^{\otimes mp} \otimes \overline{\mathfrak{b}_p^m}$$

thanks to the projection formula. Keeping in mind that $\mathfrak{b}_{mp} = \mathfrak{b}_p^m \subseteq \overline{\mathfrak{b}_p^m}$, one finds

$$\begin{aligned} H^0\big(X^+, (L^+)^{\otimes m}\big) &= H^0\big(X, \nu_*((L^+)^{\otimes m})\big) \\ &= H^0\big(X, L^{\otimes mp} \otimes \overline{\mathfrak{b}_p^m}\big) \\ &= H^0\big(X, L^{\otimes mp}\big), \end{aligned}$$

as required.) □

Remark 2.1.32. (Theorem of Zariski–Fujita). In his fundamental paper [623], Zariski proved a surprising fact about linear series having finite base loci:

> **Theorem.** *Let L be a line bundle on a projective variety X with the property that the base locus $\mathrm{Bs}\big(|L|\big)$ is a finite set. Then L is semiample, i.e. $L^{\otimes m}$ is free for some $m > 0$.*

Fujita [194] generalized this by showing that the same conclusion holds assuming that L restricts to an ample line bundle on the base locus $\mathrm{Bs}\big(|L|\big)$. We refer to [22, Theorem 9.17] for a very simple proof of Zariski's theorem, based on an extension of 2.1.29 to the multi-graded ring defined by several linear series. See [147] for a clean formulation of the proof of Fujita's theorem. Keel [329], [330] studies some strengthenings of these facts that hold for varieties defined over the algebraic closure of a finite field. □

2.1.C Iitaka Fibration

We now give the asymptotic analysis of the rational mappings

$$\phi_k = \phi_{|L^{\otimes k}|} \,:\, X \dashrightarrow Y_k \;\subseteq\; \mathbf{P} \;=\; \mathbf{P}H^0\big(X, L^{\otimes k}\big)$$

in the general setting. As above, we ignore the ambient projective space $\mathbf{P}$ and view ϕ_k as a rational mapping to Y_k. The main result is a birational analogue of Theorem 2.1.27:

Theorem 2.1.33. (Iitaka fibrations). *Let X be a normal projective variety, and L a line bundle on X such that $\kappa(X, L) > 0$. Then for all sufficiently large $k \in \mathbf{N}(X, L)$, the rational mappings $\phi_k : X \dashrightarrow Y_k$ are birationally equivalent to a fixed algebraic fibre space*

$$\phi_\infty \,:\, X_\infty \longrightarrow Y_\infty$$

of normal varieties, and the restriction of L to a very general fibre of ϕ_∞ has Iitaka dimension $= 0$. More specifically, there exists for large $k \in \mathbf{N}(X, L)$ a commutative diagram

$$\begin{array}{ccc} X & \xleftarrow{\;\;u_\infty\;\;} & X_\infty \\ {\scriptstyle \phi_k}\Big\downarrow & & \Big\downarrow{\scriptstyle \phi_\infty} \\ Y_k & \underset{\nu_k}{\dashleftarrow} & Y_\infty \end{array}$$

of rational maps and morphisms, where the horizontal maps are birational and u_∞ is a morphism. One has $\dim Y_\infty = \kappa(X, L)$. Moreover, if we set $L_\infty = u_\infty^* L$, and take $F \subseteq X_\infty$ to be a very general fibre of ϕ_∞, then

$$\kappa\big(F, L_\infty \,|\, F\big) \;=\; 0. \tag{2.3}$$

More precisely, the assertion is that the last displayed formula holds for the fibres of ϕ_∞ over all points in the complement of the union of countably many proper subvarieties of Y_∞.

Definition 2.1.34. (Iitaka fibration associated to a line bundle). In the setting of the theorem, $\phi_\infty : X_\infty \longrightarrow Y_\infty$ is the *Iitaka fibration* associated to L. It is unique up to birational equivalence. The Iitaka fibration of a divisor D is defined by passing to $\mathcal{O}_X(D)$. □

Remark 2.1.35. Conversely, if

$$\lambda : X \dashrightarrow W$$

is a rational fibre space of normal varieties (in the sense of Example 2.1.12), and if the restriction of L to a very general fibre F of λ has Iitaka dimension zero, i.e. $\kappa(F, L|F) = 0$, then λ factors through the Iitaka fibration of L. (See [439, §1].) □

Definition 2.1.36. (Iitaka fibration of a variety). The Iitaka fibration of an irreducible variety X is by definition the Iitaka fibration associated to the canonical bundle on any non-singular model of X. A very general fibre F of the Iitaka fibration satisfies $\kappa(F) = 0$ (cf. [439, §2]). □

Example 2.1.37. (Subvarieties of abelian varieties). Let $X \subseteq A$ be an n-dimensional irreducible subvariety of an abelian variety, and denote by $\mathbf{G}$ the Grassmannian of n-dimensional subspaces of the tangent space T_0A. Then the Gauss mapping determines a rational map

$$\gamma : X \dashrightarrow Y = \gamma(X) \subseteq \mathbf{G},$$

which can be identified with the Iitaka fibration of X. (This is a result of Ueno–Kawamata–Kollár: see [439, (3.10)].) □

The proof of Theorem 2.1.33 is straight-forward but a bit messy: the idea is simply to follow the outline of the argument establishing the corresponding statement for free fibrations. Mori presents a quicker approach from a more algebraic point of view in [439].

Proof of Theorem 2.1.33. Fix any $m \in \mathbf{N}(X, L)$ such that $\dim Y_m = \kappa(X, L)$. *Step 1.* We first argue that if $k \gg 0$ then all the mappings $\phi_{km} : X \dashrightarrow Y_{km}$ are birationally equivalent to a fixed algebraic fibre space $\psi_{(m)} : X_{(m)} \longrightarrow Y_{(m)}$ of normal varieties.

In fact, let $u_m : X_{(m)} \longrightarrow X$ be a resolution of indeterminacies of ϕ_m, i.e. a birational morphism with

$$u_m^* \,|\, L^{\otimes m} \,| = |\, M_m \,| + F_m,$$

where M_m is a globally generated line bundle on the normal variety $X_{(m)}$, F_m is the fixed divisor of $|\, u_m^* L^{\otimes m} \,|$, and

$$\psi_m : X_{(m)} \longrightarrow Y_m \subseteq \mathbf{P}H^0\big(X_{(m)}, M_m\big) = \mathbf{P}H^0\big(X, L^{\otimes m}\big)$$

is the morphism defined by $|\, M_m \,|$. The situation is summarized by the top triangle of the following commutative diagram:

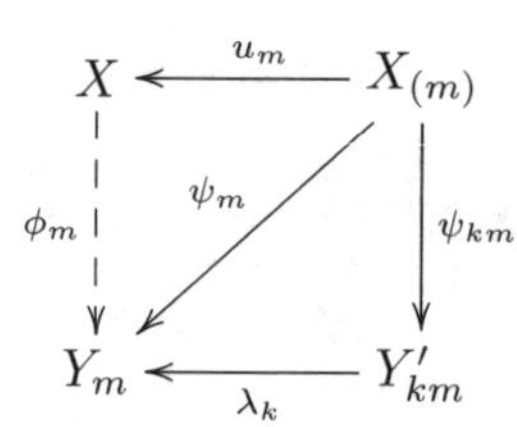

Consider now the morphism $\psi_{km} : X_{(m)} \longrightarrow Y'_{km}$ defined by $|M_m^{\otimes k}|$. Then, as indicated in the diagram, ψ_m factors as the composition

$$X_{(m)} \xrightarrow{\psi_{km}} Y'_{km} \xrightarrow{\lambda_k} Y_m,$$

with λ_k finite. It follows from Theorem 2.1.27 that for $k \gg 0$ the ψ_{km} stabilize to a fixed fibre space $\psi_{(m)} : X_{(m)} \longrightarrow Y_{(m)}$. On the other hand, $|M_m^{\otimes k}|$ defines a linear subseries of $|u_m^* L^{\otimes km}|$, and hence also of $|L^{\otimes km}|$. Since $X_{(m)}$ is birational to X we can view $\psi_{(m)}$ as a rational mapping from X to $Y_{(m)}$, and then we get a factorization

$$\begin{array}{ccc} X_{(m)} \sim X & & \\ & \overset{\phi_{km}}{\searrow} & \\ \Big\downarrow \psi_{(m)} & & Y_{km} \\ & \overset{\mu_k}{\swarrow} & \\ Y_{(m)} & & \end{array}$$

with μ_k generically finite. But $\mathbf{C}(Y_{(m)})$ is algebraically closed in $\mathbf{C}(X)$ (Example 2.1.12), and hence μ_k is birational. So ϕ_{km} coincides birationally with $\psi_{(m)} : X_{(m)} \longrightarrow Y_{(m)}$, as asserted.

Step 2. The fibre spaces $X_{(m)} \longrightarrow Y_{(m)}$ just constructed are all birationally isomorphic, and the next step is to choose a common model.

Specifically, replacing L by $L^{\otimes e}$ we may assume first of all that L has exponent $e(L) = 1$. Fix then two large relatively prime integers p and q such that $\dim Y_p = \dim Y_q = \kappa(X, L)$ (Example 2.1.10). With notation as in Step 1 there exists by construction an integer $m \gg 0$ such that $Y_{(p)}$ and $Y_{(q)}$ are the images of $X_{(p)}$ and $X_{(q)}$ under the morphisms defined by the complete linear series $|M_p^{\otimes(p^{m-1})}|$ and $|M_q^{\otimes(q^{m-1})}|$ respectively. Fix next a normal variety X_∞ together with birational morphisms

$$v_p : X_\infty \longrightarrow X_{(p)} \quad , \quad v_q : X_\infty \longrightarrow X_{(q)}$$

such that $u_p \circ v_p = u_q \circ v_q$. Write $u_\infty : X_\infty \longrightarrow X$ for the resulting birational morphism determined by u_p and u_q. Now consider on X_∞ the globally generated line bundle

$$M_{p,q} = v_p^* M_p^{\otimes(p^{m-1})} \otimes v_q^* M_q^{\otimes(q^{m-1})}.$$

Denote by Y_∞ the normalization of the image of X_∞ under the morphism defined by the complete linear series $|M_{p,q}|$, and by $\phi_\infty : X_\infty \longrightarrow Y_\infty$ the corresponding mapping. Then Y_∞ maps finitely to the Segre embedding of the product $Y_{(p)} \times Y_{(q)}$, and so one has morphisms

$$w_p : Y_\infty \longrightarrow Y_{(p)} \quad , \quad w_q : Y_\infty \longrightarrow Y_{(q)},$$

sitting in the commutative diagram

$$\begin{array}{ccccc}
X_{(p)} & \xleftarrow{\;v_p\;} & X_\infty & \xrightarrow{\;v_q\;} & X_{(q)} \\
\Big\downarrow{\scriptstyle \psi_{(p)}} & & \Big\downarrow{\scriptstyle \phi_\infty} & & \Big\downarrow{\scriptstyle \psi_{(q)}} \\
Y_{(p)} & \xleftarrow[\;w_p\;]{} & Y_\infty & \xrightarrow[\;w_q\;]{} & Y_{(q)}
\end{array}$$

Then $\dim Y_\infty = \kappa(X, L)$, and therefore w_p and w_q are generically finite; since they factor the algebraic fibre spaces $\psi_{(p)} \circ v_p$ and $\psi_{(q)} \circ v_q$, they are actually birational. By the same token, ϕ_∞ is an algebraic fibre space since the composition $w_p \circ \phi_\infty$ is. Note that by construction Y_∞ carries an ample and globally generated line bundle $A_{p,q}$ such that $\phi_\infty^* A_{p,q} = M_{p,q}$ (viz. the pull-back of the hyperplane line bundle defining the appropriate Segre embedding of $Y_{(p)} \times Y_{(q)}$).

Now fix positive integers $c, d \geq 1$. Then one has

$$\begin{aligned}
H^0\big(X_\infty, M_{p,q}\big) \;&\subseteq\; H^0\big(X_\infty, u_p^* M_p^{\otimes(cp^{m-1})} \otimes u_q^* M_q^{\otimes(dq^{m-1})}\big) \\
&\subseteq\; H^0\big(X_\infty, u_\infty^* L^{\otimes(cp^m + dq^m)}\big),
\end{aligned}$$

the first inclusion being defined via multiplication by a fixed section of $u_p^* M_p^{\otimes(c-1)p^{m-1}} \otimes u_q^* M_q^{\otimes(d-1)q^{m-1}}$. This gives rise in the usual way to a factorization

$$\begin{array}{ccc}
X_\infty & \xrightarrow{\;\;u_\infty\;\;} & X \\
\Big\downarrow{\scriptstyle \phi_\infty} & & \Big\downarrow{\scriptstyle \phi_{cp^m+dq^m}} \\
Y_\infty & \xleftarrow[\;\mu_{cp^m+dq^m}\;]{} & Y_{cp^m+dq^m}
\end{array}$$

with $\mu_{cp^m+dq^m}$ generically finite. It follows as before from Example 2.1.12 that $\mu_{cp^m+dq^m}$ must be birational, and since every $k \gg 0$ is of the form $cp^m + dq^m$, taking $\nu_k = \mu_k^{-1}$ gives the diagram appearing in the statement of the Theorem.

Step 3. It remains to show that a very general fibre F of ϕ_∞ has L-dimension zero.

Set $L_\infty = u_\infty^* L$. Since evidently $\kappa(F, L_\infty | F) \geq 0$, the issue is to establish the reverse inequality. To this end, fix a very general point $y \in Y_\infty$, and let

$$F \;=\; F_y \;=\; \phi_\infty^{-1}(y) \;\subseteq\; X_\infty.$$

We can assume that ν_k is defined and regular at y for every $k \gg 0$, and that $u_\infty(F)$ is not contained in the locus of indeterminacy of any of the ϕ_k.[4] Then $\phi_k \circ u_\infty$ maps F to a point, and this means that the restriction morphism

[4] We are imposing here countably many open conditions on y, explaining why we require y to be very general.

$$\rho_k : H^0\big(X_\infty, L_\infty^{\otimes k}\big) \longrightarrow H^0\big(F, L_\infty^{\otimes k} \,|\, F\big)$$

has rank one for every $k \gg 0$. We will show that in fact ρ_k is surjective for every $k \gg 0$.

Fix next a very ample line bundle B on Y_∞. We assert that there is a large positive integer $m_0 > 0$ such that

$$H^0\big(X_\infty, L_\infty^{\otimes m_0} \otimes \phi_\infty^*(B^*)\big) \;\neq\; 0. \tag{2.4}$$

Granting this for the time being, we get for fixed k and any $r > 0$ a diagram

$$\begin{array}{ccc}
H^0\big(X_\infty, L_\infty^{\otimes k} \otimes \phi_\infty^* B^{\otimes r}\big) & \hookrightarrow & H^0\big(X_\infty, L_\infty^{\otimes(k+rm_0)}\big) \\
\Big\downarrow{\scriptstyle \beta_{k,r}} & & \Big\downarrow{\scriptstyle \rho_{k+rm_0}} \\
H^0\big(F, \big(L_\infty^{\otimes k} \otimes \phi_\infty^* B^{\otimes r}\big) \,|\, F\big) & \hookrightarrow & H^0\big(F, L_\infty^{\otimes(k+rm_0)} \,|\, F\big)
\end{array}$$

the vertical maps arising via restriction to F. Now for general $F = F_y$ the homomorphism $\beta_{k,r}$ on the left is identified with the map

$$H^0\big(Y_\infty, \phi_{\infty,*}\big(L_\infty^{\otimes k}\big) \otimes B^{\otimes r}\big) \longrightarrow \Big(\phi_{\infty,*}\big(L_\infty^{\otimes k}\big) \otimes B^{\otimes r}\Big) \otimes \mathbf{C}(y)$$

obtained by evaluating sections of the indicated direct image sheaf at the point $y \in Y$; here $\mathbf{C}(y) = \mathcal{O}_y Y_\infty / \mathfrak{m}_y$ is the residue field of Y_∞ at y. But since B is ample, for fixed k the sheaf $\phi_{\infty,*}\big(L_\infty^{\otimes k}\big) \otimes B^{\otimes r}$ is globally generated for $r \gg 0$. Therefore $\beta_{k,r}$ is surjective for fixed k and $r \gg 0$. On the other hand, we've seen that ρ_{k+rm_0} has rank one. It follows from the diagram that rank $\beta_{k,r} = 1$, and therefore

$$\dim H^0\big(F, \big(L_\infty^{\otimes k} \otimes \phi_\infty^* B^{\otimes r}\big) \,|\, F\big) \;=\; 1.$$

But since $\phi_\infty^* B$ is trivial along F, this implies that

$$h^0\big(F, L_\infty^{\otimes k} \,|\, F\big) \;=\; 1,$$

as required.

It remains only to verify the non-vanishing (2.4). Recall that Y_∞ carries an ample and globally generated line bundle $A = A_{p,q}$ pulling back to $M_{p,q}$ on X_∞. Then $A^{\otimes m_1} \otimes B^*$ has a non-zero section for some $m_1 \gg 0$. On the other hand, $M_{p,q}^{\otimes m_1}$ is by construction a subsheaf of $L_\infty^{\otimes(p^m+q^m)m_1}$, and (2.4) follows. □

We conclude with an alternative characterization of the Iitaka dimension of a line bundle.

Corollary 2.1.38. *Let L be a line bundle on an irreducible normal projective variety X, and set $\kappa = \kappa(X, L)$. Then there are constants $a, A > 0$ such that*

$$a \cdot m^{\kappa} \ \le\ h^0\big(X, L^{\otimes m}\big) \ \le\ A \cdot m^{\kappa}$$

for all sufficiently large $m \in \mathbf{N}(X, L)$. □

Remark 2.1.39. It seems to be unknown whether or not

$$\liminf_{m \in \mathbf{N}(L)} \frac{h^0\big(X, L^{\otimes m}\big)}{m^{\kappa}} \ =\ \limsup_{m \in \mathbf{N}(L)} \frac{h^0\big(X, L^{\otimes m}\big)}{m^{\kappa}}.$$

However when $\kappa(X, L) = \dim X$ — i.e. when L is big (Definition 2.2.1) — then in fact $h^0(X, L^{\otimes m})/m^{\dim X}$ has a limit: see Example 11.4.7. □

Sketch of Proof of Corollary 2.1.38. The assertion is invariant under replacing X by a birational modification, so we may assume that $\phi : X \longrightarrow Y$ is the Iitaka fibration associated to L.[5] By resolving singularities, we may suppose also that X is non-singular. It follows from equation (2.4) in the proof of Theorem 2.1.33 that there is a large positive integer m_0 together with an ample bundle B on Y such that $L^{m_0} \otimes \phi^* B^*$ has a non-zero section. This implies that

$$h^0\big(X, L^{\otimes(\ell \cdot m_0)}\big) \ \ge\ h^0\big(Y, B^{\otimes \ell}\big) \ =\ \frac{\ell^{\kappa}}{\kappa!} \int_Y c_1(B)^{\kappa} + o(\ell^{\kappa}).$$

This gives a lower bound on $h^0\big(X, L^{\otimes m}\big)$ of the appropriate shape for all sufficiently divisible $m \in \mathbf{N}(L)$, and we leave it to the reader to extend the bound to all $m \in \mathbf{N}(L)$.

For the reverse inequality we will suppose for simplicity that $H^0\big(X, L\big) \neq 0$, so that $\mathbf{N}(X, L) = \mathbf{N}$: then take D to be an effective divisor on X with $\mathcal{O}_X(D) = L$. Now decompose D as a sum

$$D \ =\ D_{\text{vert}} + D_{\text{horiz}},$$

where D_{vert} consists of those components of D that map via ϕ to a proper subvariety of Y, while every component of D_{horiz} maps onto Y. Thus every fibre F of ϕ meets D_{horiz} but the general fibre is disjoint from D_{vert}. We assert that

$$H^0\big(X, \mathcal{O}_X(mD_{\text{vert}})\big) \ =\ H^0\big(X, \mathcal{O}_X(mD)\big) \quad \text{for all } \ m \ge 0. \qquad (*)$$

Grant this for the moment. We can find an ample divisor H on Y such that $D_{\text{vert}} \preccurlyeq \phi^* H$, and then by (*),

$$h^0\big(X, \mathcal{O}_X(mD)\big) \ =\ h^0\big(X, \mathcal{O}_X(mD_{\text{vert}})\big) \ \le\ h^0\big(Y, \mathcal{O}_Y(mH)\big).$$

The required upper bound follows as before from asymptotic Riemann–Roch on Y. As for (*), it is equivalent to show that every section of $\mathcal{O}_X(mD)$ contains mD_{horiz} as a base-divisor. But if this fails, then

[5] I.e. we take $X = X_{\infty}$, $Y = Y_{\infty}$, and $L = L_{\infty}$.

$$h^0\big(F, \mathcal{O}_F(mD)\big) \;=\; h^0\big(F, \mathcal{O}_F(mD_{\text{horiz}})\big) \;\geq\; 2$$

for a general fibre F of ϕ, and this contradicts the fact (Theorem 2.1.33) that $\kappa(F, L \,|\, F) = 0$ for a very general fibre F. □

2.2 Big Line Bundles and Divisors

In this section we study a particularly important class of line bundles, namely those of maximal Iitaka dimension. The basic facts are presented in the first subsection, and we discuss the big and pseudoeffective cones in Section 2.2.B. Finally, in the third subsection we define and study the volume of a big divisor.

Throughout this section, X continues to denote unless otherwise stated an irreducible projective variety of dimension n.

2.2.A Basic Properties of Big Divisors

Definition 2.2.1. (Big). A line bundle L on the irreducible projective variety X is *big* if $\kappa(X, L) = \dim X$. A Cartier divisor D on X is *big* if $\mathcal{O}_X(D)$ is so. □

For example, the pullback of an ample line bundle under a generically finite morphism is big. When X is normal, the Iitaka fibration theorem implies that L is big if and only if the mapping $\phi_m : X \dashrightarrow \mathbf{P}H^0\big(X, L^{\otimes m}\big)$ defined by $L^{\otimes m}$ is birational onto its image for some $m > 0$.[6]

Example 2.2.2. (Varieties of general type). A smooth projective variety X is (by definition) of general type if and only if K_X is big. □

Recall that the Iitaka dimension of a divisor on a variety X is computed by passing to the normalization of X. We start by observing that in the present situation this is unnecessary.

Lemma 2.2.3. *Assume that X is a projective variety of dimension n. A divisor D on X is big if and only if there is a constant $C > 0$ such that*

$$h^0\big(X, \mathcal{O}_X(mD)\big) \;\geq\; C \cdot m^n \tag{2.5}$$

for all sufficiently large $m \in \mathbf{N}(X, D)$.

Proof. If X is normal, this follows from the definition and Theorem 2.1.33. In general, let $\nu : X' \longrightarrow X$ be the normalization of X, and put $D' = \nu^* D$. We will show that D' has maximal Iitaka dimension on X' if and only if (2.5) holds on X. Using the projection formula, one has

[6] This also follows directly from Corollary 2.2.7 below.

$$h^0\big(X', \mathcal{O}_{X'}(mD')\big) = h^0\big(X, \nu_*\mathcal{O}_{X'} \otimes \mathcal{O}_X(mD)\big)$$

for all m. On the other hand, there is an exact sequence

$$0 \longrightarrow \mathcal{O}_X \longrightarrow \nu_*\mathcal{O}_{X'} \longrightarrow \eta \longrightarrow 0$$

of sheaves on X where η is supported on a scheme of dimension $\leq n-1$. Therefore

$$\begin{aligned} h^0\big(X, \mathcal{O}_X(mD)\big) &\leq h^0\big(X, \nu_*\mathcal{O}_{X'} \otimes \mathcal{O}_X(mD)\big) \\ &\leq h^0\big(X, \mathcal{O}_X(mD)\big) + h^0\big(X, \eta \otimes \mathcal{O}_X(mD)\big). \end{aligned}$$

But $h^0\big(X, \eta \otimes \mathcal{O}_X(mD)\big) = O(m^{n-1})$ since η is supported on a proper subscheme (Example 1.2.33). It follows that $h^0\big(X, \mathcal{O}_X(mD)\big)$ grows like m^n if and only if $h^0\big(X', \mathcal{O}_{X'}(mD')\big)$ does. □

Example 2.2.4. (Blow-ups of projective space). The blow-up of projective space at a finite set of points provides some interesting examples. Let $S \subseteq \mathbf{P}^n$ be a finite set, viewed as a reduced scheme, and let

$$\mu : X = \mathrm{Bl}_S(\mathbf{P}^n) \longrightarrow \mathbf{P}^n$$

be the blowing up of $\mathbf{P}^n$ along S. Denote by $E = E_S$ and $H = H_S$ respectively the exceptional divisor and the pullback of a hyperplane class, and put $D = dH - rE$. One has

$$\mu_*\mathcal{O}_X(mD) = \mathcal{O}_{\mathbf{P}^n}(md) \otimes \mathcal{I}^{mr},$$

$\mathcal{I} = \mathcal{I}_{S/\mathbf{P}}$ being the ideal sheaf of S in $\mathbf{P}^n$, and therefore $H^0\big(X, \mathcal{O}_X(mD)\big)$ is identified with the space of hypersurfaces of degree md that vanish to order $\geq mr$ at each point of S. Assume that $\#S = s$. Then D is big provided that $d^n > s \cdot r^n$. On the other hand, it is quite possible that D fails to be nef: for example, if $d < 2r$ then the proper transform of a line through two points in S has negative intersection with D. □

Remark 2.2.5. (Hyperplane sections). Note that we deal with bigness only for line bundles on an irreducible projective variety. On the other hand, it is sometimes convenient to be able to work inductively with hyperplane sections. Although with a little more care one could avoid this, it is useful to recall that if X is an irreducible projective variety of dimension ≥ 2, and if H is a very ample divisor on X that is general in its linear series, then H is itself irreducible and reduced. In fact, being a general hyperplane section of a reduced scheme, H is reduced (cf. [186, Corollary 3.4.9] or [308, Theorem 6.3 (3)]), and it is irreducible by Theorem 3.3.1. □

The first essential result is due to Kodaira.

Proposition 2.2.6. (Kodaira's lemma). *Let D be a big Cartier divisor and F an arbitrary effective Cartier divisor on X. Then*

$$H^0\big(X, \mathcal{O}_X(mD - F)\big) \neq 0$$

for all sufficiently large $m \in \mathbf{N}(X, D)$.

Proof. Suppose that $\dim X = n$, and consider the exact sequence

$$0 \longrightarrow \mathcal{O}_X(mD - F) \longrightarrow \mathcal{O}_X(mD) \longrightarrow \mathcal{O}_F(mD) \longrightarrow 0.$$

Since D is big, there is a constant $c > 0$ such that $h^0\big(X, \mathcal{O}_X(mD)\big) \geq c \cdot m^n$ for sufficiently large $m \in \mathbf{N}(X, D)$. On the other hand, F being a scheme of dimension $n-1$, $h^0\big(F, \mathcal{O}_F(mD)\big)$ grows at most like m^{n-1} (Example 1.2.20 or 1.2.33). Therefore

$$h^0\big(X, \mathcal{O}_X(mD)\big) > h^0\big(F, \mathcal{O}_F(mD)\big)$$

for large $m \in \mathbf{N}(X, D)$, and the assertion follows from the displayed sequence. □

Kodaira's lemma has several important consequences. First, it leads to a useful characterization of big divisors:

Corollary 2.2.7. (Characterization of big divisors). *Let D be a divisor on an irreducible projective variety X. Then the following are equivalent:*

(i) *D is big.*

(ii). *For any ample integer divisor A on X, there exists a positive integer $m > 0$ and an effective divisor N on X such that $mD \equiv_{\text{lin}} A + N$.*

(iii). *Same as in (ii) for some ample divisor A.*

(iv). *There exists an ample divisor A, a positive integer $m > 0$, and an effective divisor N such that $mD \equiv_{\text{num}} A + N$.*

Proof. Assuming that D is big, take $r \gg 0$ so that $rA \equiv_{\text{lin}} H_r$ and $(r+1)A \equiv_{\text{lin}} H_{r+1}$ are both effective. Apply 2.2.6 with $F = H_{r+1}$ to find a positive integer m and an effective divisor N' with

$$mD \equiv_{\text{lin}} H_{r+1} + N' \equiv_{\text{lin}} A + \big(H_r + N'\big).$$

Taking $N = H_r + N'$ gives (ii). The implications (ii) $\Longrightarrow$ (iii) $\Longrightarrow$ (iv) being trivial, we assume (iv) and deduce (i). If $mD \equiv_{\text{num}} A + N$, then $mD - N$ is numerically equivalent to an ample divisor, and hence ample. So after possibly passing to an even larger multiple of D we can assume that $mD \equiv_{\text{lin}} H + N'$, where H is very ample and N' is effective. But then

$$\kappa(X, D) \geq \kappa(X, H) = \dim X,$$

so D is big. □

Corollary 2.2.8. (Numerical nature of bigness). *The bigness of a divisor D depends only on its numerical equivalence class.*

Proof. This follows from statement (iv) in the previous corollary. □

Example 2.2.9. (Cohomological characterization of big divisors). Let D be a big divisor on a projective variety X. Then D is big if and only if the following holds:

> For any coherent sheaf $\mathcal{F}$ on X, there exists a positive integer $m = m(\mathcal{F})$ such that $\mathcal{F} \otimes \mathcal{O}_X(mD)$ is generically globally generated, i.e. such that the natural map
>
> $$H^0\big(X, \mathcal{F} \otimes \mathcal{O}_X(mD)\big) \otimes_{\mathbf{C}} \mathcal{O}_X \longrightarrow \mathcal{F} \otimes \mathcal{O}_X(mD)$$
>
> is generically surjective.

(Combine 2.2.7 (ii) with the theorem of Cartan–Serre–Grothendieck.) This statement is given by Mori in [439]. □

Concerning the exponent of a big divisor, one finds:

Corollary 2.2.10. (Exponent of big divisors). *If D is big then $e(D) = 1$, i.e. every sufficiently large multiple of D is effective.*

Proof. Given D, choose a very ample divisor H on X such that $H - D \equiv_{\text{lin}} H_1$ is effective. By Corollary 2.2.7 there is an integer $m \in \mathbf{N}(X, D)$ such that $mD \equiv_{\text{lin}} H + N$ for some effective divisor N. Then

$$(m-1)D \equiv_{\text{lin}} (H - D) + N \equiv_{\text{lin}} H_1 + N$$

is also effective. In other words, the two consecutive integers m and $m - 1$ both lie in the semigroup $\mathbf{N}(X, D)$ of D, and hence $e(D) = 1$. □

Finally, Kodaira's lemma implies that the restriction of a big line bundle to a "sufficiently general" subvariety remains big:

Corollary 2.2.11. (Restrictions of big divisors). *Let L be a big line bundle on a projective variety X. There is a proper Zariski-closed subset $V \subseteq X$ having the property that if $Y \subseteq X$ is any subvariety of X not contained in V then the restriction $L_Y = L \mid Y$ of L to Y is a big line bundle on Y. In particular, if H is a general member of a very ample linear series, then L_H is big.*

Proof. Say $L = \mathcal{O}_X(D)$, and using Corollary 2.2.7 write $mD \equiv_{\text{lin}} H + N$ where N is effective and H is very ample. Take V to be the support of N. If $Y \not\subseteq V$, then the restriction mD_Y of mD to Y is again the sum of a very ample and an effective divisor, and hence is big. □

Example 2.2.12. One should not expect that an arbitrary restriction of a big divisor is big. For instance let $X = \mathrm{Bl}_P \mathbf{P}^2$ be the blowing up of the plane at a point P, and let E and H denote respectively the exceptional divisor and the pullback of a line. Then $\mathcal{O}_X(H+E)$ is big, but $\mathcal{O}_E(H+E) = \mathcal{O}_{\mathbf{P}^1}(-1)$ is not. □

Example 2.2.13. (Variational pathologies). It is possible to have a family of line bundles L_t on varieties X_t, parameterized say by a curve T, such that $\kappa(X_t, L_t) = -\infty$ for general $t \in T$, but with L_0 big for some special point $0 \in T$. For example, starting with a finite set $S \subset \mathbf{P}^2$ of points in the plane, consider as in 2.2.4 the blow-up

$$X_S = \mathrm{Bl}_S(\mathbf{P}^2) \longrightarrow \mathbf{P}^2$$

of $\mathbf{P}^2$ along S, and put $D_S = 2H_S - E_S$. By taking S to be a large number of sufficiently general points, one can assume that there are no curves of degree $2m$ in $\mathbf{P}^2$ having multiplicity $\geq m$ at each of the points of S. This means that

$$H^0\big(X_S, \mathcal{O}_{X_S}(mD_S)\big) = 0 \quad \text{for all} \;\; m > 0.$$

On the other hand, if the points of S are collinear, then $D_S - H_S$ is effective, and hence D_S is big. The family (X_t, L_t) is obtained by letting S vary in a suitable one-parameter family S_t. Observe however that on a fixed variety, bigness is invariant under deformation thanks to Corollary 2.2.8. □

Note that an integral divisor A is big if and only if every (or equivalently some) positive multiple of A is big. This leads to a natural notion of bigness for a $\mathbf{Q}$-divisor:

Definition 2.2.14. (Big Q-divisors). A $\mathbf{Q}$-divisor D is *big* if there is a positive integer $m > 0$ such that mD is integral and big. □

As in 2.2.8, bigness is a numerical property of $\mathbf{Q}$-divisors. In the next subsection we extend the definition to $\mathbf{R}$-divisors and discuss the corresponding cone in $N^1(X)_{\mathbf{R}}$.

We now come to a basic result of Siu [537] giving a numerical condition for a difference of nef divisors to be big.

Theorem 2.2.15. (Numerical criterion for bigness). *Let X be a projective variety of dimension n, and let D and E be nef $\mathbf{Q}$-divisors on X. Assume that*

$$\big(D^n\big) \; > \; n \cdot \big(D^{n-1} \cdot E\big). \tag{2.6}$$

Then $D - E$ is big.

Proof. By continuity the hypothesis (2.6) continues to hold if we fix an ample divisor H and replace D and E respectively by $D + \varepsilon H$ and $E + \varepsilon H$ for $0 < \varepsilon \ll 1$. Therefore we may suppose that both D and E are ample. Then, since (2.6) is preserved under multiplying D and E by a large integer, we may — and do — assume also that D and E are integral and very ample.

Choose a sequence $E_1, E_2, E_3, \ldots \in |E|$ of general divisors linearly equivalent to E, and fix an integer $m \geq 1$. Then

$$\mathcal{O}_X\big(m(D-E)\big) \;\cong\; \mathcal{O}_X\Big(mD - \sum_{i=1}^{m} E_i\Big),$$

so $H^0\big(X, \mathcal{O}_X(m(D-E))\big) \cong H^0\big(X, \mathcal{O}_X(mD - \sum E_i)\big)$ is identified with the group of sections of $\mathcal{O}_X(mD)$ vanishing on each of the divisors $E_1, \ldots, E_m$. This in turn may be analyzed via the exact sequence

$$0 \longrightarrow H^0\big(X, \mathcal{O}_X(mD - \textstyle\sum E_i)\big) \longrightarrow H^0\big(X, \mathcal{O}_X(mD)\big) \longrightarrow \displaystyle\bigoplus_{i=1}^{m} H^0\big(E_i, \mathcal{O}_{E_i}(mD)\big). \quad (2.7)$$

For large m the dimensions $h^0\big(E_i, \mathcal{O}_{E_i}(mD)\big)$ are independent of i, and then asymptotic Riemann–Roch (Example 1.2.19) on X and each E_i yields

$$\begin{aligned} h^0\big(X, \mathcal{O}_X(m(D-E))\big) &\geq h^0\big(X, mD\big) - \sum_{i=1}^{m} h^0\big(E_i, \mathcal{O}_{E_i}(mD)\big) \\ &= \frac{\big(D^n\big)}{n!} m^n - \sum_{i=1}^{m} \frac{\big(D^{n-1} \cdot E_i\big)}{(n-1)!} m^{n-1} + O(m^{n-1}) \\ &= \frac{\big(D^n\big)}{n!} m^n - n\, \frac{\big(D^{n-1} \cdot E\big)}{n!} m^n + O(m^{n-1}). \end{aligned}$$

In particular, the hypothesis (2.6) implies that $h^0\big(X, \mathcal{O}_X(m(D-E))\big)$ grows like a positive multiple of m^n, and the theorem follows. □

As a first consequence, one obtains an important characterization of bigness for nef divisors.

Theorem 2.2.16. (Bigness of nef divisors). *Let D be a nef divisor on an irreducible projective variety X of dimension n. Then D is big if and only if its top self-intersection is strictly positive, i.e. $\big(D^n\big) > 0$.*

Proof. Suppose to begin with that $\big(D^n\big) > 0$. Then the hypothesis of Theorem 2.2.15 is satisfied with $E = 0$, and hence D is big. Conversely, suppose that D is nef and big. Then $mD \equiv_{\text{lin}} H + N$ for some very ample H and effective N, and suitable $m > 0$ (2.2.7). But $\big(D^{n-1} \cdot N\big) \geq 0$ by Kleiman's Theorem 1.4.9, and therefore

$$\begin{aligned} m \cdot \big(D^n\big) &= \big((H+N) \cdot D^{n-1}\big) \\ &\geq \big(H \cdot D^{n-1}\big). \end{aligned}$$

In light of Corollary 2.2.11 we may assume moreover that the restriction $D \mid H$ is a big and nef divisor on H. So by induction $\big(D^{n-1} \cdot H\big) > 0$, and the required inequality $\big(D^n\big) > 0$ follows. □

Remark 2.2.17. (Alternative approach to Theorem 2.2.16). The criterion for the bigness of nef divisors alternatively follows from asymptotic Riemann–Roch in the form of Corollary 1.4.41 (see also 1.4.42). □

Example 2.2.18. (Big line bundles with negative numerics). One should keep in mind that absent nefness, the intersection numbers associated with a big line bundle can be very negative. For example, let X be a smooth surface containing a (-1)-curve E and let A be a very ample divisor on X. Then $D_\ell = A + \ell E$ is big for all $\ell \geq 0$ but $\big(D_\ell \cdot D_\ell\big) \ll 0$ for large ℓ. □

Example 2.2.19. (Alternative characterization of big and nef divisors). Let D be a divisor on an irreducible projective variety X. Then D is nef and big if and only if there is an effective divisor N such that $D - \frac{1}{k}N$ is ample for all $k \gg 0$. (Assume that D is big and nef. Then there is a positive integer m, an effective divisor N, and an ample divisor A with $mD \equiv_{\text{num}} A + N$. Thus for $k > m$,

$$kD \ \equiv_{\text{num}} \ \big((k-m)D + A\big) \ + \ N,$$

and the term in parentheses, being the sum of a nef and an ample divisor, is ample.) □

Remark 2.2.20. (Holomorphic Morse inequalities). One can view Theorem 2.2.15 as a special case of more general inequalities on the cohomology of a difference of nef divisors. Specifically, let X be a projective variety of dimension n, and let D and E be nef divisors on X, and put $F = D - E$. Then for any integer $q \in [0, n]$ one has the two inequalities

$$h^q\big(X, \mathcal{O}_X(mF)\big) \ \leq \ m^n \, \frac{\big(D^{n-q} \cdot E^q\big)}{(n-q)!q!} \ + \ O(m^{n-1}),$$

$$\sum_{i=0}^{q} (-1)^{q-i}\, h^i\big(X, \mathcal{O}_X(mF)\big) \ \leq \ \frac{m^n}{n!} \sum_{i=0}^{q} (-1)^{q-i} \binom{n}{i} \Big(D^{n-i} \cdot E^i\Big) + O(m^{n-1}).$$

Theorem 2.2.15 follows from the case $q = 1$ of the second inequality. These were originally established analytically by Demailly [126, §12] using his holomorphic Morse inequalities [120]. Angelini [11] later gave a simple algebraic proof along the lines of the argument leading to 2.2.15. (Demailly and Angelini assume that X is smooth, but using 1.2.35 one can reduce to this case by passing to a resolution of singularities.) □

2.2.B Pseudoeffective and Big Cones

In this subsection we rephrase 2.2.7 to characterize the cone of all big divisors. As before, X is an irreducible projective variety of dimension n.

We start by extending the definitions to **R**-divisors.

Definition 2.2.21. (Big R-divisors). An **R**-divisor $D \in \mathrm{Div}_{\mathbf{R}}(X)$ is *big* if it can be written in the form

$$D = \sum a_i \cdot D_i$$

where each D_i is a big integral divisor and a_i is a positive real number. □

This is justified by the observation that if D_1 and D_2 are big **Q**-divisors, then so too is $a_1D_1 + a_2D_2$ for any positive rational numbers a_1, a_2.

The formal properties established in the previous subsection extend to the present setting:

Proposition 2.2.22. (Formal properties of big R-divisors). *Let D and D' be* **R***-divisors on X.*

(i). *If $D \equiv_{\mathrm{num}} D'$, then D is big if and only if D' is big.*

(ii). *D is big if and only if $D \equiv_{\mathrm{num}} A + N$ where A is an ample and N is an effective* **R***-divisor.*

Sketch of Proof. For (i) argue as in the proof of Proposition 1.3.13. Turning to (ii), it follows immediately from 2.2.7 (iv) that a big **R**-divisor has an expression of the indicated sort. For the converse one reduces to showing that if B and N are integral divisors, with B big and N effective, then $B + sN$ is big for any real number $s > 0$. If $s \in \mathbf{Q}$ this follows again from 2.2.7. In general, choose positive rational numbers $s_1 < s < s_2$ and $t \in [0, 1]$ such that $s = ts_1 + (1 - t)s_2$. Then

$$B + sN = t\big(B + s_1N\big) + (1 - t)\big(B + s_2N\big)$$

exhibits $B + sN$ as a positive linear combination of big **Q**-divisors. □

Example 2.2.23. (Big and nef R-divisors). Let D be a nef and big **R** divisor. Then there is an effective **R**-divisor N such that $D - \frac{1}{k}N$ is an ample **R**-divisor for every sufficiently large $k \in \mathbf{N}$. (Compare Example 2.2.19.) □

Corollary 2.2.24. *Let $D \in \mathrm{Div}_{\mathbf{R}}(X)$ be a big* **R***-divisor, and let $E_1, \ldots, E_t \in \mathrm{Div}_{\mathbf{R}}(X)$ be arbitrary* **R***-divisors. Then*

$$D + \varepsilon_1 E_1 + \ldots + \varepsilon_t E_t$$

remains big for all sufficiently small real numbers $0 \leq |\varepsilon_i| \ll 1$.

Proof. This follows from statement (ii) of the previous proposition thanks to the open nature of amplitude (Example 1.3.14). □

Note that in view of 2.2.22 (i) it makes sense to talk about a big **R**-divisor class. We can then define some additional cones in $N^1(X)_{\mathbf{R}}$.

Definition 2.2.25. (Big and pseudoeffective cones). The *big cone*

$$\mathrm{Big}(X) \subseteq N^1(X)_{\mathbf{R}}$$

is the convex cone of all big **R**-divisor classes on X. The *pseudoeffective cone*

$$\overline{\mathrm{Eff}}(X) \subseteq N^1(X)_{\mathbf{R}}$$

is the closure of the convex cone spanned by the classes of all effective **R**-divisors. A divisor $D \in \mathrm{Div}_{\mathbf{R}}(X)$ is *pseudoeffective* if its class lies in the pseudoeffective cone.[7] □

One then has

Theorem 2.2.26. *The big cone is the interior of the pseudoeffective cone and the pseudoeffective cone is the closure of the big cone:*

$$\mathrm{Big}(X) = \mathrm{int}\Big(\overline{\mathrm{Eff}}(X)\Big) \quad , \quad \overline{\mathrm{Eff}}(X) = \overline{\mathrm{Big}(X)}.$$

Proof. The pseudoeffective cone is closed by definition, the big cone is open by 2.2.24, and $\mathrm{Big}(X) \subseteq \overline{\mathrm{Eff}}(X)$ thanks to 2.2.22 (ii). It remains to establish the inclusions

$$\overline{\mathrm{Eff}}(X) \subseteq \overline{\mathrm{Big}(X)} \quad , \quad \mathrm{int}\Big(\overline{\mathrm{Eff}}(X)\Big) \subseteq \mathrm{Big}(X).$$

We focus on the first of these. Given $\eta \in \overline{\mathrm{Eff}}(X)$, one can write η as the limit $\eta = \lim_k \eta_k$ of the classes of effective divisors. Fixing an ample class $\alpha \in N^1(X)_{\mathbf{R}}$ one has

$$\eta = \lim_{k \to \infty} \big(\eta_k + \tfrac{1}{k}\alpha\big).$$

Each of the classes $\eta_k + \frac{1}{k}\alpha$ is big thanks to 2.2.22 (ii), so η is a limit of big classes. □

Remark 2.2.27. (Pseudoeffective divisors on a surface). On a smooth projective surface X, where we may identify curves with divisors, the pseudoeffective cone coincides with the closed cone of curves $\overline{\mathrm{NE}}(X)$ (Definition 1.4.26). In particular, it follows from Proposition 1.4.28 that a divisor D on a smooth projective surface X is pseudoeffective if and only if $\big(D \cdot H\big) \geq 0$ for every nef divisor H on X. □

Remark 2.2.28. (Dual of the pseudoeffective cone). We saw in Section 1.4.C that the nef cone $\mathrm{Nef}(X)$ of a projective variety X is dual to the closed cone $\overline{\mathrm{NE}}(X) \subseteq N_1(X)_{\mathbf{R}}$ of effective curves on X. In the same spirit, it is natural to ask what cone in $N_1(X)_{\mathbf{R}}$ is dual to $\overline{\mathrm{Eff}}(X)$. This has only recently been determined, by Boucksom, Demailly, Paun, and Peternell [68]. Roughly

[7] The term "pseudoeffective" (regrettably) seems to be standard: cf. [126], [472], [21].

speaking, they show that the dual of the pseudoeffective cone is spanned by "mobile" curves, i.e. classes of curves arising as complete intersections of ample divisors on a modification of X. The precise statement and proof of the theorem of BDPP appears in Section 11.4.C. □

Remark 2.2.29. (Fujita's rationality conjecture). Let X be a smooth projective variety whose canonical bundle is not pseudoeffective, and let L be an ample divisor on X. Fujita [198], [200] conjectures that the quantity

$$a(X,L) \;=\; \inf\big\{t > 0 \mid (K_X + tL) \text{ is big }\big\}$$

is always rational. This is the analogue for the effective cone of the rationality theorem of Kawamata and Shokurov (cf. [368, Chapter 3.4]). The invariant $a(X,L)$ plays an important role in work of Batyrev, Manin, Tschinkel, and others concerning the number of points of bounded height on varieties for which $-K_X$ is effective (cf. [188]). Tschinkel's survey [573] gives a very nice overview of Fujita's program and its arithmetic significance. □

Finally, although being effective is not characterized numerically, it is useful to have a notion of an effective numerical equivalence class.

Definition 2.2.30. (Effective class). A class $e \in N^1(X)_{\mathbf{Q}}$ or $e \in N^1(X)_{\mathbf{R}}$ is *effective* if e is represented by an effective divisor with the indicated coefficients. □

2.2.C Volume of a Big Divisor

We turn now to an interesting invariant of a big divisor D that measures the asymptotic growth of the linear series $|mD|$ for $m \gg 0$:

Definition 2.2.31. (Volume of a line bundle). Let X be an irreducible projective variety of dimension n, and let L be a line bundle on X. The *volume*[8] of L is defined to be the non-negative real number

$$\mathrm{vol}(L) \;=\; \mathrm{vol}_X(L) \;=\; \limsup_{m\to\infty} \frac{h^0\big(X, L^{\otimes m}\big)}{m^n/n!}. \tag{2.8}$$

The volume $\mathrm{vol}(D) = \mathrm{vol}_X(D)$ of a Cartier divisor D is defined similarly, or by passing to $\mathcal{O}_X(D)$. □

Note that $\mathrm{vol}(L) > 0$ if and only if L is big. If L is nef, then it follows from asymptotic Riemann–Roch (Corollary 1.4.41) that

$$\mathrm{vol}(L) \;=\; \int_X c_1(L)^n \tag{2.9}$$

[8] This was called the "degree" of the graded linear series $\oplus H^0\big(X, \mathcal{O}_X(mL)\big)$ in [158], but the present terminology is more natural and seems to be becoming standard.

is the top self-intersection of L.[9] We will show later that in general there is an analogous interpretation in terms of "moving self-intersection numbers," and that moreover the lim sup in (2.8) is actually a limit: see Remark 2.2.50 and Section 11.4.

Example 2.2.32. (Volume on a blow-up of projective space). In the situation of 2.2.4, let $X = \mathrm{Bl}_S(\mathbf{P}^n)$ be the blowing up of $\mathbf{P}^n$ at $s = \#S$ points, and put $D = dH - rE$. Then

$$\mathrm{vol}(D) \ \geq \ d^n - s \cdot r^n. \quad \square$$

Example 2.2.33. (Volume of difference of nef divisors, I). It follows from the proof of Theorem 2.2.15 that if D and E are nef, then

$$\mathrm{vol}\big(D - E\big) \ \geq \ \big(D^n\big) - n\,\big(D^{n-1} \cdot E\big). \quad \square$$

Remark 2.2.34. (Irrational volumes). The volume of a big line bundle can be an irrational number: see Section 2.3.B. □

The next result summarizes the first formal properties of this invariant.

Proposition 2.2.35. (Properties of the volume). *Let D be a big divisor on an irreducible projective variety X of dimension n.*

(i). *For a fixed natural number $a > 0$,*

$$\mathrm{vol}(\,aD\,) \ = \ a^n \ \mathrm{vol}(\,D\,).$$

(ii). *Fix any divisor N on X, and any $\varepsilon > 0$. Then there exists an integer $p_0 = p_0(N, \varepsilon)$ such that*

$$\frac{1}{p^n} \cdot \big|\mathrm{vol}\big(pD - N\big) \ - \ \mathrm{vol}\big(pD\big)\big| \ < \ \varepsilon$$

for every $p > p_0$.

Remark 2.2.36. One can view the second statement as hinting at the continuity properties of the volume. A stronger result along these lines appears below as Theorem 2.2.44. □

We start with two lemmas:

Lemma 2.2.37. *Let L be a big line bundle and A a very ample divisor on X. If $E, E' \in |A|$ are very general divisors, then*

$$\mathrm{vol}_E(L \mid E) \ = \ \mathrm{vol}_{E'}(L \mid E').$$

[9] When X is smooth and L is ample this integral computes (up to constants) the volume of X in the metric determined by a Kähler form representing $c_1(L)$: hence the term "volume."

Proof. In fact, for every $m > 0$ there exists by semi-continuity a Zariski-open dense subset $U_m \subseteq |A|$ such that

$$h^0\big(E, L^{\otimes m} \mid E\big) \;=\; h^0\big(E', L^{\otimes m} \mid E'\big)$$

for every $E, E' \in U_m$. □

Lemma 2.2.38. *Let D be any divisor on X, and $a \in \mathbf{N}$ a fixed positive integer. Then*

$$\limsup_m \frac{h^0\big(X, \mathcal{O}_X(mD)\big)}{m^n\,/\,n!} \;=\; \limsup_k \frac{h^0\big(X, \mathcal{O}_X(akD)\big)}{(ak)^n\,/\,n!}. \tag{2.10}$$

Proof. If $\kappa(X, D) < n = \dim X$ then both sides vanish, so we suppose that D is big. Given $r \in \mathbf{N}$, put

$$v_r \;=\; \limsup_k \frac{h^0\big(X, \mathcal{O}_X\big((ak+r)D\big)\big)}{(ak+r)^n\,/\,n!},$$

so that the assertion to be established is that $\mathrm{vol}(D) = v_0$. For any fixed natural number $r_0 \in \mathbf{N}$ the definition of $\mathrm{vol}(D)$ implies that

$$\mathrm{vol}(D) \;=\; \max\big\{v_{r_0+1}, \ldots, v_{r_0+a}\big\}.$$

So it is enough to produce any $r_0 > 0$ for which we can show that

$$v_0 \;=\; v_r \quad \text{for all } \; r \in \big[r_0+1, r_0+a\big].$$

To this end, fix $r_0 \gg 0$ such that $h^0\big(X, \mathcal{O}_X(rD)\big) \neq 0$ for $r \geq r_0$. Then choose $q \gg 0$ such that $qa - (r_0+a) \geq r_0$. For $r \in \big[r_0+1, r_0+a\big]$ we can then fix divisors

$$D_r \;\in\; |rD| \;\;,\;\; D'_r \;\in\; |(qa-r)D|.$$

These in turn give rise to inclusions

$$\mathcal{O}_X\big(kaD\big) \;\subseteq\; \mathcal{O}_X\big((ka+r)D\big) \;\subseteq\; \mathcal{O}_X\big((k+q)aD\big)$$

for every $k \in \mathbf{N}$ and $r \in \big[r_0+1, r_0+a\big]$. Hence

$$h^0\big(X, \mathcal{O}_X\big(kaD\big)\big) \;\leq\; h^0\big(X, \mathcal{O}_X\big((ka+r)D\big)\big) \;\leq\; h^0\big(X, \mathcal{O}_X\big((k+q)aD\big)\big). \tag{*}$$

But

$$\limsup_k \frac{h^0\big(X, \mathcal{O}_X\big((ka+r)D\big)\big)}{(ka)^n\,/\,n!} \;=\; \limsup_k \frac{h^0\big(X, \mathcal{O}_X\big((ka+r)D\big)\big)}{(ka+r)^n\,/\,n!}$$

(multiply on the left by $\frac{(ka)^n}{(ka+r)^n}$), and similarly $\limsup_k \frac{h^0(X, \mathcal{O}_X((k+q)aD)}{(ka)^n\,/\,n!} = v_0$. Therefore (*) implies that $v_0 = v_r$ for $r \in \big[r_0+1, r_0+a\big]$, as required. □

Proof of Proposition 2.2.35. The first statement follows immediately from Lemma 2.2.38, which yields

$$\begin{aligned}\operatorname{vol}(aD) &= \limsup_k \frac{h^0\big(X, \mathcal{O}_X(akD)\big)}{k^n / n!} \\ &= a^n \limsup_k \frac{h^0\big(X, \mathcal{O}_X(akD)\big)}{(ak)^n / n!} = a^n \operatorname{vol}(D).\end{aligned}$$

For (ii) we start with some reductions. Write $N \equiv_{\text{lin}} A - B$ as the difference of two effective divisors. Since D is big we can write $rD - B \equiv_{\text{lin}} B_1$ for some effective divisor B_1 and suitable integer $r > 0$, so that

$$pD - N \equiv_{\text{lin}} (p+r)D - \big(A + B_1\big).$$

Moreover by (i),

$$\lim_{p \to \infty} \frac{\operatorname{vol}\big((p+r)D\big)}{p^n} = \lim_{p \to \infty} \frac{\operatorname{vol}(pD)}{p^n} = \operatorname{vol}(D).$$

Thus we can assume without loss of generality that N is effective. If N' is a second effective divisor, then evidently

$$\operatorname{vol}\big(pD - (N + N')\big) \le \operatorname{vol}\big(pD - N\big) \le \operatorname{vol}\big(pD\big).$$

So it is enough to prove the statement for $N + N'$ in place of N. In particular, by taking N' to be a sufficiently large multiple of an ample divisor we are reduced to the case in which N is very ample.

Assuming that N is very ample (and in particular effective), the argument used in the proof of Theorem 2.2.15, together with Lemma 2.2.37, shows that if $E \in |N|$ is very general then

$$\begin{aligned}&h^0\Big(X, \mathcal{O}_X\big(m(pD - N)\big)\Big) \ge \\ &\qquad h^0\Big(X, \mathcal{O}_X\big(mpD\big)\Big) - m \cdot h^0\Big(E, \mathcal{O}_E\big(mpD\big)\Big).\end{aligned}$$

This implies that

$$\operatorname{vol}_X(pD - N) \ge \operatorname{vol}_X(pD) - n \cdot \operatorname{vol}_E(pD_E), \qquad (*)$$

where $D_E = D|E$ is the restriction of D to E. But by (i),

$$\operatorname{vol}_E(pD_E) = p^{n-1} \operatorname{vol}_E(D_E).$$

Statement (ii) follows upon dividing through by p^n in (*). □

Remark 2.2.39. (Volume of a Q-divisor). When D is a **Q**-divisor, one can define the volume $\operatorname{vol}(D)$ of D as in (2.8), taking the lim sup over those m for which mD is integral. However it is perhaps quicker to choose some $a \in \mathbf{N}$ for which aD is integral, and then set $\operatorname{vol}(D) = \frac{1}{a^n}\operatorname{vol}(aD)$. It follows from 2.2.35 (i) that this is independent of the choice of a. □

Example 2.2.40. (Volume of a finitely generated divisor). Let D be a finitely generated divisor (Definition 2.1.19) on a normal projective variety X. Then $\mathrm{vol}(D)$ is rational. (Via Example 2.1.31 this reduces to the case of semiample divisors, where (2.9) applies.) □

Using Fujita's vanishing theorem (Section 1.4.D), we next deduce from 2.2.35 that the volume of a line bundle depends only upon its numerical equivalence class:

Proposition 2.2.41 (Numerical nature of volume). *If D, D' are numerically equivalent divisors on X, then*

$$\mathrm{vol}\big(D\big) \ = \ \mathrm{vol}\big(D'\big).$$

In particular, it makes sense to talk of the volume $\mathrm{vol}(\xi)$ of a class $\xi \in N^1(X)_{\mathbf{Q}}$.

For 2.2.41, the essential point is the following:

Lemma 2.2.42. *There exists a fixed divisor N having the property that*

$$H^0\big(X, \mathcal{O}_X(N+P)\big) \neq 0$$

for every numerically trivial divisor P. In particular, if P_0 is any numerically trivial divisor then for every $m \in \mathbf{Z}$, $N \pm mP_0$ is linearly equivalent to an effective divisor.

Proof. Fix a very ample divisor B on X. Fujita's vanishing theorem guarantees that if F is a sufficiently large multiple of B, then $H^i\big(X, \mathcal{O}_X(F+A)\big) = 0$ for $i > 0$ and every nef divisor A. In particular, this applies to $A = P + (n-i)B$, where as always $n = \dim X$. It follows by Castelnuovo–Mumford regularity (Theorem 1.8.5) that $F + P + nB$ is globally generated for every numerically trivial P. In particular, it suffices to take $N = F + nB$. □

Proof of Proposition 2.2.41. We can assume that D and D' are big, and the issue is to show that $\mathrm{vol}(D) = \mathrm{vol}(D+P)$ for any numerically trivial divisor P. To this end, fix any integer $p > 0$, and take N as in the preceding lemma. Then $H^0\big(X, \mathcal{O}_X(N - pP)\big) \neq 0$: choosing a non-zero section in this group gives rise for every $m \geq 0$ to an inclusion

$$\mathcal{O}_X\big(m(p(D+P) - N)\big) \hookrightarrow \mathcal{O}_X\big(m(pD)\big).$$

Therefore $\mathrm{vol}\big(p(D+P) - N\big) \leq \mathrm{vol}\big(pD\big) = p^n \,\mathrm{vol}(D)$. On the other hand,

$$\frac{1}{p^n} \cdot \mathrm{vol}\big(p(D+P) - N\big) \to \mathrm{vol}\big(D+P\big)$$

as $p \to \infty$ thanks to 2.2.35 (ii). We conclude that $\mathrm{vol}(D+P) \leq \mathrm{vol}(D)$. The reverse inequality follows upon replacing P by $-P$ and D by $D+P$. □

The volume of a divisor is also a birational invariant.

Proposition 2.2.43. (Birational invariance of volume, I). *Consider a birational projective mapping*

$$\nu : X' \longrightarrow X$$

of irreducible varieties of dimension n. Given an integral or $\mathbf{Q}$-divisor D on X, put $D' = \nu^* D$. Then

$$\mathrm{vol}_{X'}(D') \;=\; \mathrm{vol}_X(D).$$

Proof. This follows from Example 1.2.35, but it is more elementary to argue directly. It suffices to treat the case of integral D, and then by the projection formula one has

$$\begin{aligned} H^0\big(X', \mathcal{O}_{X'}(mD')\big) \;&=\; H^0\big(X, \nu_*\big(\mathcal{O}_{X'}(mD')\big)\big) \\ &=\; H^0\big(X, (\nu_*\mathcal{O}_{X'}) \otimes \mathcal{O}_X(mD)\big). \end{aligned}$$

To analyze the sheaf $\nu_*\mathcal{O}_{X'}$ we proceed as in the proof of Proposition 2.2.3. Specifically, since ν is birational there is an exact sequence

$$0 \longrightarrow \mathcal{O}_X \longrightarrow \nu_*\mathcal{O}_{X'} \longrightarrow \eta \longrightarrow 0,$$

where η is supported on a scheme of dimension $\leq n-1$. By what we said above, this leads to the inequalities

$$\begin{aligned} h^0\big(X, \mathcal{O}_X(mD)\big) \;&\leq\; h^0\big(X', \mathcal{O}_{X'}(mD')\big) \\ &=\; h^0\big(X, \mathcal{O}_X(mD)\big) \;+\; h^0\big(X, \eta \otimes \mathcal{O}_X(mD)\big) \\ &=\; h^0\big(X, \mathcal{O}_X(mD)\big) \;+\; O(m^{n-1}), \end{aligned}$$

and the assertion follows. □

Finally, we study the continuity properties of the volume.

Theorem 2.2.44. (Continuity of volume). *Let X be an irreducible projective variety of dimension n, and fix any norm $\| \ \|$ on $N^1(X)_{\mathbf{R}}$ inducing the usual topology on that finite-dimensional vector space. Then there is a positive constant $C > 0$ such that*

$$\big|\, \mathrm{vol}(\xi) \;-\; \mathrm{vol}(\xi') \,\big| \;\leq\; C \cdot \Big(\max\big(\|\xi\| \,,\, \|\xi'\| \big) \Big)^{n-1} \cdot \| \, \xi \;-\; \xi' \, \| \qquad (2.11)$$

for any two classes $\xi,\ \xi' \in N^1(X)_{\mathbf{Q}}$.

Corollary 2.2.45. (Volume of real classes). *The function $\xi \mapsto \mathrm{vol}(\xi)$ on $N^1(X)_{\mathbf{Q}}$ extends uniquely to a continuous function*

$$\mathrm{vol} \,:\, N^1(X)_{\mathbf{R}} \longrightarrow \mathbf{R}.$$

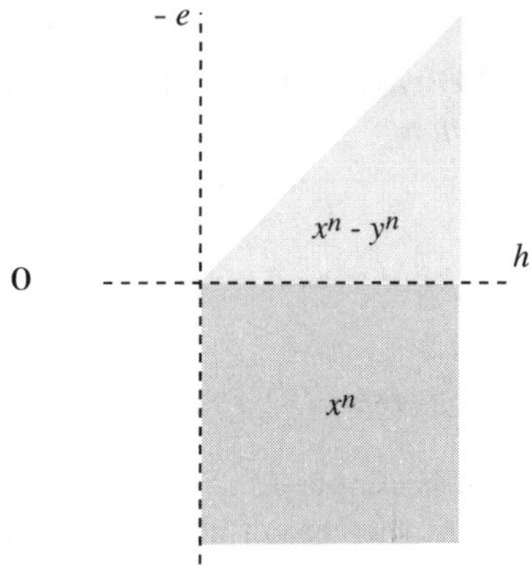

Figure 2.1. Volume on blow-up of projective space

Proof of Corollary 2.2.45. Equation (2.11) guarantees that if we choose a sequence $\xi_i \in N^1(X)_{\mathbf{Q}}$ converging to a given real class $\xi \in N^1(X)_{\mathbf{R}}$, then $\lim_{i\to\infty} \mathrm{vol}(\xi_i)$ exists and is independent of the choice of $\{\xi_i\}$. □

Example 2.2.46. (Volume on blow-up of projective space). We work out the volume function in a simple case. Let $X = \mathrm{Bl}_P(\mathbf{P}^n)$ be the blowing up of $\mathbf{P}^n$ at a point P, and denote by $e, h \in N^1(X)_{\mathbf{R}} = \mathbf{R}^2$ respectively the classes of the exceptional divisor E and the pullback H of a hyperplane. The nef cone of X is generated by h and $h - e$. In the sector of the plane they span, the volume is given by

$$\mathrm{vol}(x \cdot h \; - \; y \cdot e) \;=\; \big((x \cdot h \; - \; y \cdot e)^n \big) \;=\; x^n \; - \; y^n.$$

On the other hand, when $k, \ell \geq 0$ the linear series $|k\,H + \ell\,E|$ contains ℓE as a fixed component. So in the region spanned by h and e — which corresponds to effective divisors that are not nef — the volume is given by

$$\mathrm{vol}(x \cdot h - y \cdot e) \;=\; \big((x \cdot h)^n \big) \;=\; x^n.$$

Elsewhere the volume is zero. The situation is illustrated in Figure 2.1, which shows the volume $\mathrm{vol}(x\,h - y\,e)$ as a function of $(x, y) \in \mathbf{R}^2$. □

Proof of Theorem 2.2.44. Since both sides of (2.11) are homogeneous of degree n in (ξ, ξ'), we are free to prove the inequality after replacing the given classes by a common large multiple. Having said this, fix very ample divisors $A_1, \ldots, A_r$ on X whose classes form a basis of $N^1(X)_{\mathbf{Q}}$. It is enough to prove that (2.11) holds for divisors

$$\begin{aligned} D &= a_1 \cdot A_1 + \ldots + a_r \cdot A_r, \\ D' &= a'_1 \cdot A_1 + \ldots + a'_r \cdot A_r, \end{aligned}$$

with $a_i, a_i' \in \mathbf{Z}$. It will be convenient to work with the norm given by $\|D\| = \max\{|a_i|\}$. We will assume to begin with that $a_i - a_i' = b_i \geq 0$, so write

$$D' = D - B \quad \text{with} \quad B = \sum b_i \cdot A_i \succcurlyeq 0.$$

Let $E_j \in |A_j|$ be a very general divisor, so that in particular E_j meets each A_i properly. We claim that

$$\operatorname{vol}(D - B) \geq \operatorname{vol}(D) - n \cdot \sum_{j=1}^{r} b_j \cdot \operatorname{vol}_{E_j}(D_j), \tag{2.12}$$

where $D_j = D|E_j$ is the restriction of D to E_j. In fact, it suffices by induction to prove this in the special case $B = b_1 A_1$, where we argue as in the proof of 2.2.15. Specifically, given $m \gg 0$, choose mb_1 very general divisors $F_\alpha \in |A_1|$ and consider the exact sequence

$$0 \longrightarrow \mathcal{O}_X\big(mD - mb_1 A_1\big) \longrightarrow \mathcal{O}_X\big(mD\big) \longrightarrow \bigoplus_{\alpha=1}^{mb_1} \mathcal{O}_{F_\alpha}\big(mD\big).$$

Using Lemma 2.2.37, the dimension of H^0 of each of the terms on the right is approximated by $\operatorname{vol}_{E_1}(D_1) \cdot \frac{m^{n-1}}{(n-1)!}$, and the assertion follows.

Now let $D^+ = \sum |a_i| \cdot A_i$ be the divisor obtained by making all the coefficients of D non-negative, and let $D_j^+ = D^+ \mid E_j$. Then $D_j^+ - D_j$ is effective, so

$$\operatorname{vol}_{E_j}(D_j) \leq \operatorname{vol}_{E_j}(D_j^+).$$

On the other hand, D_j^+ is nef, and consequently

$$\begin{aligned} \operatorname{vol}_{E_j}(D_j^+) &= \Big(\big(\sum |a_i| \cdot A_i \big)^{n-1} \cdot A_j \Big) \\ &\leq C_j \cdot \Big(\max\{ |a_i| \} \Big)^{n-1} \\ &= C_j \cdot \big(\|D\| \big)^{n-1} \end{aligned}$$

for a suitable constant $C_j > 0$ depending on j and the divisors A_i. So taking $C = 2rn \cdot \max\{ C_j \}$, and noting that $\sum b_i \leq r \max\{b_j\} = r\,\|B\|$, (2.12) leads to the bound

$$\operatorname{vol}\big(D\big) - \operatorname{vol}\big(D - B\big) \leq \frac{C}{2} \cdot \big(\|D\|\big)^{n-1} \cdot \|B\|. \tag{2.13}$$

Now consider the general case

$$D' = D + E - F,$$

where E and F are effective integral linear combinations of the A_i, with no non-zero "coordinates" in common. Then (2.13) gives

$$\operatorname{vol}(D) \;-\; \operatorname{vol}(D-F) \;\leq\; \frac{C}{2} \cdot \big(\|D\|\big)^{n-1} \cdot \|F\|,$$
$$\operatorname{vol}(D+E-F) \;-\; \operatorname{vol}\big((D+E-F)-E\big) \;\leq\; \frac{C}{2} \cdot \big(\|D+E-F\|\big)^{n-1} \cdot \|E\|.$$

Observing that $\max\big(\|E\|, \|F\|\big) = \|E-F\|$, the triangle inequality then yields

$$\begin{aligned} &|\operatorname{vol}(D) \;-\; \operatorname{vol}(D+E-F)| \\ &\qquad\qquad \leq C \cdot \Big(\max\big(\|D\|, \|D+E-F\|\big)\Big)^{n-1} \cdot \|E-F\|, \end{aligned}$$

as required. □

Example 2.2.47. (Volume of difference of nef divisors, II). Let $\xi, \eta \in \operatorname{Nef}(X)_{\mathbf{R}}$ be real nef classes on an n-dimensional projective variety X. Then

$$\operatorname{vol}(\xi-\eta) \;\geq\; (\xi^n) \;-\; n \cdot (\xi^{n-1} \cdot \eta).$$

(For rational classes this follows by homogeneity from Example 2.2.33, and for $\mathbf{R}$-divisors one argues by continuity.) □

Example 2.2.48. (Volume increases in effective directions). If $\xi \in N^1(X)_{\mathbf{R}}$ is big and $e \in N^1(X)_{\mathbf{R}}$ is effective, then

$$\operatorname{vol}(\xi) \;\leq\; \operatorname{vol}(\xi+e).$$

(If ξ and e are rational this is clear from the fact that $\operatorname{vol}(D+E) \geq \operatorname{vol}(D)$ for integral divisors D and E with E effective. The statement follows in general by perturbing the classes in question and invoking the continuity of volume.) □

Example 2.2.49. (Birational invariance of volume, II). Let

$$\nu : X' \longrightarrow X$$

be a birational morphism of irreducible projective varieties. Then

$$\operatorname{vol}_X(\xi) \;=\; \operatorname{vol}_{X'}(\nu^*\xi)$$

for any class $\xi \in N^1(X)_{\mathbf{R}}$. (By continuity this follows from Proposition 2.2.43.) □

Remark 2.2.50. (Further properties of the volume). In Section 11.4 we prove a result of Fujita to the effect that the volume of a big line bundle on X can be approximated arbitrarily closely by that of a nef $\mathbf{Q}$-divisor on a modification of X. This in turn will allow us to deduce several additional facts about this invariant:

- The lim sup in (2.8) is actually a limit.
- The volume of a big divisor can be interpreted geometrically in terms of "moving self-intersection numbers", generalizing (2.9).

- The volume function satisfies the log-concavity relation

$$\operatorname{vol}(\xi+\xi')^{1/n} \;\geq\; \operatorname{vol}(\xi)^{1/n} + \operatorname{vol}(\xi')^{1/n}$$

for any two big classes $\xi, \xi' \in N^1(X)_{\mathbf{R}}$, generalizing the Hodge-type inequality 1.6.3 (iii) for nef classes. □

Remark 2.2.51. (Volume on toric varieties). It is established in [157] that if X is a smooth projective toric variety, then $\overline{\mathrm{Eff}}(X)$ is a polyhedral cone, and carries a decomposition into finitely many closed polyhedral subcones on which vol_X is given by a polynomial of degree $n = \dim X$. The piecewise polynomial nature of vol_X in the toric setting is illustrated by Example 2.2.46. □

Remark 2.2.52. (Nature of volume function). Bauer, Küronya, and Szemberg [40] have used Zariski decompositions (Section 2.3.E) to study the volume function when X is a smooth projective surface. They show that in this case $\mathrm{Big}(X)$ admits a decomposition into locally polyhedral chambers on which $\operatorname{vol}(\xi)$ is a quadratic polynomial function of ξ. On the other hand, they also exhibit a threefold X for which $\operatorname{vol}(\xi)$ is not piecewise polynomial (although it is locally analytic in this example). It remains a very interesting open question to say something about the nature of the function $\operatorname{vol}_X : N^1(X)_{\mathbf{R}} \longrightarrow \mathbf{R}$ in general. The natural expectation here is that there exists a "large" open set $U \subseteq N^1(X)_{\mathbf{R}}$ such that vol_X is real analytic on each connected component of U: one could hope that U is actually dense, but this might run into trouble with clustering phenomena. □

Remark 2.2.53. (Volume on a Kähler manifold). Boucksom [66] defines and studies the volume of a big class $\alpha \in H^{1,1}(X, \mathbf{R})$ on an arbitrary compact Kähler manifold X. Among other things, he proves that the continuity of volume remains true in this setting as well. □

2.3 Examples and Complements

This section is devoted to some concrete examples and further results concerning the phenomena that can occur for linear series. We start by explaining two methods for exhibiting interesting examples, one originating with Zariski and the other a very powerful construction introduced and exploited by Cutkosky. We then discuss the base loci of big and nef linear series, and present a criterion for the corresponding graded ring to be finitely generated. Next, as an application of Siu's numerical criterion 2.2.15 for bigness, we prove a theorem of Campana and Peternell to the effect that the Nakai inequalities characterize ample $\mathbf{R}$-divisor classes. Finally, we discuss Zariski decompositions on surfaces.

2.3.A Zariski's Construction

Zariski observed that one could use the geometry of algebraic curves to produce examples of linear series on surfaces with various sorts of noteworthy behavior. In order to illustrate the method, we reproduce from [623] an example of a big and nef divisor D on a surface X with the property that the graded ring $R(D) = \oplus H^0\big(X, \mathcal{O}_X(mD)\big)$ is not finitely generated. Mumford's paper [451] gives a very nice account of the history of Zariski's example and its relation with Hilbert's fourteenth problem. See also 2.3.3 for another example and Section 2.3.C for some related general results.

Start with a smooth cubic curve $C_0 \subset \mathbf{P}^2$, and denote by ℓ the hyperplane class on $\mathbf{P}^2$. Choose twelve general points $P_1, \ldots, P_{12} \in C_0$ in such a way that the line bundle

$$\mathcal{O}_{C_0}\big(P_1 + \ldots + P_{12} - 4\ell\big) =_{\text{def}} \eta \in \operatorname{Pic}^0(C_0)$$

is a non-torsion class. Now take

$$\mu : X \ = \ \mathrm{Bl}_{\{P_1,\ldots,P_{12}\}}\mathbf{P}^2 \longrightarrow \mathbf{P}^2$$

to be the blowing up of $\mathbf{P}^2$ at the chosen points, with exceptional divisor $E = \sum_{i=1}^{12} E_i$ (E_i being the component over P_i). Write $H = \mu^*\ell$ for the pullback of the hyperplane class on $\mathbf{P}^2$, and let $C \in |3H - E|$ be the proper transform of C_0: thus $C \cong C_0$. We consider the divisor

$$D \ = \ 4H - E.$$

Note that $D \equiv_{\text{lin}} H + C$, and consequently D is big. Observing that $\mathcal{O}_C(D) = \eta^*$ is nef, and that $D - C$ is free, one sees that D is also nef.

We assert that for every $m \geq 1$ the linear series $|mD|$ contains C in its base locus, whereas $|mD - C|$ is free. This easily implies that $R(D)$ cannot be finitely generated: in fact, finite generation requires that the multiplicity of a base-curve in $|mD|$ must go to infinity with m. (See Section 2.3.C.)

The fact that $C \subseteq \operatorname{Bs}|mD|$ is clear, since $\mathcal{O}_C(mD) = \eta^{\otimes -m}$ has no sections. On the other hand, $mD - C \equiv_{\text{lin}} (m-1)D + H$, and using the exact sequence

$$0 \longrightarrow \mathcal{O}_X((m-1)D + H - C) \stackrel{\cdot C}{\longrightarrow} \mathcal{O}_X((m-1)D + H) \\ \longrightarrow \mathcal{O}_C((m-1)D + H) \longrightarrow 0$$

one shows by induction on m that $mD - C$ is free. (Observe that the term on the right is a line bundle of degree three on the elliptic curve C, hence free.)

Example 2.3.1. (A generalization). As Mumford [451] points out, a similar construction works in a more general setting. Start with an algebraic surface X and a curve $C \subseteq X$ such that $\big(C^2\big) < 0$ and such that moreover the map $\mathrm{Pic}(X) \longrightarrow \mathrm{Pic}(C)$ is injective. (As above, examples arise by blowing up sufficiently many general points on a smooth curve sitting on a surface having Picard number one.) Fix a very ample divisor A on X and put

$$a = (C \cdot A) \ , \ \ -b = (C \cdot C) \ , \ \ D \ = \ bA + aC.$$

As before D is nef and big, and C appears in the base locus of $|mD|$ for every $m \geq 1$. On the other hand, it follows using Fujita's vanishing theorem 1.4.35 and Castelnuovo–Mumford regularity 1.8.5 that there is a fixed integer $p_0 > 0$ such that $mD - p_0C$ is free for all $m \gg 0$. □

2.3.B Cutkosky's Construction

Cutkosky starts with a direct sum of line bundles on a base variety V and then passes to the corresponding projective bundle. The philosophy is that one can exploit in this manner the geometry of the various cones sitting inside the Néron–Severi space $N^1(V)_{\mathbf{R}}$.

Specifically, let V be an irreducible projective variety, say of dimension $\dim V = v$, and fix integral (Cartier) divisors $A_0, A_1, \ldots, A_r$ on V. Put

$$\mathcal{E} \ = \ \mathcal{O}_V(A_0) \oplus \ldots \oplus \mathcal{O}_V(A_r),$$

so that $\mathcal{E}$ is a vector bundle of rank $r+1$ on V. The examples arise by taking

$$X \ = \ \mathbf{P}(\mathcal{E}) \quad \text{and} \quad L = \mathcal{O}_{\mathbf{P}(\mathcal{E})}(1).$$

Thus X is an irreducible projective variety of dimension $n = v + r$. Cutkosky's idea is that judicious choices of the A_i lead to interesting behavior of the linear series associated to L.

Lemma 2.3.2. *Let $X = \mathbf{P}(\mathcal{E})$ be the variety just introduced, and for simplicity of notation write $\mathcal{O}_X(k)$ to denote the bundle $\mathcal{O}_{\mathbf{P}(\mathcal{E})}(k)$ on X.*

(i). *One has*

$$H^0\big(X, \mathcal{O}_X(k)\big) \ = \ \bigoplus_{a_0 + \ldots + a_r = k} H^0\big(V, \mathcal{O}_V\big(a_0A_0 + \ldots + a_rA_r\big)\big).$$

(ii). *$\mathcal{O}_X(1)$ is ample if and only if each of the divisors A_i is ample on V.*

(iii). *$\mathcal{O}_X(1)$ is nef if and only if each of the A_i is a nef divisor on V.*

(iv). *$\mathcal{O}_X(1)$ is big if only only if some non-negative $\mathbf{Z}$-linear combination of the A_i is a big divisor on V.*

(v). *Given $m \in \mathbf{N}$, $\mathcal{O}_X(m)$ is free if and only if mA_i moves in a free linear series on V for each $0 \leq i \leq r$.*

Sketch of Proof. The first assertion is a consequence of the isomorphism

$$H^0\big(X, \mathcal{O}_X(k)\big) \;=\; H^0\big(V, S^k E\big).$$

Statements (ii) and (iii) follow most naturally from the theory of ample vector bundles: see 6.1.13 and 6.2.12. Statement (iv) follows from Example 6.1.23, but for a direct argument we assert first that if the condition on the A_i appearing in (iv) holds, then

$$\#\Big\{ (a_0, \ldots, a_r) \;\Big|\; \sum a_i = k \text{ and } \sum a_i A_i \text{ is big} \Big\} \;\geq\; C \cdot k^r \qquad (*)$$

for some $C > 0$ and all $k \gg 0$. To see this, consider the linear mapping

$$\phi : \mathbf{R}^{r+1} \longrightarrow N^1(V)_{\mathbf{R}} \ , \quad (b_0, \ldots, b_r) \mapsto \sum b_i A_i.$$

The set $\mathrm{Big}(V) \subseteq N^1(V)_{\mathbf{R}}$ of big classes is an open convex cone (Section 2.2.B), and hence so too is its inverse image $\phi^{-1}\mathrm{Big}(V)$. On the other hand, the condition in (iv) guarantees that this cone has non-empty intersection with the first orthant. Therefore each of the r-simplices

$$\Delta_t \;=\; \Big\{ \sum x_i = t \ , \ x_i \geq 0 \Big\} \;\subseteq\; \mathbf{R}^{r+1}$$

meets $\phi^{-1}\mathrm{Big}(V)$ in a set of positive Lebesgue measure, and this implies (*). It follows from (*) and (i) using Corollary 2.2.7 that $h^0\big(X, \mathcal{O}_X(k)\big) \geq C' k^{v+r}$ for $k \gg 0$ and so $\mathcal{O}_X(k)$ is big. We leave the converse to the reader. Finally, if mA_i is free for every i then it is immediate that $\mathcal{O}_X(m)$ is free. Conversely, the quotient $\mathcal{E} \longrightarrow \mathcal{O}_V(A_i)$ defines an embedding $V = V_i \subseteq X$ sectioning the bundle projection $\mathbf{P}(\mathcal{E}) \longrightarrow V$. If every section of $\mathcal{O}_V(mA_i)$ vanishes at some point $y \in V$, then one sees that $\mathcal{O}_X(m)$ has a base-point at the point of V_i lying over y. □

We now give several explicit examples.

To begin with, take V to be a smooth projective curve of genus $g \geq 1$, $P \in \mathrm{Div}^0(X)$ a divisor of degree zero, and A a divisor of degree $a > 0$. We run Cutkosky's construction with $A_0 = P$ and $A_1 = A$. Thus

$$X \;=\; \mathbf{P}(\mathcal{O}_V(P) \oplus \mathcal{O}_V(A))$$

is a ruled surface, and as above we write $\mathcal{O}_X(1)$ for the Serre line bundle on X. It follows from the lemma that $\mathcal{O}_X(1)$ is big and nef. By varying the specifications on P and A, we exhibit several interesting phenomena.

Example 2.3.3. Take P non-torsion in $\mathrm{Pic}^0(V)$ and $a = \deg A \gg 0$. Then $H^0\big(V, \mathcal{O}_V(kP)\big) = 0$ for all k. So here $\mathcal{O}_X(k)$ has a base-component for every $k \geq 0$. Moreover, we can easily see that the graded ring

$$R \;=\; R_\bullet \;=\; \oplus H^0\big(X, \mathcal{O}_X(k)\big)$$

is not finitely generated. In fact, R is the space of sections of the quasi-coherent sheaf $\mathrm{Sym}\big(\mathcal{O}_V(P) \oplus \mathcal{O}_V(A)\big)$ on V, and the algebra structure on R is compatible with that on the symmetric algebra. Therefore we can realize R as a bigraded ring by setting

$$R_{i,j} \;=\; H^0\big(V, \mathcal{O}_V(i\,P + j\,A)\big),$$

so that $R_k = \oplus_{i+j=k} R_{i,j}$. But now consider for any $k \geq 1$ the subspace

$$R_{k-1,1} \;=\; H^0\big(V, \mathcal{O}_V\big((k-1)P + A\big)\big) \subseteq R_k.$$

This is non-zero since A has large degree. On the other hand,

$$R_{k-p,1} \otimes R_{p-1,0} \;=\; R_{k-1,0} \otimes R_{0,1} \;=\; 0$$

for $p \geq 2$, so $R_{k-1,1}$ consists of minimal generators of R. In particular, R is not finitely generated. (This example is an illustration of Theorem 2.3.15, which asserts that the graded ring $R(X, D)$ associated to a big and nef divisor D on a projective variety X is finitely generated if and only if $|mD|$ is free for some $m > 0$.) □

Example 2.3.4. Take P defining an e-torsion element in $\mathrm{Pic}^0(V)$, and $a = \deg A \gg 0$.

This gives an example where $\mathcal{O}_X(m)$ is free if $m \equiv 0 \pmod{e}$, but has fixed components when $m \not\equiv 0 \pmod{e}$. □

Example 2.3.5. Take V to have genus $g = 1$, P defining a non-torsion element in $\mathrm{Pic}^0(V)$, and $A = x$ a divisor of degree one.

As in Example 2.3.3, $\mathcal{O}_X(k)$ has a fixed component F for every $k \geq 1$, and we can write

$$|\mathcal{O}_X(k)| \;=\; F + |M_k|,$$

where M_k is the moving part of the indicated linear series. Then $|M_k|$ has no fixed components, but it has a basepoint $z_k \in F$ with the property that $z_k \neq z_\ell$ for $k \neq \ell$. (In fact, there is a unique point $y_k \in V$ such that

$$y_k \;\equiv_{\mathrm{lin}}\; (k-1)P + x.$$

Then z_k lies over y_k in the bundle projection $X \longrightarrow V$.) This example appears in [102]. □

Example 2.3.6. (Integral divisor with small volume). We construct an example of a big line bundle having arbitrarily small volume. As before, let V be a smooth projective curve, say of genus $g = 1$ to simplify the calculations. Fix a point $x \in V$, and take

$$A_0 \;=\; (1-a) \cdot x \quad , \quad A_1 \;=\; x$$

for some positive integer a. Writing $h^0(m \cdot x)$ for $h^0(V, \mathcal{O}_V(m \cdot x))$ one finds:

$$\begin{aligned} h^0(X, \mathcal{O}_X(k)) &= h^0(k \cdot x) + h^0((k-a) \cdot x) + \ldots + h^0((k - [\tfrac{k}{a}]a) \cdot x) \\ &= \frac{k^2}{2a} + O(k), \end{aligned}$$

and hence $\mathrm{vol}_X(\mathcal{O}_X(1)) = \frac{1}{a}$. □

We now turn to an example from [97] and [101] of a big divisor having irrational volume. Take $V = E \times E$ to be the product of an elliptic curve with itself. Recall from Section 1.5.B that $\mathrm{Nef}(V)$ is a circular cone $\mathcal{K}$, consisting of classes having non-negative self-intersection (and non-negative intersection with an ample divisor). We may then choose ample integral divisors

$$D \ , \ H \ \in \mathrm{Div}(V) \ \text{ with classes } \ \delta \ , \ h \ \in \ N^1(V)_{\mathbf{R}},$$

such that the ray in $N^1(V)_{\mathbf{R}}$ emanating from δ in the direction of $-h$ meets the boundary of $\mathcal{K}$ at an *irrational* point. In other words, we ask that

$$\sigma \ =_{\text{def}} \ \max\left\{ t \mid \delta - t \cdot h \text{ is nef} \right\} \ \notin \mathbf{Q}.$$

In fact, σ is the smallest root of the quadratic polynomial

$$s(t) \ = \ \left((\delta - t \cdot h)^2 \right),$$

so "most" choices of D and H will lead to irrational σ. (See 2.3.8 for a completely concrete calculation.)

The plan is to apply Cutkosky's construction with

$$A_0 \ = \ D \ , \ A_1 \ = \ -H.$$

Then

$$h^0(X, \mathcal{O}_X(k)) \ = \ \sum_{i+j=k} h^0(V, \mathcal{O}_V(iD - jH)). \tag{2.14}$$

Now the divisor $iD - jH$ appearing in this sum is nef precisely if $\frac{j}{i} < \sigma$, and has no sections if $\frac{j}{i} > \sigma$. Using Riemann–Roch on V, it follows that for $i, j \geq 0$,

$$h^0(V, \mathcal{O}_V(iD - jH)) \ = \ \begin{cases} \frac{1}{2}\left((i\delta - jh)^2 \right) & \text{if } \frac{j}{i} < \sigma, \\ 0 & \text{if } \frac{j}{i} > \sigma. \end{cases} \tag{2.15}$$

The situation is indicated in Figure 2.2, which shows the points (i, j) in the first quadrant corresponding to the divisor $iD - jH$. The divisors appearing in the sum (2.14) are parameterized by the integer points on the line $x + y = k$, indicated for $k = 9$ by open circles. The points lying below the line $y = \sigma x$ give a non-zero contribution to the sum, while those lying above that line correspond to divisors with no sections.

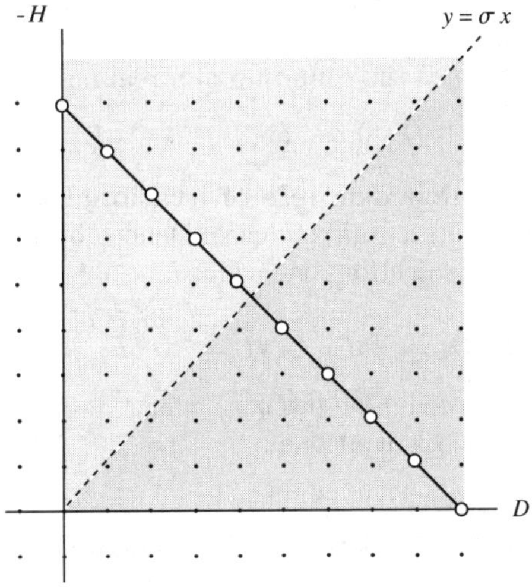

Figure 2.2. Computation of irrational volume

Now combine (2.14) and (2.15), substitute $(k-i)$ for j, and divide by $\frac{k^3}{3!}$ to obtain

$$\frac{h^0\big(X, \mathcal{O}_X(k)\big)}{k^3/3!} = \frac{3!}{2} \cdot \sum_{i \geq \frac{k}{1+\sigma}}^{k} \Big(\big(\tfrac{i}{k}\delta - (1 - \tfrac{i}{k})h \big)^2 \Big) \cdot \tfrac{1}{k} \ . \tag{2.16}$$

(Here the lower limit of summation is obtained by computing the intersection of the lines $x+y=k$ and $y=\sigma x$.) The right-hand side of (2.16) is a Riemann sum for an integral of the quadratic function

$$q(x) = \Big(\big(x\delta - (1-x)h\big)^2 \Big).$$

As $k \to \infty$ in (2.16), we therefore arrive at the expression

$$\operatorname{vol}_X\big(\mathcal{O}_X(1)\big) = \frac{3!}{2} \cdot \int_{\frac{1}{1+\sigma}}^{1} q(x)\, dx. \tag{2.17}$$

Since $\frac{1}{1+\sigma}$ is a quadratic irrationality, the value of this integral will likewise be irrational for most choices of D and H.

Example 2.3.7. (Base loci). It is instructive to analyze the base locus of $|\mathcal{O}_X(k)|$ in this example. Put $c = \frac{1}{\sigma+1}$ and let

$$E = \mathbf{P}\big(\mathcal{O}_V(-H)\big) \subseteq \mathbf{P}(\mathcal{E}) = X$$

be the "divisor at infinity." Then E is (set-theoretically) the base locus of $|\mathcal{O}_X(k)|$, and it appears in the base scheme with multiplicity exactly $\ulcorner kc \urcorner$, where as usual $\ulcorner r \urcorner$ denotes the round-up of a real number r. In other words

$$\mathfrak{b}\big(|\mathcal{O}_X(k)|\big) = \mathcal{O}_X\big(-\ulcorner kc \urcorner \cdot E\big). \quad \square$$

Example 2.3.8. (Explicit example of irrational volume). We give the computations explicitly in a concrete case. Denote by $F_1, F_2, \Delta \subseteq V$ respectively the fibres of the projections $V = E \times E \longrightarrow E$ and the diagonal. Take for instance

$$D = F_1 + F_2 \quad , \quad H = 3 \cdot (F_2 + \Delta).$$

Then $\sigma = \frac{3-\sqrt{5}}{6}$ is irrational. One has $q(x) = 38x^2 - 54x + 18$, and one checks that the integral in (2.17) is irrational. $\square$

2.3.C Base Loci of Nef and Big Linear Series

We discuss here some boundedness results for the base loci of nef and big divisors. As an application we prove that the graded ring associated to such a divisor is finitely generated if any only if the divisor is semiample.

We begin with a theorem due to Wilson [607]:

Theorem 2.3.9. (Wilson's theorem). *Let X be an irreducible projective variety of dimension n, and let D be a nef and big divisor on X. Then there exists a natural number $m_0 \in \mathbf{N}$ together with an effective divisor N such that*

$$|mD - N|$$

is free for every $m \geq m_0$.

The crucial point here is that the divisor N is independent of m.

Example 2.3.10. We gave in 2.3.5 an example of a big and nef divisor D on a surface X having the property that $|mD| = F + |M_m|$, where F is the fixed divisor of $|mD|$, and $|M_m|$ has a base point at a variable point $z_m \in X$. Here the statement of the theorem is satisfied with $N = 2F$. $\square$

Proof of Theorem 2.3.9. This is a consequence of Fujita's vanishing theorem and Castelnuovo–Mumford regularity. In fact, according to 1.4.35 one can fix a very ample divisor B on X satisfying $H^i\big(X, \mathcal{O}_X(B+P)\big) = 0$ for all $i > 0$ and every nef divisor P. Since D is big, there is an integer $m_0 > 0$ and an effective divisor N with $m_0 D \equiv_{\text{lin}} (n+1)B + N$. Then

$$mD - N \equiv_{\text{lin}} (m - m_0)D + (n+1)B = B + nB + (\text{nef}),$$

and consequently

$$H^i\big(X, \mathcal{O}_X(mD - N - iB)\big) = 0 \quad \text{for} \quad i > 0 \quad \text{and} \quad m \geq m_0.$$

Therefore $mD - N$ is free thanks to 1.8.5. $\square$

One should view 2.3.9 as reflecting the principle that base loci of nef and big divisors are bounded. The next corollary gives a more concrete statement in this direction. First, a definition:

Definition 2.3.11. (Multiplicity of a linear series at a point). Let X be a projective variety, and $x \in X$ a fixed point. Given a divisor D on X and a linear series $|V| \subseteq |D|$, denote by $\text{mult}_x |V|$ the multiplicity at x of a general divisor in $|V|$. Equivalently,

$$\text{mult}_x |V| \ = \ \min_{D' \in |V|} \ \{ \ \text{mult}_x D' \ \}.$$

We refer to this integer as the *multiplicity of $|V|$ at x*. □

Corollary 2.3.12. (Boundedness of multiplicity). *In the situation of Theorem 2.3.9, there is a fixed positive constant $C > 0$ independent of m and x such that*

$$\text{mult}_x |mD| \ \leq \ C$$

for every $x \in X$.

Proof. In fact, let N be the divisor whose existence is asserted in 2.3.9. Since $mD - N$ is free, $\text{mult}_x |mD| \leq \text{mult}_x N$ for every $m \geq m_0$. The corollary follows. □

Remark 2.3.13. (Goodman's theorem). Conversely, Goodman shows in [226, Proposition 8] that if D is any divisor on a projective variety X such that $\text{mult}_x |mD|$ is uniformly bounded above independently of x and m, then D is nef. (When X is smooth, a quick proof using multiplier ideals appears in Example 11.2.19.) □

Remark 2.3.14. (Nakamaye's theorem). An interesting theorem of Nakamaye [468] states that the stable base locus of (a small perturbation of) a big and nef divisor D is determined by the numerical class of D and its restriction to subvarieties. We present this result in Section 10.3. □

Recall (Definition 2.1.17) that one can associate to any divisor D the graded ring $R(X, D)$, and as we have stressed it is very interesting to ask when this algebra is finitely generated. For big nef divisors there is a very simple answer due to Zariski [623] and Wilson [607]:

Theorem 2.3.15. (Criterion for finite generation of big nef divisors). *Assume that X is a normal projective variety and that D is a big nef divisor on X. Then $R(X, D)$ is finitely generated if and only if D is semiample, i.e. $|\ell D|$ is free for some $\ell > 0$.*

Proof of Theorem 2.3.15. If D is semiample then $R(X, D)$ is finitely generated by Example 2.1.30. For the converse, fix any point $x \in X$. Supposing that $R(X, D)$ is finitely generated, there is an integer $\ell > 0$ such that

$H^0\big(X, \mathcal{O}_X(\ell D)\big)$ generates the Veronese subring $R(X,D)^{(\ell)}$. So when $k > 0$ every section of $\mathcal{O}_X(k\ell D)$ is a degree k polynomial in sections of $\mathcal{O}_X(\ell D)$, which implies that

$$\operatorname{mult}_x|k\ell D| \ \geq \ k \cdot \operatorname{mult}_x|\ell D|.$$

But the left-hand side remains bounded as $k \to \infty$, and therefore $\operatorname{mult}_x|\ell D| = 0$. This means that x is not in the base locus of $|\ell D|$, and so ℓD is free. □

Example 2.3.16. (Counter-examples for divisors that fail to be big). Theorem 2.3.9 and Corollary 2.3.12 can fail for nef divisors D that are not big. For example, starting with a smooth curve V of genus ≥ 2, run Cutkosky's construction with $A_0 = 0$ and $A_1 = P$ where P is a divisor of degree zero that is non-torsion in $\operatorname{Pic}^0(V)$. Then $\mathcal{O}_X(1)$ is a nef line bundle on $X = \mathbf{P}(\mathcal{O}_V \oplus \mathcal{O}_V(P))$, but the divisor at infinity $\mathbf{P}(\mathcal{O}_V(P)) \subseteq X$ appears with multiplicity m in the base locus of $\mathcal{O}_X(m)$. □

Remark 2.3.17. (Abundant nef divisors). The pathology illustrated in the previous example disappears if one restricts attention to abundant (or good) divisors. Given a nef divisor D on a normal projective variety X, the *numerical dimension* $\nu(D) = \nu(X,D)$ of D is defined to be the largest integer $e > 0$ such that $\big(D^e \cdot V\big) \neq 0$ for some irreducible $V \subseteq X$ of dimension e. One sees that $\kappa(D) \leq \nu(D)$, and D is said to be *good* (or *abundant*) if equality holds. (One makes of course similar definition for line bundles.) Using results of Kawamata [318, §2], Mourougane and Russo [443] prove that analogues of 2.3.9 and 2.3.12 continue to hold for good nef divisors. □

2.3.D The Theorem of Campana and Peternell

We prove a theorem of Campana and Peternell [78] to the effect that the Nakai inequalities characterize the amplitude of **R**-divisor classes. Some applications appeared in Section 1.5.E. We again call the reader's attention to the paper of Demailly and Paun [132] in which an analogous statement is established on an arbitrary Kähler manifold: see Remark 1.5.32.

Theorem 2.3.18. (Nakai criterion for R-divisors). *Let X be a projective scheme, and let $\delta \in N^1(X)_{\mathbf{R}}$ be a class having positive intersection with every irreducible subvariety of X. In other words, assume that*

$$\left(\delta^{\dim V} \cdot V\right) > 0$$

for every $V \subseteq X$ of positive dimension. Then δ is an ample class.

Proof. There is no loss in generality in taking X reduced and irreducible, and the assertion is clear if $\dim X = 1$. So we proceed by induction on $n = \dim X$, and assume that for every proper subvariety $Y \subset X$ the restriction

$$\delta \mid Y \in N^1(Y)_{\mathbf{R}}$$

is an ample class on Y. Note that in any event, δ is nef.

Choose any norm on $N^1(X)_{\mathbf{R}}$ inducing the standard topology on that finite-dimensional real vector space. Arguing as in the proofs of 1.4.10 or 1.4.9, one sees that there exist arbitrarily small ample classes $\alpha, \alpha' \in N^1(X)_{\mathbf{R}}$ such that

$$\delta + \alpha' \ , \ \ \alpha + \alpha' \ , \ \ \text{and} \ \ \delta - \alpha$$

are rational. Moreover since $\big(\delta^n\big) > 0$, we can suppose by taking α and α' sufficiently small that

$$\Big(\, (\delta + \alpha')^n \, \Big) \ > \ n \cdot \Big(\, (\delta + \alpha')^{n-1} \cdot (\alpha + \alpha') \, \Big).$$

It then follows from the numerical criterion for bigness (Theorem 2.2.15) that

$$\delta - \alpha \ = \ (\delta + \alpha') - (\alpha + \alpha')$$

is represented by an *effective* $\mathbf{Q}$-divisor E. Denote by $Y_1, \ldots, Y_t \subset X$ the irreducible components of a support of E.

We now claim that if $0 < \varepsilon \ll 1$ is sufficiently small, then $\big(\delta - \varepsilon \cdot \alpha\big)$ is a nef class on X. In fact $\delta \mid Y_i$ is ample by the induction hypothesis. So by taking $0 < \epsilon \ll 1$ we can arrange that each of the restrictions $\big(\delta - \varepsilon \cdot \alpha\big) \mid Y_i$ are nef (Example 1.3.14). Now let $C \subset X$ be any curve. If $C \subset Y_i$ for some i, then $\big((\delta - \varepsilon \cdot \alpha) \cdot C\big) \geq 0$ by what we have just said. On the other hand, if $C \not\subset \mathrm{Supp}(E)$, then evidently

$$\Big(E \cdot C\Big) \ = \ \Big((\delta - \alpha) \cdot C\Big) \ \geq 0,$$

and a fortiori $\big((\delta - \varepsilon \cdot \alpha) \cdot C\big) \geq 0$. Thus $(\delta - \varepsilon \cdot \alpha)$ is indeed nef. But then δ — being the sum of a nef and an ample class — is ample. □

2.3.E Zariski Decompositions

In his fundamental paper [623], Zariski established a result that yields a great deal of insight into the behavior of linear series on a smooth surface. Here we state (but do not prove) his theorem, and give some elementary applications. At the end we briefly discuss the Riemann–Roch problem on surfaces and higher dimensional varieties.

Zariski's theorem. Zariski's beautiful idea is that any effective divisor on a surface can be decomposed in a non-trivial way into a positive and negative part:

Theorem 2.3.19. (Zariski decomposition). *Let X be a smooth projective surface, and let D be a pseudoeffective integral divisor on X (Definition 2.2.25). Then D can be written uniquely as a sum*

$$D \ = \ P + N$$

of $\mathbf{Q}$-divisors with the following properties:

(i). *P is nef;*

(ii). *$N = \sum_{i=1}^{r} a_i E_i$ is effective, and if $N \neq 0$ then the intersection matrix*

$$\| (E_i \cdot E_j) \|$$

determined by the components of N is negative definite;

(iii). *P is orthogonal to each of the components of N, i.e. $\big(P \cdot E_i\big) = 0$ for every $1 \leq i \leq r$.*

P and N are called respectively the *positive* and *negative* parts of D. Historically, the theorem represents one of the earliest settings in which $\mathbf{Q}$-divisors appear in an essential way: one cannot in general take P and N to be integral (Example 2.3.20).

Zariski [623] established Theorem 2.3.19 when D is effective; the extension to pseudoeffective divisors is due to Fujita [193]. We refer to Bădescu's book [22, Chapter 14] for a nice account of the proof, and for applications of the theorem to classification theory. Here we content ourselves with some illustrations and simple consequences.

Example 2.3.20. Let X be the surface obtained by blowing up three collinear points $x_1, x_2, x_3 \in \mathbf{P}^2$. Write H for the pullback to X of a line in $\mathbf{P}^2$, E_1, E_2, E_3 for the exceptional divisors over the three points, and

$$E_{123} \equiv_{\text{lin}} H - E_1 - E_2 - E_3$$

for the proper transform of the line through the x_i. Then

$$D =_{\text{def}} 3H - 2E_1 - 2E_2 - 2E_3$$

is a big divisor on X whose negative and positive parts are

$$\begin{aligned} N &= \tfrac{3}{2} \cdot E_{123} \\ P = D - N &\equiv_{\text{num}} \tfrac{1}{2} \cdot \big(3H - E_1 - E_2 - E_3\big). \quad \square \end{aligned}$$

Returning to the situation of the theorem, a very important point is that the positive part P in effect carries all the sections of D and its multiples. In the following statement, $\ulcorner mN \urcorner$ denotes the round-up of mN, i.e. the divisor obtained from mN by rounding up the coefficients of each of its components (Definition 9.1.2).

Proposition 2.3.21. *In the situation of Theorem 2.3.19, the natural map*

$$H^0\big(X, \mathcal{O}_X(mD - \ulcorner mN \urcorner)\big) \longrightarrow H^0\big(X, \mathcal{O}_X(mD)\big)$$

is bijective for every $m \geq 1$.

In other words, at least after passing to a multiple to clear denominators — so that $mD - \ulcorner mN \urcorner = mP$ — this means that all the sections of a line bundle on a surface come from a nef divisor.

We will prove the proposition shortly, but first we note a number of interesting consequences:

Corollary 2.3.22. (Volume of divisors on a surface). *In the situation of Theorem 2.3.19,*

$$\mathrm{vol}(D) \;=\; \big(P^2\big).$$

In particular, the volume of an integral divisor on a surface is always a rational number.

Proof. In fact, the proposition implies that $\mathrm{vol}(D) = \mathrm{vol}(P) = \big(P^2\big)$. □

Corollary 2.3.23. (Criterion for finite generation). *If D is a big divisor on a smooth surface X, then the section ring $R(X,D)$ is finitely generated if and only if the positive part P of D is semiample.*

Sketch of Proof. Assume first that P is integral. Then the proposition implies that $R(X,D) = R(X,P)$, and so the assertion follows from 2.3.15. For the general case, one argues as in Example 2.1.30. □

Remark 2.3.24. (Surfaces defined over finite fields). Cutkosky and Srinivas [101, Theorem 3] show that if X is a non-singular projective surface defined over the algebraic closure of a finite field, and if D is an effective divisor on X, then the graded ring $R(X,D)$ is always finitely generated. Keel ([329], [330]) explores some related phenomena for higher-dimensional varieties defined over finite fields. □

Corollary 2.3.25. *Assume that D is big, and fix a prime divisor E on X. Given $m \gg 1$ denote by $\mathrm{ord}_E\big(|mD|\big)$ the order along E of a general divisor in $|mD|$. Then*

$$\limsup_{m\to\infty} \tfrac{1}{m}\cdot \mathrm{ord}_E\big(|mD|\big)$$

is a rational number, and it is non-zero for only finitely many E.

Proof. Let $D = P + N$ be the Zariski decomposition of D. It follows from 2.3.9 that $\mathrm{ord}_E\big(|\,[mP]\,|\big)$ is bounded independently of m. Therefore

$$\limsup_{m\to\infty} \frac{1}{m}\cdot \mathrm{ord}_E\big(|mD|\big) \;=\; \mathrm{ord}_E\big(N\big),$$

and the assertion follows. □

Proof of Proposition 2.3.21. The issue is to show that if $D' \equiv_{\mathrm{lin}} mD$ is an effective divisor, then $D' \succcurlyeq mN$. It is enough to prove this after replacing each of the divisors in question by a multiple, so without loss of generality we may assume to begin with that P and N are integral and that $m = 1$. Given $D' \equiv_{\mathrm{lin}} D$ effective, write $D' = N_1 + M_1$, where N_1 is an effective linear combination of the E_i and M_1 does not contain any of these components. We are required to prove that $N_1 \succcurlyeq N$. Since $D' - N \equiv_{\text{num}} P$ is perpendicular to each of the E_i, and since M_1 is an effective divisor meeting each E_i properly, we see that

$$(N_1 - N) \cdot E_i \ \leq \ 0 \quad \text{for all} \ \ i. \tag{*}$$

Now write $N_1 - N = N' - N''$ where N' and N'' are non-negative linear combinations of the E_i with no common components, and assume for a contradiction that $N'' \neq 0$. Then $N'' \cdot N'' < 0$ by property (ii) in 2.3.19, and so

$$(N_1 - N) \cdot N'' \ > 0.$$

But this contradicts (*). □

Higher dimensions. It is natural to ask whether analogous results hold on a smooth projective variety X of dimension $n \geq 3$. There are various different ways to state the question precisely, but one of the simplest is to focus on a higher-dimensional analogue of Proposition 2.3.21. Specifically, one says that a divisor D on X has a rational Zariski decomposition in the sense of Cutkosky, Kawamata, and Moriwaki (a CKM decomposition) if one can find a smooth birational modification $\mu : X' \longrightarrow X$ of X, together with an effective $\mathbf{Q}$-divisor N' on X', such that:

(i). $P' =_{\text{def}} \mu^* D - N'$ is a nef $\mathbf{Q}$-divisor on X'; and
(ii). The natural maps

$$H^0\big(X', \mathcal{O}_{X'}(\mu^*(mD) - \ulcorner mN' \urcorner)\big) \longrightarrow H^0\big(X', \mathcal{O}_{X'}(\mu^*(mD))\big)$$

are bijective for every $m \geq 1$.

Since in any event $H^0\big(X', \mathcal{O}_{X'}(\mu^*(mD))\big) = H^0\big(X, \mathcal{O}_X(mD)\big)$ for all m, this would again imply in effect that all the sections of D and its multiples are carried by a nef divisor. When $D = K_X$ the existence of such a decomposition has important structural consequences [319], [326, Chapter 7-3].

However Cutkosky [97] showed that such decompositions cannot exist in general. In fact, if D is a big divisor admitting a rational CKM decomposition, then it follows as in Corollary 2.3.22 that

$$\mathrm{vol}_X(D) \ = \ \mathrm{vol}_{X'}(P') \ = \ \big((P')^n\big)_{X'}$$

is rational. Therefore a big divisor D with irrational volume (Section 2.3.B) does not admit a rational CKM-decomposition. Nakayama [472] constructed a related example where it is impossible to find a decomposition of this type even if one allows N' and P' to be $\mathbf{R}$-divisors. Prokhorov's survey [507] gives an overview of this circle of ideas.

One can also try to generalize Corollary 2.3.25 by attaching asymptotically defined multiplicities to components of the stable base locus of D. This was first carried out by Nakayama [472], whose work is extended in [67] and [157]. One can in effect construct the "codimension one" part of the Zariski decomposition, but higher codimension phenomena remain difficult to attack. In yet another direction, Fujita [199] proved a very interesting theorem showing that one can capture "most" of the sections of a divisor D by a nef bundle on a modification. This result, along with some applications, appears in Section 11.4.

The Riemann–Roch problem. Zariski was led to Theorem 2.3.19 in the course of his work on the Riemann–Roch problem, which we now survey very briefly.

Specifically, let D be a divisor on a non-singular projective surface X. One wants to understand how the dimension $h^0\big(X, \mathcal{O}_X(mD)\big)$ behaves as a function of m. In this direction Zariksi [623] proves:

Theorem. *Assume that D is effective. Then there is a quadratic polynomial $P(x)$ such that for all $m \gg 0$,*

$$h^0\big(X, \mathcal{O}_X(mD)\big) \;=\; P(m) + \lambda(m)$$

where $\lambda(m)$ is a bounded function of m.

Zariski's theorem was completed by Cutkosky and Srinivas [101], who prove that in fact $\lambda(m)$ is a periodic function of m. The paper [101] also contains a nice discussion of several related questions.

It is natural to ask about the function $h^0\big(X, \mathcal{O}_X(mD)\big)$ for a divisor D on a smooth projective variety X of dimension n. This is already very interesting when D is ample or nef, in which case the dominant contribution to the dimension in question is $\frac{(D^n)\cdot m^n}{n!}$. Kollár and Matsusaka [366] prove

Theorem. *Assume that D is semiample. Then there exists a polynomial $Q(m)$ of degree $\leq n-1$ such that*

$$\left| h^0\big(X, \mathcal{O}_X(mD)\big) \;-\; \frac{(D^n)\cdot m^n}{n!} \right| \;\leq\; Q(m).$$

Moreover, the coefficients of $Q(m)$ depend only on the intersection numbers $\big(D^n\big)$ and $\big(D^{n-1}\cdot K_X\big)$.

A somewhat sharper statement was later established by Nielson [478]. Luo [410] showed that the theorem of Kollár–Matsusaka remains valid assuming only that D is big and nef. Some related results appear in Section 10.2 below. Tsai [572] gives some interesting applications of this circle of ideas to the boundedness of certain families of morphisms.

Example 2.3.26. Let D be a big and nef divisor on a projective variety X of dimension n. Then given any $\varepsilon > 0$ there is an integer m_0 such that

$$\left| h^0\big(X, \mathcal{O}_X(mD)\big) - \frac{(D^n)\cdot m^n}{n!} \right| \leq \varepsilon \cdot m^n$$

for $m \geq m_0$. (This is an elementary consequence of Asymptotic Riemann–Roch III, Theorem 1.4.41.) □

2.4 Graded Linear Series and Families of Ideals

We have devoted considerable attention in the previous sections to the asymptotic behavior of the complete linear series $|mD|$ determined by a divisor D. However it is sometimes important to be able to discuss analogous properties for possibly incomplete series, and we introduce in this section a formalism for doing so. We also outline a related algebraic concept that allows one to build simple local models of many phenomena that can occur for global linear series. Both of these constructions will play a central role when we develop the theory of asymptotic multiplier ideals in Chapter 11. Here we focus principally on definitions and examples.

2.4.A Graded Linear Series

The essential idea of a graded linear series goes at least as far back as Zariski's paper [623]. More recently the formalism was developed and applied in the paper [158].

We begin with the definition.

Definition 2.4.1. (Graded linear series). Let X be an irreducible variety. A *graded linear series* on X consists of a collection

$$V_\bullet = \{ V_m \}_{m \geq 0}$$

of finite dimensional vector subspaces $V_m \subseteq H^0\big(X, \mathcal{O}_X(mD)\big)$, D being some integer divisor on X. These subspaces are required to satisfy the property

$$V_k \cdot V_\ell \subseteq V_{k+\ell} \quad \text{for all} \quad k, \ell \geq 0, \tag{2.18}$$

where $V_k \cdot V_\ell$ denotes the image of $V_k \otimes V_\ell$ under the homomorphism

$$H^0\big(X, \mathcal{O}_X(kD)\big) \otimes H^0\big(X, \mathcal{O}_X(\ell D)\big) \longrightarrow H^0\big(X, \mathcal{O}_X((k+\ell)D)\big)$$

determined by multiplication. It is also required that V_0 contain all constant functions. □

One can of course replace the divisor D by a line bundle L. Note however that giving either D or L is an essential part of the data, since they are needed to determine the multiplication in (2.18). In practice this ambient complete linear series is usually clear from context, but when it is not we will specify $V_\bullet$ as belonging to (or being associated with) the divisor D or bundle L.

Given a graded linear series $V_\bullet = \{V_m\}$, set

$$R(V_\bullet) \;=\; \bigoplus_{m=0}^{\infty} V_m. \tag{2.19}$$

Then (2.18) is equivalent to the condition that $R(V_\bullet)$ be a graded $\mathbf{C}$-subalgebra of the section ring $R(D) = R(X,D)$ of D (Definition 2.1.17). We call $R(V_\bullet)$ the *section ring* or *graded ring* associated to $V_\bullet$.

Definition 2.4.2. (Finite generation of a graded linear series). A graded linear series $V_\bullet$ on a projective variety X is *finitely generated* if $R(V_\bullet)$ is finitely generated as a $\mathbf{C}$-algebra. □

We next give some examples and describe some operations on graded linear series.

Example 2.4.3. (Examples of graded linear series). Let X be an irreducible projective variety, and D a divisor on X.

(i). The full space of sections $V_m = H^0\big(X, \mathcal{O}_X(mD)\big)$ defines a graded linear series, which we write $\Gamma_\bullet\big(X, \mathcal{O}_X(D)\big)$.

(ii). Let $\mathfrak{b} \subseteq \mathcal{O}_X$ be an ideal sheaf, and $c > 0$ a positive real number. Set

$$V_m \;=\; H^0\big(X, \mathcal{O}_X(mD) \otimes \mathfrak{b}^{\ulcorner mc \urcorner}\big)$$

where as usual $\ulcorner r \urcorner$ is the round-up (i.e. least integer greater than or equal to) a real number r. For example, when $\mathfrak{a} = \mathfrak{m}$ is the maximal ideal of a smooth point $x \in X$, V_m is the subspace of sections of $\mathcal{O}_X(mD)$ vanishing to order $\geq mc$ at x. Then $V_\bullet = \{V_m\}$ is a graded linear series (belonging to D), which we denote by[10]

$$\Gamma_\bullet\big(X, \mathcal{O}_X(D) \otimes \mathfrak{b}^c\big).$$

Even when $R(D)$ is finitely generated, this graded linear series may fail to be so.

(iii). Fix a point $x \in \mathbf{P}^n$ and for $m \geq 1$ define $V_m \subseteq H^0\big(\mathbf{P}^n, \mathcal{O}_{\mathbf{P}^n}(m)\big)$ to be the space of homogeneous polynomials of degree m that vanish at x. This determines a graded linear series $V_\bullet$ on $\mathbf{P}^n$ that is not finitely generated.

(iv). Suppose $Y \supseteq X$ is a variety that contains X as a subvariety, and $D = D'|X$ is the restriction of a divisor D' on Y. Then the images

[10] The notation is purely formal: in particular, we do not try to attach any meaning to $\mathfrak{b}^c$.

$$V_m = \operatorname{im}\Big(H^0\big(Y, \mathcal{O}_Y(mD')\big) \longrightarrow H^0\big(X, \mathcal{O}_X(mD)\big) \Big)$$

form a graded linear series on X, which we denote by $\mathrm{Tr}_X(D')_\bullet$.

(v). Keeping the notation of (iv), suppose more generally that $W_\bullet$ is a graded linear series on Y belonging to D'. Then the restriction $V_\bullet = W_\bullet \mid X$ of $W_\bullet$ to X is the graded linear series given by

$$V_m = \Big(\text{image of } W_m \text{ in } H^0\big(X, \mathcal{O}_X(mD)\big) \Big). \quad \square$$

Example 2.4.4. (Product of two graded linear series). Let $V_\bullet$ and $W_\bullet$ be graded linear series on X associated to divisors D and E respectively. Then one gets a graded linear series $V_\bullet \cdot W_\bullet$ belonging to $D + E$ by putting

$$\big(V_\bullet \cdot W_\bullet\big)_m = V_m \cdot W_m \subseteq H^0\big(X, \mathcal{O}_X\big(m(D+E)\big)\big).$$

Here $V_m \cdot W_m$ denotes (somewhat abusively) the image of $V_m \otimes W_m$ under the multiplication

$$H^0\big(X, \mathcal{O}_X(mD)\big) \otimes H^0\big(X, \mathcal{O}_X(mE)\big) \longrightarrow H^0\big(X, \mathcal{O}_X\big(m(D+E)\big)\big). \quad \square$$

Example 2.4.5. (Intersection and span of graded linear series). Let $V_\bullet$ and $W_\bullet$ be graded linear series on a variety X belonging to the same divisor D. Then we get two new graded linear series $V_\bullet \cap W_\bullet$ and $\mathrm{Span}(V_\bullet, W_\bullet)$ by the rules

$$\big(V_\bullet \cap W_\bullet\big)_m = V_m \cap W_m,$$
$$\mathrm{Span}(V_\bullet\, , W_\bullet)_m = \sum_{k+\ell=m} V_k \cdot W_\ell,$$

these both being subspaces of $H^0\big(X, \mathcal{O}_X(mD)\big)$. $\square$

Example 2.4.6. (Veronese of a graded linear series). If $V_\bullet$ is a graded linear series belonging to D and $p \geq 1$ is a positive integer, then $V_\bullet^{(p)}$ is the graded series associated to pD whose component of degree m is

$$\big(V_\bullet^{(p)}\big)_m = V_{mp} \subseteq H^0\big(X, \mathcal{O}_X(mpD)\big). \quad \square$$

The basic asymptotic invariants of complete linear series extend without difficulty to the present setting. In the following X is an irreducible projective variety and $V_\bullet$ is a graded linear series belonging to a divisor D.

Definition 2.4.7. (Semigroup and exponent of a graded linear series). Let $V_\bullet$ be a graded linear series belonging to a divisor D. The semigroup of $V_\bullet$ is the set

$$\mathbf{N}(V_\bullet) = \big\{ m \geq 0 \mid V_m \neq 0 \big\}.$$

The exponent $e(V_\bullet)$ of $V_\bullet$ is the greatest common divisor of all the elements in $\mathbf{N}(V_\bullet)$. $\square$

Similarly:

Definition 2.4.8. (Iitaka dimension of $\mathbf{V}_\bullet$). If $V_m = 0$ for all m, put $\kappa(V_\bullet) = -\infty$. Otherwise, given a positive integer $m \in \mathbf{N}(V_\bullet)$, write

$$\phi_m : X \dashrightarrow \mathbf{P}H^0\big(X, V_m\big)$$

for the rational mapping determined by V_m, and denote by $\phi_m(X)$ the closure of its image. Assuming that X is normal, the *Iitaka dimension* $\kappa(V_\bullet)$ of $V_\bullet$ is defined to be

$$\kappa(V_\bullet) \;=\; \max_{m \in \mathbf{N}(V_\bullet)} \{\dim \phi_m(X)\}. \quad \square$$

Definition 2.4.9. (Stable base locus of a graded linear series). The *stable base locus* $\mathbf{B}\big(V_\bullet\big)$ of $V_\bullet$ is the algebraic subset

$$\mathbf{B}\big(V_\bullet\big) \;=\; \bigcap_{m \geq 0} \mathrm{Bs}\big(|V_m|\big). \quad \square$$

The next example shows however that in every case the invariants for a graded linear series can differ from those of the divisor to which it belongs.

Example 2.4.10. Let D be an ample divisor on a smooth projective variety X.

(i). Put $V_m = H^0\big(X, \mathcal{O}_X(mD)\big)$ when m is even and $V_m = 0$ when m is odd. Then $e(V_\bullet) = 2$ but $e(D) = 1$.

(ii). Assume that there is an effective integral divisor E on X such that $0 < \kappa(D - E) < \dim X$. Take $\mathfrak{a} = \mathcal{O}_X(-E)$ to be the ideal of E, and apply the construction of Example 2.4.3 (ii) to define $V_\bullet = \Gamma_\bullet\big(X, \mathcal{O}_X(D) \otimes \mathfrak{a}\big)$. Then $\kappa(V_\bullet) = \kappa(D - E) < \kappa(D) = \dim X$.

(iii). Take $V_\bullet$ to be the graded linear series belonging to $\mathcal{O}_{\mathbf{P}^n}(1)$ on $\mathbf{P}^n$ constructed in 2.4.3 (iii). Then $\mathbf{B}\big(V_\bullet\big) = \{x\}$, whereas of course $\mathbf{B}\big(\mathcal{O}_{\mathbf{P}^n}(1)\big)$ is empty. $\square$

Example 2.4.11. Let $V_\bullet$ be a graded linear series on X, and let $\mathfrak{b}_m = \mathfrak{b}\big(\,|V_m|\,\big)$ be the base ideal of V_m (Definition 1.1.8). Then

$$\mathfrak{b}_k \cdot \mathfrak{b}_\ell \;\subseteq\; \mathfrak{b}_{k+\ell} \tag{2.20}$$

for all $k, \ell \geq 0$. $\square$

Finally, one can discuss the volume of a graded linear series:

Definition 2.4.12. Let X be a projective variety of dimension n, and $V_\bullet$ a graded linear series on X belonging to a divisor D. Then the volume of $V_\bullet$ is the non-negative real number

$$\mathrm{vol}(V_\bullet) \;=\; \limsup_{m\in \mathbf{N}(V_\bullet)} \frac{\dim V_m}{m^n/n!}. \quad \square$$

As in the case of complete linear series, this invariant can easily be irrational:

Example 2.4.13. Let D be the hyperplane divisor on $X = \mathbf{P}^n$, let $\mathfrak{m}$ be the ideal of a point $x \in \mathbf{P}^n$, and for $0 < c < 1$ let $V_\bullet = \Gamma_\bullet\big(\mathbf{P}^n, \mathcal{O}_{\mathbf{P}^n}(1) \otimes \mathfrak{m}^c\big)$ be the graded linear series given by forms of degree m vanishing to order $\geq cm$ at x (Example 2.4.3.(ii)). Then

$$\mathrm{vol}(V_\bullet) \;=\; 1 - c.$$

In particular, this volume is irrational if c is. $\square$

2.4.B Graded Families of Ideals

Much of the geometry of a graded linear series $V_\bullet$ is captured by the sequence $\mathfrak{b}_m = \mathfrak{b}\big(|V_m|\big)$ of its base ideals. The most important property of these ideals is the multiplicative behavior summarized in the inclusion (2.20). Taking this relation as an axiom leads to an essentially local construction that can be used to model some of the complexity of global linear series. These graded systems of ideals are also very natural from an algebraic perspective, and one can conversely use globally inspired techniques to study them. This was the viewpoint taken in the papers [159], [160] of Ein, Smith, and the author, from which much of the present subsection is adapted.

Definition 2.4.14. (Graded family of ideals). Let X be an irreducible variety. A *graded family* or *graded system* of ideals $\mathfrak{a}_\bullet = \{\mathfrak{a}_m\}$ on X is a collection of ideal sheaves $\mathfrak{a}_m \subseteq \mathcal{O}_X$ $(m \geq 1)$ such that

$$\mathfrak{a}_k \cdot \mathfrak{a}_\ell \;\subseteq\; \mathfrak{a}_{k+\ell} \tag{2.21}$$

for all $k, \ell \geq 1$. It is convenient to set $\mathfrak{a}_0 = \mathcal{O}_X$. $\square$

Definition 2.4.15. (Rees algebra and finite generation). The *Rees algebra* of $\mathfrak{a}_\bullet$ is the graded $\mathcal{O}_X$-algebra

$$\mathrm{Rees}(\mathfrak{a}_\bullet) \;=\; \bigoplus_{m\geq 0} \mathfrak{a}_m.$$

We say that $\mathfrak{a}_\bullet$ is *finitely generated* if $\mathrm{Rees}(\mathfrak{a}_\bullet)$ is finitely generated as an $\mathcal{O}_X$-algebra in the sense that there is an integer m_0 such that $\mathrm{Rees}(\mathfrak{a}_\bullet)$ is generated as an $\mathcal{O}_X$-algebra by its terms of degrees $\leq m_0$. $\square$

In order to give a feeling for the definition, we present an extended collection of examples. Most of the constructions that follow yield graded families that fail in general to be finitely generated (Remark 2.4.17).

Example 2.4.16. (Examples of graded families of ideals). Let X be an irreducible variety.

(i). If $\mathfrak{b} \subseteq \mathcal{O}_X$ is any ideal, then the powers $\mathfrak{a}_m = \mathfrak{b}^m$ of $\mathfrak{b}$ form a graded family, which we denote by $\mathrm{Pow}_\bullet(\mathfrak{b})$. One should view this as a trivial example, and we will see shortly (Proposition 2.4.27) that any finitely generated graded system is essentially of this form.

(ii). Let $V_\bullet$ be a graded linear series on X. Then the base ideals $\mathfrak{b}_m = \mathfrak{b}(\,|V_m|\,)$ form a graded family $\mathfrak{b}(V_\bullet)$. As just suggested, this is one of the motivating examples.

(iii). Fix an ideal $\mathfrak{b} \subseteq \mathcal{O}_X$ and a real number $c > 0$. Then we get a graded family $\mathfrak{a}_\bullet$ of ideals by putting $\mathfrak{a}_m = \mathfrak{b}^{\ulcorner mc \urcorner}$.

(iv). Assume that X is smooth, and let $Z \subseteq X$ be a reduced algebraic subset. Thus the ideal sheaf

$$\mathfrak{q} =_{\text{def}} \mathcal{I}_Z \subseteq \mathcal{O}_X$$

of Z is radical. Given $m \geq 1$, recall that the mth *symbolic power* $\mathfrak{q}^{<m>} \subseteq \mathcal{O}_X$ of $\mathfrak{q}$ is the ideal consisting of germs of functions that vanish to order $\geq m$ at a general point (and hence also at every point) of each component of Z.[11] These constitute a graded system $\{\mathfrak{q}^{<\bullet>}\}$: the required inclusion

$$\mathfrak{q}^{<k>} \cdot \mathfrak{q}^{<\ell>} \subseteq \mathfrak{q}^{<k+\ell>}$$

follows from the observation that if $f \in \mathcal{O}_x X$ has multiplicity $\geq k$ and $g \in \mathcal{O}_x X$ has multiplicity $\geq \ell$ at a point $x \in Z$, then fg vanishes to order $\geq k + \ell$ at x.

(v). Assume that X is normal. Let $\mu : X' \longrightarrow X$ be a projective (or proper) birational mapping, and let $D \subset X$ be an effective μ-exceptional Cartier divisor. Then the ideals

$$\mathfrak{a}_m \;=\; \mu_* \mathcal{O}_{X'}(-mD) \;\subseteq\; \mathcal{O}_X$$

form a graded family. The inclusion $\mathfrak{a}_k \cdot \mathfrak{a}_\ell \subseteq \mathfrak{a}_{k+\ell}$ is deduced from the natural mapping

$$\mu_* \mathcal{O}_{X'}(-kD) \otimes \mu_* \mathcal{O}_{X'}(-\ell D) \longrightarrow \mu_* \mathcal{O}_{X'}\big(-(k+\ell)D\big).$$

Note that the previous example of symbolic powers can be realized in this fashion. (Take X' to dominate the blow-up of $\mathcal{I}_Z$, and let D be the union of the irreducible exceptional divisors corresponding to the valuations determined by order of vanishing along the components of Z.)

---

[11] In commutative algebra, one defines the symbolic power somewhat differently. It is a theorem of Nagata and Zariski that in our setting this coincides with the stated definition. Eisenbud's book [164, Chapter 3.9] contains a nice discussion.

(vi). Assume that X is affine and non-singular, and identify ideal sheaves with ideals in the coordinate ring $\mathbf{C}[X]$ of X. Given any non-zero ideal $\mathfrak{b} \subseteq \mathbf{C}[X]$, set

$$\mathfrak{a}_m = \left\{ f \in \mathbf{C}[X] \;\middle|\; \begin{array}{c} Df \in \mathfrak{b} \\ \forall \text{ differential operators } D \text{ on } X \text{ of order } < m \end{array} \right\}.$$

This graded family reduces to the symbolic powers $\{\mathfrak{q}^{<m>}\}$ when $\mathfrak{b} = \mathfrak{q}$ is radical.

(vii). Let $\mathfrak{a}_\bullet = \{\mathfrak{a}_m\}$ be a graded family, and $\mathfrak{b} \subseteq \mathcal{O}_X$ a fixed ideal. Then the colon ideals

$$\mathfrak{r}_m = \left(\mathfrak{a}_m : \mathfrak{b}^m\right) =_{\text{def}} \left\{ f \in \mathcal{O}_X \mid f \cdot \mathfrak{b}^m \in \mathfrak{a}_m \right\}$$

form a graded family.

(viii). In the setting of (vii), observe that for fixed m the ideals $(\mathfrak{a}_m : \mathfrak{b}^\ell)$ are increasing in ℓ, and hence stabilize to an ideal denoted by $(\mathfrak{a}_m : \mathfrak{b}^\infty)$. These stable colon ideals form yet another graded family indexed by m.

(ix). Let $<$ be a term order on $\mathbf{C}[x_1, \ldots, x_n]$, and let $\mathfrak{b} \subseteq \mathbf{C}[x_1, \ldots, x_n]$ be an ideal. Then the initial ideals $\mathfrak{a}_m = \text{in}_<(\mathfrak{b}^m)$ of powers of $\mathfrak{b}$ determine a graded family of monomial ideals. □

Remark 2.4.17. (Failure of finite generation). Except for (i), most of the graded families appearing in the previous example can fail to be finitely generated. For instance, the system $\mathfrak{b}^{\ulcorner mc \urcorner}$ in (ii) is finitely generated if and only if c is rational. Goto, Nishida, and Watanabe gave an example of a monomial curve $Z \subseteq \mathbf{C}^3$ for which the symbolic power system in (iv) is not finitely generated: see [585, §7.3]. It is observed in [585, Corollary 7.2.5] that the stable colon construction in (viii) contains (iv) as a special case, as do (v) and (vi). □

Example 2.4.18. (Graded systems from valuations). Suppose v is a valuation on the function field $\mathbf{C}(X)$ taking values in $\mathbf{R}$ and that is non-negative on regular functions. Then the valuation ideals

$$\mathfrak{a}_m =_{\text{def}} \{ f \in \mathcal{O}_X \mid v(f) \geq m \}$$

form a graded system of ideal sheaves. When v is a divisorial valuation — i.e. v is given by order of vanishing along a prime divisor on a birational model of X — one can realize $\mathfrak{a}_\bullet$ as in Example 2.4.16 (iv). However already when $X = \mathbf{C}^2$, so that $\mathbf{C}(X) = \mathbf{C}(x, y)$, one also gets very interesting examples from non-divisorial valuations.

(i). (*Order of vanishing along a transcendental arc*). Let $p(t) = \sum_{i=1}^{\infty} \frac{t^i}{i!} \in \mathbf{C}[[t]]$ be the power series of the function $e^t - 1$. Given $f \in \mathbf{C}[x, y]$ define

$$v(f) = \operatorname{ord}_t f\big(t\,,\,p(t)\big).$$

The corresponding valuation ideals are given explicitly by

$$\mathfrak{a}_m = \big(\ x^m\ ,\ y - p_{m-1}(x)\ \big),$$

where $p_\ell(t) = \sum_{i=1}^{\ell} \frac{1}{i!} t^i$ is the ℓth Taylor polynomial of $e^t - 1$. Note that each $\mathfrak{a}_k$ contains a polynomial (viz. $y - p_{k-1}(x)$) defining a smooth curve in $\mathbf{C}^2$: this implies that $\mathfrak{a}_\bullet$ cannot be finitely generated.

(ii). (*Valuation by irrational weighted degree*). Let $\mathfrak{a}_m \subseteq \mathbf{C}[x,y]$ be the monomial ideal spanned by all monomials $x^i y^j$ with

$$i + j\sqrt{2}\ \geq\ m.$$

Then $\mathfrak{a}_\bullet$ is the graded family of ideals corresponding to the valuation by weighted degree with $v(x) = 1$ and $v(y) = \sqrt{2}$. Here again $\mathfrak{a}_\bullet$ fails to be finitely generated. □

Example 2.4.19. ("Diagonal" graded system). Generalizing the previous example, pick any real numbers $\delta_1, \ldots, \delta_n > 0$ and consider the monomial ideal $\mathfrak{a}_m \subseteq \mathbf{C}[x_1, \ldots, x_n]$ given by

$$\mathfrak{a}_m\ =\ \big\langle\ x_1^{i_1} \cdot \ldots \cdot x_n^{i_n}\ \big|\ \textstyle\sum \frac{i_k}{\delta_k} \geq m\ \big\rangle.$$

When the δ_i are integers, $\mathfrak{a}_m$ is just (the integral closure of) the mth power of the "diagonal ideal" $(x_1^{\delta_1}, \ldots, x_n^{\delta_n})$. In general, one can view the resulting graded system $\mathfrak{a}_\bullet = \{\mathfrak{a}_m\}$ as being an analogue defined for arbitrary real numbers $\delta_i > 0$. □

Example 2.4.20. (Models of base loci). We illustrate the philosophy that graded systems of ideals can be used to simulate the behavior of base loci of global linear series. As above, X is an irreducible variety.

(i). Let $\mathfrak{b} \subseteq \mathcal{O}_X$ be a fixed ideal, and take $\mathfrak{a}_m = \mathfrak{b}$ for every $m \geq 1$. The resulting graded family $\mathfrak{a}_\bullet$ mirrors the behavior of Zariski's example in Section 2.3.A of a divisor D for which the base locus of each of the linear series $|mD|$ consists of a fixed curve independent of m. Note that if $\mathfrak{b}$ is non-trivial then $\mathfrak{a}_\bullet$ is not finitely generated.

(ii). Remaining in the setting of (i), choose for each $m \geq 1$ an ideal $\mathfrak{c}_m$ with $\mathfrak{b} \subseteq \mathfrak{c}_m \subseteq \mathcal{O}_X$ for every m. Then $\mathfrak{a}_m = \mathfrak{b} \cdot \mathfrak{c}_m$ forms a graded family that models Example 2.3.5: there the base loci of $|mD|$ consisted of a fixed curve having an embedded point that varies with m.

(iii). Referring to Example 2.3.7, we see that the base-loci of the linear series in question are modeled by an instance of Example 2.4.16 with irrational $c > 0$. □

Remark 2.4.21. Conversely, in Chapter 11 we use global methods to study graded systems of ideals. Specifically, we will attach to any graded system $\mathfrak{a}_\bullet$ and coefficient $c > 0$ a multiplier ideal $\mathcal{J}(c \cdot \mathfrak{a}_\bullet)$ that carries interesting information about $\mathfrak{a}_\bullet$ and can be used to prove some rather surprising results. (See for instance Theorem 11.3.4.) □

Graded families of ideals give rise to graded linear series:

Example 2.4.22. Let D be a divisor on a projective variety X, and let $\mathfrak{a}_\bullet = \{\mathfrak{a}_m\}$ be a graded family of ideals on X. Then the spaces

$$V_m = H^0\big(X, \mathcal{O}_X(mD) \otimes \mathfrak{a}_m\big)$$

form a graded linear series (associated to D). Example 2.4.3.(ii) is a special case of this construction. □

Example 2.4.23. (Realizing ideals as base loci). Keeping the notation of the previous example, suppose that we are able to choose D so that each of the sheaves $\mathcal{O}_X(mD) \otimes \mathfrak{a}_m$ is globally generated. Then we have realized $\mathfrak{a}_\bullet$ as the family of base ideals of a graded linear series $V_\bullet$. This applies, for example, to the two families appearing in Example 2.4.18 provided that one compactifies and views each $\mathfrak{a}_m$ in the evident way as a sheaf on $\mathbf{P}^2$. However, there are non-trivial global obstructions to realizing a given family of ideals as the base-ideals of a *complete* linear series. The theorem of Zariski-Fujita (Remark 2.1.32) provides one such. Another, coming from the theory of multiplier ideals, appears in Corollary 11.2.22. □

We next spell out some additional definitions, starting with natural algebraic operations on graded systems of ideals.

Definition 2.4.24. (Intersection, product, and sum). Let $\mathfrak{a}_\bullet = \{\mathfrak{a}_m\}$ and $\mathfrak{b}_\bullet = \{\mathfrak{b}_m\}$ be graded families of ideals on a variety X. The intersection, product, and sum of $\mathfrak{a}_\bullet$ and $\mathfrak{b}_\bullet$ are the graded systems given by the rules

$$\begin{aligned} \big(\mathfrak{a}_\bullet \cap \mathfrak{b}_\bullet\big)_m &= \mathfrak{a}_m \cap \mathfrak{b}_m, \\ \big(\mathfrak{a}_\bullet \cdot \mathfrak{b}_\bullet\big)_m &= \mathfrak{a}_m \cdot \mathfrak{b}_m, \\ \big(\mathfrak{a}_\bullet + \mathfrak{b}_\bullet\big)_m &= \sum_{k+\ell=m} \mathfrak{a}_k \cdot \mathfrak{b}_\ell. \quad \square \end{aligned}$$

As before, we can define the semigroup and Veronese of a graded system.

Definition 2.4.25. (Semigroup and exponent of a graded system of ideals). Let $\mathfrak{a}_\bullet = \{\mathfrak{a}_m\}$ be a graded system of ideals on an irreducible variety X. The *semigroup* of $\mathfrak{a}_\bullet$ is the set

$$\mathbf{N}(\mathfrak{a}_\bullet) = \{ m \geq 0 \mid \mathfrak{a}_m \neq (0) \}.$$

The *exponent* $e(\mathfrak{a}_\bullet)$ of $\mathfrak{a}_\bullet$ is the greatest common divisor of the elements in $\mathbf{N}(\mathfrak{a}_\bullet)$. □

Definition 2.4.26. (Veronese of a graded family). Given a graded family $\mathfrak{a}_\bullet = \{\mathfrak{a}_m\}$ of ideals on X and a positive integer $p > 0$, the pth Veronese of $\mathfrak{a}_\bullet$ is the graded system $\mathfrak{a}_\bullet^{(p)}$ defined by $(\mathfrak{a}_\bullet^{(p)})_m = \mathfrak{a}_{pm}$. □

The next proposition states that a finitely generated graded system essentially looks like the trivial family (Example 2.4.16 (i)) consisting of powers of a fixed ideal:

Proposition 2.4.27. (Finitely generated graded systems). *Let $\mathfrak{a}_\bullet = \{\mathfrak{a}_m\}$ be a finitely generated graded family of ideals on a variety X. Then there is an integer $p > 0$ such that*

$$\mathfrak{a}_\bullet^{(p)} = \mathrm{Pow}_\bullet(\mathfrak{a}_p).$$

Proof. At the expense of eventually replacing p by a multiple, we can assume that X is affine. Then by definition the graded ring

$$\mathrm{Rees}(\mathfrak{a}_\bullet) = \mathbf{C}[X] \oplus \mathfrak{a}_1 \oplus \mathfrak{a}_2 \oplus \ldots$$

is finitely generated as a $\mathbf{C}[X]$-algebra. But quite generally, if a graded ring $R_\bullet = R_0 \oplus R_1 \oplus R_2 \oplus \ldots$ is finitely generated as an algebra over R_0, then for suitable $p > 0$ the Veronese subring $R_\bullet^{(p)}$ is generated over R_0 by R_p [69, Chapter III, §1, Proposition 3]. In our situation this means exactly that $\mathfrak{a}_{pm} = \mathfrak{a}_p^m$ for all $m \geq 0$. □

Example 2.4.28. (Finite generation of base ideals). Let D be a divisor on a normal projective variety X, and let $\mathfrak{b}_\bullet = \{\mathfrak{b}_m\}$ be the graded system of base ideals $\mathfrak{b}_m = \mathfrak{b}\big(|mD|\big)$. If $\mathfrak{b}_\bullet$ is a finitely generated graded system of ideals, then there is an integer $p > 0$ such that the pth Veronese subring $R^{(p)}$ of the section ring $R = R(X, D)$ of D is finitely generated. (Use the previous proposition to fix $p > 0$ so that $\mathfrak{b}_{mp} = \mathfrak{b}_p^m$ for all $m \geq 0$, and let $\nu : X^+ \longrightarrow X$ be the normalized blowing-up of X along $\mathfrak{b}_p$, with exceptional divisor N^+. Put

$$D^+ = \nu^*(pD) - N^+.$$

Then D^+ is free, and arguing as in Example 2.1.31 the equality $\mathfrak{b}_{mp} = \mathfrak{b}_p^m$ implies that

$$R(X^+, D^+) = R(X, D)^{(p)}.$$

Then it follows from 2.1.30 that $R(X, D)^{(p)}$ is finitely generated.) □

Example 2.4.29. (Finitely generated graded linear series). Let $V_\bullet$ be a graded linear series on a variety X. If $V_\bullet$ is finitely generated, then there exists a positive integer $p > 0$ such that

$$V_{pk} \cdot V_{p\ell} = V_{p(k+\ell)}$$

for every $k, \ell \geq 1$. □

Finally, we discuss an interesting invariant of a family of ideals, analogous to the volume of a graded linear series. In somewhat different language, this notion goes back at least to Cutkosky and Srinivas [101].

Let $\mathfrak{a}_\bullet = \{\mathfrak{a}_m\}$ be a graded system of ideals on a variety X with $\dim X = n$. Assume that all the $\mathfrak{a}_m$ vanish only at one point $x \in X$, independent of m. Writing $\mathfrak{m}$ for the maximal ideal of x, this means that each $\mathfrak{a}_m$ is $\mathfrak{m}$-primary, and hence of finite colength (i.e. complex codimension) in $\mathcal{O}_x X$.

Definition 2.4.30. (Multiplicity of a graded system). Given a graded system $\mathfrak{a}_\bullet = \{\mathfrak{a}_m\}$ of $\mathfrak{m}$-primary ideals, the *multiplicity* of $\mathfrak{a}_\bullet$ is defined to be

$$\operatorname{mult}(\mathfrak{a}_\bullet) \ = \ \limsup_{m \to \infty} \frac{\operatorname{colength}_{\mathcal{O}_x X}(\mathfrak{a}_m)}{m^n/n!}. \quad \square$$

This was called the volume of $\mathfrak{a}_\bullet$ in [160] and [374], but the present terminology seems preferable.[12]

Example 2.4.31. Suppose that $\mathfrak{a}_\bullet = \operatorname{Pow}_\bullet(\mathfrak{b})$ is the graded system consisting of powers of a fixed $\mathfrak{m}$-primary ideal $\mathfrak{b}$. Then $\operatorname{mult}(\mathfrak{a}_\bullet) = e(\mathfrak{b})$ is the Samuel multiplicity of $\mathfrak{b}$. (The limit defining $e(\mathfrak{a}_\bullet)$ is one of the several possible definitions of $e(\mathfrak{b})$.) $\square$

Example 2.4.32. (i). Let $\mathfrak{a}_\bullet$ be the graded family in 2.4.18 (i) coming from valuation along a transcendental arc. Then the codimension of $\mathfrak{a}_m$ in $\mathbf{C}[x,y]$ grows linearly in m, and hence $\operatorname{mult}(\mathfrak{a}_\bullet) = 0$.

(ii). Let $\mathfrak{a}_\bullet$ be the "diagonal" graded system of monomial ideals from Example 2.4.19. Then $\operatorname{mult}(\mathfrak{a}_\bullet) = \delta_1 \cdot \ldots \cdot \delta_n$. $\square$

Example 2.4.33. Let $\mathfrak{a}_\bullet = \{\mathfrak{a}_m\}$ be an $\mathfrak{m}$-primary graded system. If $\mathfrak{a}_\bullet$ is finitely generated, then $\operatorname{mult}(\mathfrak{a}_\bullet)$ is rational. (After verifying that $\operatorname{mult}(\mathfrak{a}_\bullet^{(p)}) = p^n \cdot \operatorname{mult}(\mathfrak{a}_\bullet)$, this follows from 2.4.27 and 2.4.31.) $\square$

Remark 2.4.34. (Divisorial valuation with irrational multiplicity). Küronya [374] constructs an example of a discrete divisorial valuation centered at $0 \in \mathbf{C}^4$ such that the corresponding sequence of ideals has irrational multiplicity. $\square$

Example 2.4.31 raises the question of how the multiplicity of a graded system $\mathfrak{a}_\bullet$ relates to the Samuel multiplicities $e(\mathfrak{a}_m)$ of each of the terms $\mathfrak{a}_m$. We conclude by stating without proof a theorem showing that — at least when X is smooth — the answer is as simple as one could hope.

Theorem 2.4.35. *Let $\mathfrak{a}_\bullet = \{\mathfrak{a}_m\}$ be an $\mathfrak{m}$-primary graded family of ideals on a smooth variety X of dimension n. Then*

[12] The notation $\operatorname{mult}(\mathfrak{a}_\bullet)$ is intended to avoid conflicts with the exponent $e(\mathfrak{a}_\bullet)$ of $\mathfrak{a}_\bullet$.

$$\mathrm{mult}(\mathfrak{a}_\bullet) = \lim_{m\to\infty} \frac{e(\mathfrak{a}_m)}{m^n}. \quad \square$$

This was established in [160] for the graded families associated to certain valuations, and by Mustaţă [457] in general. One would expect the statement to hold also at a singular point of X, but this isn't known.

Notes

For Section 2.1, we have drawn on the presentations in [578] and [439].

The basic numerical criterion for bigness (Theorem 2.2.15) was originally proved by Siu [537] (in a slightly weaker formulation) using Demailly's holomorphic Morse inequalities. Elementary algebraic arguments were subsequently given by Ein and the author and by Catanese: here we follow Catanese's formulation (as reported in [126]). The use of this criterion to analyze bigness for nef divisors (Theorem 2.2.16) simplifies earlier approaches. Proposition 2.2.35 appears in [158] and [130]. Theorem 2.2.44 is new, although Boucksom [66] independently obtained analogous results in the analytic setting.

Section 2.3.B is abstracted from a number of papers of Cutkosky and coauthors, notably [97], [101], and [102]. The proof of the Campana–Peternell theorem (Theorem 2.3.18) is new.

3

Geometric Manifestations of Positivity

This chapter focuses on a number of results that in one way or another express geometric consequences of positivity. In the first section we prove the Lefschetz hyperplane theorem following the Morse-theoretic approach of Andreotti–Frankel [7]. Section 3.2 deals with subvarieties of small codimension in projective space: we prove Barth's theorem and give an introduction to the conjectures of Hartshorne. The connectedness theorems of Bertini and Fulton–Hansen are established in Section 3.3, while applications of the Fulton–Hansen theorem occupy Section 3.4. Like the results from 3.2, these reflect the positivity of projective space itself. Finally, some extensions and variants are presented in Section 3.5.

A good deal of this material is classical, and much of it has been surveyed elsewhere (e.g. in [424], [279], [211], and [203], on which we have drawn closely). However many of the results appearing here play a role in later chapters, and we felt it important to include an account.

3.1 The Lefschetz Theorems

The theme of the Lefschetz hyperplane theorem is that a projective variety passes on many of its topological properties to any effective ample divisor sitting inside it. As we shall see in Section 4.2, it is very closely related to the Kodaira–Nakano vanishing theorem. Lefschetz originally proved his result by studying the topology associated with a fibration of the given variety by hypersurfaces: nice accounts of this approach appear in [376] and [407]. However Thom and Andreotti–Frankel observed that Morse theory leads to a very quick proof, and this is the path we initially follow here. Thus in the first subsection we prove the theorem of Andreotti–Frankel that the cohomology of an affine variety vanishes in degrees above its complex dimension; we also sketch the alternative approach of Artin–Grothendieck via constructible sheaves. This result is used in the second subsection to deduce the theorem on hyperplane sections. In the third we briefly discuss (without proof) the hard Lefschetz theorem.

These results constitute the very beginning of a vast and important body of research concerning the topology of algebraic varieties. A survey of this field lies far beyond the scope of the present volume. The reader interested in further information might start by consulting [227] or [203] and the references cited there. In particular, the Introduction to Part II of [227] contains a very nice historical survey of work on the topology of algebraic varieties.

3.1.A Topology of Affine Varieties

We start with an elementary and classical theorem of Andreotti and Frankel [7] concerning the topology of smooth affine varieties. We then briefly survey a related approach to these questions involving constructible sheaves.

The theorem of Andreotti and Frankel. Consider a non-singular complex variety V of (complex) dimension n. Then V is a $\mathcal{C}^\infty$ manifold of real dimension $2n$, and hence has the homotopy type of a CW complex of dimension $\leq 2n$. The theorem of Andreotti and Frankel states that if V is affine — or more generally, if V is any Stein manifold of dimension n — then in fact V is homotopically a CW complex of real dimension $\leq n$. In other words, V has only "half as much topology" as one might expect:

Theorem 3.1.1. (Submanifolds of affine space). *Let $V \subseteq \mathbf{C}^r$ be a closed connected complex submanifold of (complex) dimension n. Then V has the homotopy type of a CW complex of real dimension $\leq n$. Consequently,*

$$H^i\big(V\,;\mathbf{Z}\big) \;=\; 0 \quad \text{for} \quad i > n, \tag{3.1}$$

$$H_i\big(V\,;\mathbf{Z}\big) \;=\; 0 \quad \text{for} \quad i > n. \tag{3.2}$$

We emphasize that the theorem applies in particular to any non-singular affine variety of dimension n.

Example 3.1.2. It is not the case in general that V is homotopic to a finite CW complex. For example, let $V \subseteq \mathbf{C}^2$ be the analytic curve defined by the equation

$$w \cdot \sin(\pi z) \;-\; 1 \;=\; 0.$$

Then V is isomorphic as a complex manifold to $\mathbf{C} - \mathbf{Z}$, which of course cannot be realized as a finite CW complex. However, if $V \subseteq \mathbf{C}^r$ is a smooth *algebraic* subvariety, then V does in fact have the homotopy type of a finite CW complex of dimension $\leq \dim_{\mathbf{C}} V$ (Example 3.1.9). □

Example 3.1.3. (Middle-dimensional homology). With $V \subseteq \mathbf{C}^r$ as in Theorem 3.1.1, the homology group $H_n\big(V\,;\mathbf{Z}\big)$ is torsion-free. (Theorem 3.1.1 implies that $H_i\big(V\,;F\big)$ vanishes when $i > n$ for any coefficient field F. But by the universal coefficient theorem, any torsion in $H_n\big(V\,;\mathbf{Z}\big)$ would show up in $H_{n+1}\big(V\,;F\big)$ for suitable F.) □

Remark 3.1.4. (Singular varieties). It was established by Karčjauskas [311], [312] (for algebraic varieties) and by Hamm [264] (for Stein spaces) that Theorem 3.1.1 remains true for arbitrarily singular irreducible V. However, for the most part the elementary case of non-singular varieties will suffice for our purposes. The vanishing (3.1) for possibly singular varieties had been verified earlier by Kaup [314] and Narasimhan [475]. At least in the algebraic setting this statement also follows from general results about constructible sheaves, which we discuss briefly below.

These results for singular varieties are in turn generalized by Goresky and MacPherson using their stratified Morse theory. Among other things, they extend Theorem 3.1.1 to the setting of a possibly singular variety mapping properly to an affine variety [227, Theorem II.1.1*]. We recommend Chapter 1 of Part II of [227] for an overview of how the classical results generalize to the singular (or non-compact) setting. Fulton's survey [203] gives a very nice introduction to this circle of ideas. □

Theorem 3.1.1 is a beautiful application of elementary Morse theory, and we start by recalling the basic facts of this machine. The reader may consult Milnor's classical book [424, Part I] for details.

Consider then a $\mathcal{C}^\infty$ manifold M of real dimension n, and a smooth function

$$f : M \longrightarrow \mathbf{R}.$$

Suppose that $p \in M$ is a critical point of f, i.e. a point at which $df_p = 0$. Then the Hessian $\mathrm{Hess}_p(f)$ of f at p is an intrinsically defined symmetric bilinear form on the tangent space T_pX: in local coordinates $x_1, \ldots, x_n$ it is represented by the Hessian matrix $\big(\frac{\partial^2 f}{\partial x_i \partial x_j}(p)\big)$ of second partials of f. One says that p is a *non-degenerate* critical point if $\mathrm{Hess}_p(f)$ is non-degenerate. In this case, the *index* $\lambda = \mathrm{index}_p(f)$ of f at p is defined to be the number of negative eigenvalues of the Hessian. Very concretely, Morse's lemma ([424, Lemma 2.2]) states that in suitable local coordinates centered at a non-degenerate critical point of index λ, f is given by the quadratic function

$$f(x_1, \ldots, x_n) \;=\; -x_1^2 - \ldots - x_\lambda^2 + x_{\lambda+1}^2 + \ldots + x_n^2.$$

The proof of Theorem 3.1.1 will call on two basic results from (finite-dimensional) Morse theory. The first states that under suitable compactness hypotheses, a function on a manifold having only non-degenerate critical points allows one to reconstruct the given manifold as a CW complex:

Theorem 3.1.5. (cf. [424, Theorem 3.5]). *Let*

$$f : M \longrightarrow \mathbf{R}$$

be a smooth function having the property that $f^{-1}(-\infty, a]$ is compact for every $a \in \mathbf{R}$. Assume that f has only non-degenerate critical points. Then M has *the homotopy type of a CW complex with one cell of dimension λ for every critical point of index λ.* □

The second fact we need is that for submanifolds of Euclidean space, distance (squared) from a general point gives rise to a Morse function:

Theorem 3.1.6. (cf. [424, Theorem 6.6]). *Let $M \subseteq \mathbf{R}^N$ be a closed submanifold. Fix a point $c \in \mathbf{R}^N$ and let*

$$\phi_c : M \longrightarrow \mathbf{R} \quad , \quad x \mapsto \|x - c\|^2$$

be the function given by squared distance from c, where $\| \ \|$ denotes the standard Euclidean norm on $\mathbf{R}^N$. Then for almost all $c \in \mathbf{R}^N$, ϕ_c has only non-degenerate critical points on M. □

Observe that the compactness of $\phi_c^{-1}(-\infty, a]$ for every a is evident.

In view of 3.1.5 and 3.1.6, Theorem 3.1.1 reduces to estimating the index of the squared distance function:

Proposition 3.1.7. (Index estimate). *Let $V \subseteq \mathbf{C}^r$ be a closed complex submanifold of complex dimension n. Fix a point $c \in \mathbf{C}^r = \mathbf{R}^{2r}$, and consider as in 3.1.6 the squared distance $\phi_c(x) = \|x - c\|^2$. If $p \in V$ is any critical point of ϕ_c, then*

$$\operatorname{index}_p(\phi_c) \ \leq \ n.$$

Elegant proofs appear e.g. in [424, §7], [248, p. 158], [227, II.4.A] and [203, p. 26]. We will give a down–to–earth direct computation in the spirit of [7].

Proof of Proposition 3.1.7. Writing $r = n + k$, we can choose complex coordinates on $\mathbf{C}^r$ in such a way that $p = 0$ is the origin, V is locally the graph of a holomorphic function $f : \mathbf{C}^n \longrightarrow \mathbf{C}^k$ with $df_0 = 0$, and

$$c \ = \ (0, \ldots, 0, 1, 0, \ldots, 0)$$

is the unit vector with all but the $(n+1)^{\text{st}}$ coordinate zero (Figure 3.1). Thus V is described in local coordinates $z = (z_1, \ldots, z_n)$ at p as the set

$$V \ = \ \Big\{ \big(z_1, \ldots, z_n, f_1(z), \ldots, f_k(z)\big) \Big\},$$

with $f_i(z)$ holomorphic and $\operatorname{ord}_0(f_i) \geq 2$. Then $\phi = \phi_c$ is given by

$$\begin{aligned} \phi(z) \ &= \ |z_1|^2 + \ldots + |z_n|^2 + |f_1(z) - 1|^2 + |f_2(z)^2| + |f_k(z)|^2 \\ &= \ \big(1 - 2 \cdot \operatorname{Re}(f_1(z)\big) + \sum_{i=1}^{n} |z_i|^2 + \sum_{i=1}^{k} |f_i(z)|^2. \qquad (*) \end{aligned}$$

Since each f_i vanishes to order ≥ 2 at 0, the last term in (*) will not affect the Hessian $\operatorname{Hess}_0(\phi)$. Similarly, if we write

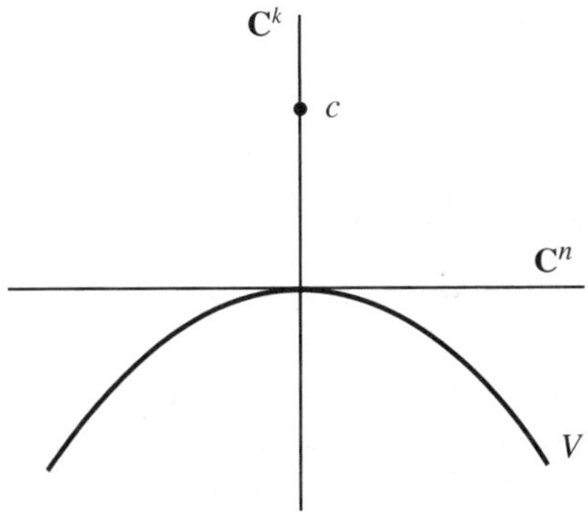

Figure 3.1. Setup for proof of Proposition 3.1.7

$$f_1(z) \;=\; Q(z) + \Big(\text{higher order terms}\Big)$$

with $Q(z)$ homogeneous of degree 2, then the contribution of the first term in (*) to $\mathrm{Hess}_0(\phi)$ depends only on Q. Therefore

$$\begin{aligned} \mathrm{Hess}_0(\phi) \;&=\; -2 \cdot \mathrm{Hess}_0\Big(\mathrm{Re}\big(\,Q(z)\,\big)\Big) \;+\; \mathrm{Hess}_0\Big(\sum |z_i|^2\Big) \\ &=\; -2 \cdot \mathrm{Hess}_0\Big(\mathrm{Re}\big(\,Q(z)\,\big)\Big) \;+\; 2 \cdot \mathrm{Id}. \qquad (**) \end{aligned}$$

The second term in (**) being positive definite, the only potential source of negative eigenvalues is from the Hessian of $\mathrm{Re}(Q)$. Therefore the proposition — and with it Theorem 3.1.1 — is a consequence of the next lemma. □

Lemma 3.1.8. *Let $Q(z)$ be a complex homogeneous form of degree 2 in n variables $z = (z_1, \ldots, z_n)$. Then*

$$\mathrm{Hess}_0\Big(\mathrm{Re}\big(\,Q(z)\,\big)\Big)$$

has $\leq n$ positive and $\leq n$ negative eigenvalues.

Proof. To start with an illustrative special case, imagine that Q is the polynomial $Q(z) \;=\; z_1^2 + \ldots + z_s^2$ for some $s \leq n$. Writing $z_j = x_j + \sqrt{-1}y_j$, one has

$$\mathrm{Re}\, Q(z) \;=\; \big(x_1^2 - y_1^2\big) \;+\; \ldots \;+\; \big(x_s^2 - y_s^2\big),$$

from which the assertion is clear. In general, as in [7, p. 716], suppose that

$$Q(z) \;=\; \sum c_{i,j} z_i z_j$$

for complex numbers $c_{i,j} \;=\; c_{j,i} \in \mathbf{C}$. Taking real and imaginary parts as above, one finds that the real quadratic form $H(x,y) = \mathrm{Re}\, Q(x + \sqrt{-1}y)$ is represented by a block matrix of the shape $\begin{pmatrix} A & -B \\ -B & -A \end{pmatrix}$, where A and B are $n \times n$ real symmetric matrices. But such a matrix has as many positive as negative eigenvalues: if $\begin{pmatrix} u \\ v \end{pmatrix}$ is an eigenvector with eigenvalue λ, then $\begin{pmatrix} -v \\ u \end{pmatrix}$ is an eigenvector with eigenvalue $-\lambda$. □

Example 3.1.9. (Finiteness of homotopy type of affine variety). Let $V \subseteq \mathbf{C}^r$ be a smooth affine algebraic variety of dimension n. Then V has the homotopy type of a *finite* CW complex of real dimension $\leq n$. (The function $\phi_c : V \longrightarrow \mathbf{R}$ appearing above is real algebraic, and hence its critical locus is a real algebraic subset of V. But the critical points of ϕ_c are discrete, and therefore there are only finitely many such.) As observed in 3.1.2, the corresponding statement can fail for Stein manifolds. □

Constructible sheaves. We now discuss briefly another approach to some of these questions, via constructible sheaves. The technique originated with Artin and Grothendieck as a mechanism for proving Lefschetz–type theorems in the setting of ℓ-adic cohomology.

Let X be a complex algebraic variety.

Definition 3.1.10. (Constructible sheaf). A constructible sheaf on X is a sheaf $\mathcal{F}$ of abelian groups having the property that X admits a finite stratification $X = \coprod_{i \in I} X_i$ into disjoint Zariski–locally closed subsets $\{X_i\}$ in such a way that the sheaf-theoretic restriction $\mathcal{F} \,|\, X_i$ to each X_i is a local system on X_i. □

By a local system we mean as usual a locally constant sheaf with respect to the classical topology on X_i. Note that then $\mathcal{F}$ will be locally constant on some union of the X_i, i.e. on a constructible subset of X. When we talk about the cohomology (or direct images) of $\mathcal{F}$, we always refer to the classical topology on X.

Example 3.1.11. (Direct images under finite maps). Let $f : Y \longrightarrow X$ be a finite morphism. Then the direct image $f_*\mathbf{Z}$ of the constant sheaf on Y is a constructible sheaf on X. More generally, the direct image $f_*\mathcal{F}$ of any constructible sheaf $\mathcal{F}$ on Y is a constructible sheaf on X. (For the first statement observe that f induces a locally closed stratification $X = \coprod X_i$ of X having the property that f is a covering space over each X_i.) □

Remark 3.1.12. (Direct images under arbitrary maps). Quite generally if $f : Y \longrightarrow X$ is any (finite-type) morphism of algebraic varieties, and if $\mathcal{F}$ is a constructible sheaf on Y, then all the direct images $R^j f_* \mathcal{F}$ are constructible sheaves on X: see [117, Theorem 1.1, p. 233]. When $\mathcal{F} = \mathbf{Z}$ is the constant sheaf on Y and f is proper, this follows from the fact [586] that X is stratified by locally closed sets on which the topological type of the fibres of f is constant. □

The theorem of Artin and Grothendieck is a cohomological analogue of 3.1.1:

Theorem 3.1.13. (Artin–Grothendieck theorem). *Let V be an affine variety of dimension n, and let $\mathcal{F}$ be a constructible sheaf on V. Then*

$$H^i\big(V,\mathcal{F}\big) \;=\; 0 \quad \text{for } \; i > n.$$

This of course implies in particular that $H^i\big(V,\mathbf{Z}\big) = 0$ for $i > n$. Note that we do not assume here that V is smooth.

Remark 3.1.14. It is not necessary in 3.1.13 to assume that V is irreducible. In fact the statement and proof of this result remain valid for any affine algebraic set V of dimension n. □

Sketch of Proof of Theorem 3.1.13. We closely follow the presentations in [595] and [15, Chapter VII, §3]. Choose to begin with a finite map $f : V \longrightarrow \mathbf{C}^n$. Then

$$H^i\big(V,\mathcal{F}\big) \;=\; H^i\big(\mathbf{C}^n, f_*\mathcal{F}\big)$$

so it is enough thanks to 3.1.11 to prove the statement for $V = \mathbf{C}^n$. Thus we suppose that $\mathcal{F}$ is a constructible sheaf on $\mathbf{C}^n$, and we prove by induction on n that

$$H^i\big(\mathbf{C}^n,\mathcal{F}\big) \;=\; 0 \quad \text{for } \; i > n. \tag{3.3}$$

This is elementary if $n = 1$, so we assume that $n \geq 2$ and that (3.3) is known for all constructible sheaves on $\mathbf{C}^{n-1}$.

By definition there exists a proper closed algebraic subset $Z \subsetneq \mathbf{C}^n$ such that $\mathcal{F}$ is locally constant on $\mathbf{C}^n - Z$. Fix a linear projection $p : \mathbf{C}^n \longrightarrow \mathbf{C}^{n-1}$ such that the restriction of p to Z is finite (and in particular proper). We assert that

$$\big(R^i p_* \mathcal{F}\big)_x \;=\; H^i\big(p^{-1}(x), \mathcal{F}\,|\,p^{-1}(x)\big) \tag{3.4}$$

for any point $x \in \mathbf{C}^{n-1}$. Grant this for the time being. Then in the first place, $\mathcal{F}\,|\,p^{-1}(x)$ being a constructible sheaf on $p^{-1}(x) = \mathbf{C}$, it follows from the induction hypothesis that $R^i p_* \mathcal{F} = 0$ for $i > 1$. On the other hand, the direct images $p_*\mathcal{F}$ and $R^1 p_* \mathcal{F}$ are constructible sheaves on $\mathbf{C}^{n-1}$: this is a consequence of the general theorem quoted in Remark 3.1.12, but could also be verified directly in the case at hand using (3.4). The required vanishing (3.3) then follows from the induction hypothesis and the Leray spectral sequence.

It remains to establish the isomorphism (3.4). For this, recall that

$$H^i\big(p^{-1}(x)\,,\,\mathcal{F}\,|\,p^{-1}(x)\big) \;=\; \varinjlim_{N} H^i\big(N,\mathcal{F}\big),$$

the limit running over all open neighborhoods $N \supseteq p^{-1}(x)$. Therefore it suffices to show that given N there are open neighborhoods $U \ni x$ of x in $\mathbf{C}^{n-1}$ and $N' \subseteq N \cap p^{-1}(U)$ of $p^{-1}(x)$ in $\mathbf{C}^n$ such that

$$H^i\big(p^{-1}(U), \mathcal{F}\big) \;=\; H^i\big(N', \mathcal{F}\big). \tag{*}$$

To this end write $\mathbf{C}^n$ as the product $\mathbf{C}^{n-1} \times \mathbf{C}$, with p the first projection. One starts by constructing a small ball $U \subseteq \mathbf{C}^{n-1}$ centered at x, and a disk $A \subseteq \mathbf{C}$ such that

$$\begin{aligned} Z \cap p^{-1}(U) \;&\subseteq\; U \times A, \\ \overline{U} \times \overline{A} \;&\subseteq\; N. \end{aligned}$$

Here $Z \subseteq \mathbf{C}^n$ is the proper subvariety chosen above off of which $\mathcal{F}$ is locally constant, and in the first inclusion we are using the properness of $p \,|\, Z$. One now uses a "tapering" process to extend $U \times A$ to a neighborhood $N' \subseteq N$ of $p^{-1}(x)$ having the property that $p^{-1}(U)$ deformation retracts onto N' via a homotopy leaving

$$Z \cap p^{-1}(U) \;=\; Z \cap N'$$

fixed. In fact, fix coordinates on $\mathbf{C}^{n-1}$ in such a way that x is the origin and U is the ball of radius 1. Now choose a continuous function $\phi : \mathbf{C} \longrightarrow (0,1]$, with $\phi \,|\, A \equiv 1$, such that

$$N' \;=_{\text{def}}\; \big\{\, (z,t) \;|\; \|z\|^2 < \phi(t) \,\big\} \;\subseteq\; N.$$

Then N' has the required properties, and (*) follows. We refer to [15, p. 316] for details. □

Remark 3.1.15. In his interesting recent paper [483], Nori studies the homological properties of constructible sheaves. He uses these to give a new approach to the theorem of Artin–Grothendieck and some related questions. □

Remark 3.1.16. (Analytic constructible sheaves). With more care one can extend these definitions and results to the setting of analytically constructible sheaves. We refer to [268] and the references therein for details. □

3.1.B The Theorem on Hyperplane Sections

We now come to the Lefschetz hyperplane theorem, which compares the topology of a smooth projective variety X with that of an effective ample divisor D on X. When we speak of the topological properties of D, we are referring of course to the subspace of X (with its classical topology) determined by the support of D.

Theorem 3.1.17. (Lefschetz hyperplane theorem). *Let X be a nonsingular irreducible complex projective variety of dimension n, and let D be any effective ample divisor on X. Then*

$$H^i\big(X\,,\, D; \mathbf{Z}\big) \;=0 \quad \text{for } i < n. \tag{3.5}$$

Equivalently, the restriction

$$H^i(X;\mathbf{Z}) \longrightarrow H^i(D;\mathbf{Z}) \tag{3.6}$$

is an isomorphism for $i \leq n-2$ and injective when $i = n-1$.

Note that we do not assume that D is non-singular. However, one cannot dispense entirely with the non-singularity of X: see 3.1.33 and 3.1.34.

Proof of Theorem 3.1.17. Since D is ample, mD is very ample for some $m \gg 0$. So we can find a projective embedding $X \subseteq \mathbf{P}^r$ together with a hyperplane $H \subseteq \mathbf{P}^r$ such that $X \cap H = mD$. In particular, $X - D = X - mD$ is affine. Therefore $H_j(X - D;\mathbf{Z}) = 0$ for $j \geq n+1$ thanks to 3.1.1. The assertion then follows from Lefschetz duality $H^i(X, D) = H_{2n-i}(X - D)$. □

Example 3.1.18. In the setting of Theorem 3.1.17, the cokernel of the restriction

$$H^{n-1}(X;\mathbf{Z}) \longrightarrow H^{n-1}(D;\mathbf{Z})$$

is torsion-free. (Note that the cokernel in question sits as a subgroup inside $H^n(X, D; \mathbf{Z})$. Then argue as above using Example 3.1.3 to see that this group is torsion-free.) □

Corollary 3.1.19. (Connectedness of ample divisors). *Let X be a smooth irreducible projective variety of dimension $n \geq 2$ and $D \subseteq X$ an ample divisor. Then D is connected.*

Proof. In fact, $H^0(D;\mathbf{Z}) = \mathbf{Z}$ thanks to 3.1.17. □

Remark 3.1.20. (Lemma of Enriques–Severi–Zariski). The conclusion of the preceding corollary is actually valid without assuming that X is non-singular. In fact, one can suppose X is normal and then $H^1(X, \mathcal{O}_X(-mD)) = 0$ for $m \gg 0$, from which the assertion follows. See [280, III.7.7 and III.7.8] for details. The connectedness of D also follows from Theorem 3.3.3 below. □

As pointed out in [424, Theorem 7.4] one can get a somewhat stronger statement on the level of homotopy by going back to the proof of Theorem 3.1.1:

Theorem 3.1.21. (Lefschetz theorem for homotopy groups). *In the situation of Theorem 3.1.17,*

$$\pi_i(X, D) = 0 \quad \text{for} \quad i < n.$$

In particular, the homomorphism

$$\pi_i(D) \longrightarrow \pi_i(X)$$

induced by inclusion is bijective if $i \leq n-2$ and surjective if $i = n-1$.

Note that except in the trivial case $n = 1$ — where the present statement is vacuous — the spaces in question are path-connected by virtue of 3.1.19. Therefore we will ignore base points.

Remark 3.1.22. Goresky and MacPherson [227, II.1.1] have proven a deep generalization of this result allowing for the possibility that X is non-compact. Their result is stated below as Theorem 3.5.10. □

Proof of Theorem 3.1.21. Choose as above a projective embedding $X \subseteq \mathbf{P}^r$ and a hyperplane H such that $X \cap H = mD$ for some $m \gg 0$. Then we can view $V = X - D$ as being embedded in the affine space $\mathbf{C}^r = \mathbf{P}^r - H$. As in the previous subsection, squared distance from a general point $c \in \mathbf{C}^r$ not lying on V determines a Morse function $\phi_c : V \longrightarrow \mathbf{R}$. Consider now the function $f : X \longrightarrow \mathbf{R}$ given by

$$f(x) \;=\; \begin{cases} \frac{1}{\phi_c(x)} & \text{for } x \in V = X - D, \\ 0 & \text{for } x \in D. \end{cases}$$

The critical points of f off D coincide with those of ϕ_c, but the signs of the eigenvalues of the Hessian are reversed. Therefore f has only non-degenerate critical points away from D, each of index $\geq n$. So for any $\varepsilon > 0$, X has the homotopy type of $X^\varepsilon =_{\text{def}} f^{-1}[0, \varepsilon]$ with finitely many cells of dimensions $\geq n$ attached. In particular, the homomorphism

$$\pi_i(X^\varepsilon, D) \longrightarrow \pi_i(X, D)$$

is bijective for $i \leq n - 1$.

Now since one can triangulate X with D as a subcomplex, we may fix a neighborhood U of D in X that deformation retracts onto D. Take $\varepsilon > 0$ sufficiently small so that $X^\varepsilon \subseteq U$, and consider the commutative diagram

$$\begin{array}{ccc} \pi_i(X^\varepsilon, D) & \longrightarrow & \pi_i(U, D) \\ & {\scriptstyle\cong}\searrow \quad \swarrow & \\ & \pi_i(X, D). & \end{array}$$

The diagonal map on the left is an isomorphism when $i \leq n - 1$ whereas the group on the top right vanishes by construction. The assertion follows. □

Example 3.1.23. Let X be a non-singular simply connected projective variety of dimension ≥ 3, and let $D \subseteq X$ be an ample divisor. Then D is simply connected. This applies in particular to hypersurfaces in $\mathbf{P}^n$ for $n \geq 3$.

Example 3.1.24. (Lefschetz theorem for Hodge groups). Let X be a smooth projective variety of dimension n, and D a non-singular ample effective divisor on X. Denote by

$$r_{p,q} : H^q\big(X, \Omega^p_X\big) \longrightarrow H^q\big(D, \Omega^p_D\big)$$

the natural maps determined by restriction. Then

$$r_{p,q} \text{ is } \begin{cases} \text{an isomorphism} & \text{for } p+q \leq n-2, \\ \text{injective} & \text{for } p+q = n-1. \end{cases}$$

(Apply the Hodge decomposition on X and on D to the complexification of the restriction maps (3.6): see Lemma 4.2.2.) The case $p = 0$ of this statement will lead very quickly to the Kodaira vanishing theorem in Section 4.2. □

Example 3.1.25. (Lefschetz theorem for Picard groups). Let X be a smooth projective variety of dimension ≥ 4, and $D \subseteq X$ a reduced effective ample divisor. Then the map

$$\text{Pic}(X) \longrightarrow \text{Pic}(D)$$

is an isomorphism. (One compares the exponential sequences on X and on D: the reducedness of D guarantees that the exponential sheaf sequence on D is exact. In view of the isomorphisms (3.6) for $i = 1, 2$ it is enough to show that the restrictions

$$H^i\big(X, \mathcal{O}_X\big) \longrightarrow H^i\big(D, \mathcal{O}_D\big) \tag{*}$$

are bijective when $i \leq 2$. But this follows from Kodaira vanishing (Theorem 4.2.1), which implies that $H^i\big(X, \mathcal{O}_X(-D)\big) = 0$ for $i \leq 3$. If D is non-singular, then (*) alternatively follows from 3.1.24.) □

Remark 3.1.26. (Grothendieck's Lefschetz-type theorem for Picard groups). Using different methods, Grothendieck proves in [258, Exposé XII, Corollary 3.6] that the statement in 3.1.25 remains true even if the ample divisor D is non-reduced: see [276, Corollary IV.3.3].[1] □

Example 3.1.27. (Factoriality of homogeneous coordinate rings of hypersurfaces). Let

$$F \in \mathbf{C}[X_0, \ldots, X_n]$$

be a non-zero homogeneous polynomial in $n + 1 \geq 5$ variables, and assume that the hypersurface $D \subseteq \mathbf{P}^n$ defined by F is non-singular. Then the quotient ring

$$R = \mathbf{C}[X_0, \ldots, X_n] \,/\, (F)$$

is a unique factorization domain. (Since D is in any event projectively normal, it is well known that R is factorial if and only if $\text{Pic}(D) = \text{Pic}(\mathbf{P})$: see [276, Chapter IV, Exercise 3.5].) □

---

[1] Hartshorne assumes that D is non-singular in order to invoke Kodaira vanishing on D to establish the vanishings $H^i\big(D, \mathcal{O}_D(mD)\big) = 0$ for $i = 1, 2$ and $m < 0$. However, as in Example 3.1.25 these follow from Kodaira on X.

Remark 3.1.28. (Further applications to local algebra). The previous example hints at the fact that this circle of ideas has interesting applications to local algebra, in particular to questions of factoriality. This viewpoint was greatly developed by Grothendieck in his seminar [258], especially Exposé XI. Among many other things, Grothendieck proves a conjecture of Samuel to the effect that if A is a local Noetherian ring arising as a complete intersection, and if A is factorial in codimension ≤ 3, then A is factorial [258, Corollary XI.3.14]. Example 3.1.27 is a very special case of this result. We recommend Lipman's survey [402] for a nice overview of this and related work. □

Example 3.1.29. (Lefschetz theorems and nef cones). Let X be a smooth projective variety of dimension ≥ 4 and $D \subseteq X$ a smooth ample divisor. Then it follows from Example 3.1.24 and the Lefschetz $(1,1)$-theorem (Remark 1.1.21) that restriction determines an isomorphism

$$N^1(X)_{\mathbf{R}} \xrightarrow{\cong} N^1(D)_{\mathbf{R}}$$

of the Néron–Severi spaces of X and D. One can ask how this isomorphism behaves with respect to the natural cones of divisors sitting inside these vector spaces. Since the restriction of a nef divisor is nef, the nef cone of X evidently maps into the nef cone of D. However, simple examples show that the latter cone can be strictly bigger. For instance, following [282] consider $X = \mathbf{P}^1 \times \mathbf{P}^{n-1}$, and let $D \subseteq X$ be a general divisor of type (a,b), with $a \geq n+1$. Then the projection $D \longrightarrow \mathbf{P}^{n-1}$ is finite, and hence $\mathcal{O}_{\mathbf{P}^{n-1}}(1)$ pulls back to an ample line bundle on D. It follows that if $k \gg 0$ then the restriction of $\mathrm{pr}_1^*\mathcal{O}_{\mathbf{P}^1}(-1) \otimes \mathrm{pr}_2^*\mathcal{O}_{\mathbf{P}^{n-1}}(k)$ to D is ample, but of course this is not nef — or even big — on X itself. However Kollár [65, Appendix], Wiśniewski [610], and Hassett–Lin–Wang [282] obtain some positive results by taking into account the Mori cone of X. □

Remark 3.1.30. (Noether–Lefschetz theorem). Simple examples — e.g. a quadric surface in $\mathbf{P}^3$ — show that if $D \subseteq X$ is an ample divisor, then the restriction $\mathrm{Pic}(X) \longrightarrow \mathrm{Pic}(D)$ can fail to be an isomorphism when $\dim X \leq 3$. However a classical theorem of Noether and Lefschetz states that if $D \subseteq \mathbf{P}^3$ is a "very general" hypersurface of degree ≥ 4, then indeed $\mathrm{Pic}(D) = \mathrm{Pic}(\mathbf{P}^3)$ (although for special D — e.g. those containing a line — this can fail). See [82], [236], or [83, Chapter 7.5] for a modern discussion of this and related results. □

Example 3.1.31. (Lefschetz theorem for complete intersections). Suppose that $Y \subseteq \mathbf{P}^n$ is the transversal complete intersection of hypersurfaces $D_1, \ldots, D_e \subseteq \mathbf{P}^r$ of degrees $d_1 \geq d_2 \geq \ldots \geq d_e$. Then

$$H^i\big(\mathbf{P}^n, Y; \mathbf{Z}\big) \ = \ 0 \quad \text{for } \ i \leq n-e.$$

(Argue that after possibly modifying the D_i, one can take each of the intermediate intersections $D_1 \cap \ldots \cap D_{i+1}$ to be smooth. Then apply the Lefschetz theorem inductively.) □

Remark 3.1.32. (Arbitrary intersections of ample divisors). More generally, let X be a smooth projective variety of dimension n, and let $D_1, \ldots, D_e \subseteq X$ be any effective ample divisors on X. Consider the intersection

$$Z = D_1 \cap \ldots \cap D_e$$

of the given divisors (which we view as a closed subspace of X). Then

$$H^i\big(X, Z; \mathbf{Z}\big) = 0 \quad \text{for } i \leq n - e. \tag{3.7}$$

This is a special case of a result of Sommese, which we prove as Theorem 7.1.1. Under additional hypotheses on Z and the D_i one can establish (3.7) inductively as in the previous example. However in general this approach runs into problems if the D_i are singular or Z has larger than expected dimension. □

Example 3.1.33. (Counterexamples for singular varieties). A construction due to Landman shows that Theorem 3.1.17 need not be valid if X is singular. Let Y be a non-singular simply connected variety of dimension ≥ 3, and let $p, q \in Y$ be distinct points. Then one can construct a new projective variety X that is isomorphic to Y with p and q identified to a singular point $o \in X$.[2] Thus $H^1\big(X\big) \neq 0$, but if $D \subset X$ is an ample divisor whose support does not pass through o, then D — being isomorphic to an ample divisor on Y — is simply connected. In particular, $H^1\big(X, D\big) \neq 0$. □

Remark 3.1.34. (Local complete intersections). Using local results of Hamm ([264], [265]), Goresky and MacPherson [227, p. 24 and II.5.2] show that 3.1.17 and 3.1.21 remain true as stated if X is locally a complete intersection in some smooth variety. For arbitrarily singular X they introduce a numerical invariant $s = s(X)$ that measures how far X is from being a local complete intersection, and they prove that $\pi_i\big(X, D\big) = 0$ for $i < n - s$. We again refer to [203, §2] for a very readable introduction. □

Remark 3.1.35. (Local theorems of Lefschetz type). Grothendieck stressed in [258, Exposé XIII] that the classical results of Lefschetz should arise from more general local results, and he proposed a number of problems in this direction. Grothendieck's program was substantially carried out by Hamm and Lê [266], [267] and Goresky–MacPherson [227]. For example, Goresky–MacPherson [227, p. 26, II.1.3, and II.5.3] prove that if $X \subseteq \mathbf{C}^r$ is an analytic subvariety of pure dimension $n+1$ having an isolated singularity at the origin 0, and if H is a general hyperplane through 0, then

$$\pi_i\big(X \cap \partial B_\varepsilon\,,\, X \cap H \cap \partial B_\varepsilon\big) = 0 \quad \text{for } i < n,$$

[2] Under a sufficiently positive projective embedding $Y \subseteq \mathbf{P}$, a general point on the secant line joining p and q will not meet any other secant lines of Y. Then take X to be the image of Y under projection from this point.

B_ε being a ball of radius $\varepsilon \ll 1$ centered at 0. They deduce the classical Lefschetz theorem from this, at least when D is a general divisor in a very ample linear series on X. □

Remark 3.1.36. (Nori's connectedness theorem). Nori [482] established an interesting variant of 3.1.17 and 3.1.32, which he applied in a surprising manner to study algebraic cycles. Specifically, let X be a non-singular complex projective variety of dimension n, and D a very ample divisor on X. Given $e \le n$ positive integers $a_1, \dots, a_e$, denote by

$$S \subseteq |a_1 D| \times \ldots \times |a_e D|$$

the variety parameterizing e-tuples of divisors in the corresponding linear series. Then the universal family $\mathcal{D}$ of complete intersections of type $a_1, \dots, a_e$ on X sits naturally inside $X_S = X \times S$:

$$\mathcal{D} = \Big\{ (x, D_1, \dots, D_e) \mid x \in D_1 \cap \ldots \cap D_e \Big\} \subseteq X_S.$$

For any morphism $T \longrightarrow S$, let $\mathcal{D}_T \subseteq X_T = X \times T$ denote the pullback of $\mathcal{D}$ to T. Nori proves that if the a_i are sufficiently large, then for any *smooth* morphism $T \longrightarrow S$ one has the vanishing

$$H^i\big(X_T, \mathcal{D}_T; \mathbf{Q}\big) = 0 \quad \text{for } i \le 2(n-e). \tag{*}$$

In the range $i \le n - e$ this follows — for any morphism $T \longrightarrow S$ — from (3.7) and the Leray spectral sequence. So the essential assertion is that one has a stronger statement for locally complete families. The proof of (*) is Hodge-theoretic in nature. Besides Nori's original paper [482], we recommend [239], [600, Chapter 20] and [83, Chapter 10] for expositions of the theorem and its application to the study of algebraic cycles. □

Remark 3.1.37. (Lefschetz-type statements for fundamental groups of subvarieties). In his book [362], Kollár considers an irreducible subvariety $Z \subseteq X$ of a smooth projective variety X and tries to find geometric explanations for the possibility that $\pi_1(Z)$ has small image in $\pi_1(X)$. By studying families of cycles on X, he proves for example that if $\rho(X) = 1$ and $\pi_1(X)$ is infinite, and if Z passes through a very general point of X, then in fact the image of

$$\pi_1(Z') \longrightarrow \pi_1(X)$$

is infinite, $Z' \longrightarrow Z$ being the normalization of Z. See [362, Chapter 1] for more information. □

Remark 3.1.38. (Lefschetz theorem for CR manifolds). In their paper [476], Ni and Wolfson prove an analogue of the Lefschetz hyperplane theorem for certain compact CR submanifolds of $\mathbf{P}^r$. This is an example of the emerging principle that the Lefschetz hyperplane theorem and related results discussed in the present chapter have analogues in non-algebraic settings: see Remarks 3.2.7 and 3.3.12. □

3.1.C Hard Lefschetz Theorem

On several occasions we will require a second (deeper) theorem of Lefschetz. Here we briefly state this result without proof.

Let X be a smooth projective variety of dimension n, which we view as a complex manifold. Recall (Definition 1.2.39) that a Kähler form ω on X is a closed positive $\mathcal{C}^\infty$ $(1,1)$-form. Wedge product with ω determines a homomorphism

$$L = L_\omega \, : \, H^i\big(X;\mathbf{C}\big) \longrightarrow H^{i+2}\big(X;\mathbf{C}\big),$$

which of course depends only on the cohomology class $[\omega] \in H^2\big(X;\mathbf{C}\big)$ of ω.

The result in question is the following:

Theorem 3.1.39. (Hard Lefschetz theorem). *For any Kähler form ω on X, the k-fold iterate*

$$L^k : H^{n-k}\big(X;\mathbf{C}\big) \xrightarrow{\cong} H^{n+k}\big(X;\mathbf{C}\big) \tag{3.8}$$

of $L = L_\omega$ is an isomorphism on the indicated cohomology groups.

Note that this applies in particular when ω is a Kähler form representing the first Chern class of any ample line bundle on X. See [600, §6.2.3] for a very nice account of the proof via harmonic theory.

Hard Lefschetz is often applied via

Corollary 3.1.40. *In the situation of the theorem, the mapping*

$$L^k : H^i\big(X;\mathbf{C}\big) \longrightarrow H^{i+2k}\big(X;\mathbf{C}\big)$$

is injective when $i \leq n-k$ and surjective when $i \geq n-k$.

Proof. If $i \leq n-k$, the composition

$$H^i\big(X;\mathbf{C}\big) \xrightarrow{\quad L^k \quad} H^{i+2k}\big(X;\mathbf{C}\big) \xrightarrow{\quad L^{n-k-i} \quad} H^{2n-i}\big(X;\mathbf{C}\big)$$

is an isomorphism thanks to 3.1.39, and so the first mapping L^k is injective. The second statement is similar. □

We conclude by outlining an interesting application of hard Lefschetz due to Wiśniewski [611], and some subsequent developments thereof.

Example 3.1.41. (Semiample deformations of ample line bundles). Let

$$f : X \longrightarrow T$$

be a smooth projective mapping of algebraic varieties, and let L be a line bundle on X. Assume that L_t is ample on the fibre X_t for general $t \in T$, and semiample (Definition 2.1.26) on X_0 for some special point $0 \in T$. Choose an integer $m > 0$ such that $L_0^{\otimes m}$ is free, and consider the corresponding morphism

$$\phi_0 : X_0 \longrightarrow Y_0 \subseteq \mathbf{P}.$$

Then for any irreducible subvariety $Z_0 \subseteq X_0$, one has the inequality

$$2 \cdot \dim Z_0 - \dim X_0 \leq \dim \phi(Z_0). \tag{3.9}$$

In other words, a semiample deformation of an ample line bundle cannot contract too much. (Set $n = \dim X$, $a = \dim Z_0$, and $b - 1 = \dim \phi(Z_0)$. Writing $\eta \in H^{2(n-a)}(X; \mathbf{R})$ for the cohomology class of Z_0, one evidently has

$$c_1(L_0)^b \cdot \eta = 0 \in H^{2n-2a+2b}(X_0; \mathbf{C}).$$

On the other hand, the homomorphisms

$$H^{2n-2a}(X_t; \mathbf{C}) \longrightarrow H^{2n-2a+2b}(X_t; \mathbf{C})$$

given by cup product with $c_1(L_t)^b$, being defined over $\mathbf{Z}$ and topologically determined, are independent of $t \in T$. So by Corollary 3.1.40 they are injective provided that $b \leq 2a - n$.) □

Remark 3.1.42. (Topology of semismall maps). A morphism $f : X \longrightarrow Y$ of irreducible projective varieties is said to be *semismall* if each of the locally closed sets

$$Y_k =_{\text{def}} \{ y \in Y \mid \dim f^{-1}(y) = k \}$$

has codimension $\geq 2k$ in Y. This is equivalent to asking that

$$2 \cdot \dim Z - \dim f(Z) \leq \dim X$$

for every irreducible subvariety $Z \subseteq X$. So one can rephrase the conclusion (3.9) of the previous example as asserting that the morphism ϕ_0 appearing there is semismall. In their paper [104], de Cataldo and Migliorini establish a number of interesting facts about the topology of such maps. For instance they show that a semiample line bundle L on a smooth projective variety X of dimension n is the pullback of an ample line bundle under a semismall map if and only if the hard Lefschetz theorem holds for L in the sense that k-fold cup product with $c_1(L)$ determines an isomorphism

$$H^{n-k}(X; \mathbf{C}) \xrightarrow{\cong} H^{n+k}(X; \mathbf{C})$$

for every $k \geq 0$. In their subsequent paper [105], de Cataldo and Migliorini study the Hodge-theoretic properties of arbitrary morphisms $f : X \longrightarrow Y$ of projective varieties, with X smooth. Among other things, they arrive at new proofs of the fundamental theorems of Beilinson, Bernstein, Deligne, and Gabber [48] concerning the topology of such maps. □

3.2 Projective Subvarieties of Small Codimension

An elementary classical theorem states that any smooth projective variety of dimension n can be embedded in $\mathbf{P}^{2n+1}$ (cf. [532, II.5.4]). Therefore subvarieties $X \subseteq \mathbf{P}^r$ with $\operatorname{codim} X > \dim X$ cannot be expected to exhibit any special properties. On the other hand, starting with a theorem of Barth in 1970 it became clear that if $\operatorname{codim} X < \dim X$, then X must be quite special. This viewpoint was developed in Hartshorne's influential article [279]. The present section is devoted to some of the properties of projective subvarieties of small codimension: we start by discussing Barth's theorem, and then give a brief account of Hartshorne's conjecture on complete intersections. The Fulton–Hansen connectedness theorem leads to some related results appearing in Section 3.4.

3.2.A Barth's Theorem

Let $X \subseteq \mathbf{P}^r$ be a smooth irreducible subvariety of dimension n and codimension $e = r - n$. It follows inductively from the Lefschetz hyperplane theorem (Theorem 3.1.17 and Example 3.1.31) that if X is the transversal complete intersection of e hypersurfaces, then there are isomorphisms

$$H^i(\mathbf{P}^r; \mathbf{Z}) \xrightarrow{\cong} H^i(X; \mathbf{Z}) \quad \text{for } i \leq n-1.$$

In 1970 Barth made the remarkable discovery that a similar statement holds (in a smaller range of degrees) for any smooth variety X having small codimension in projective space. Barth's result sparked a great deal of interest in the geometry of such subvarieties.

We start by stating the result.

Theorem 3.2.1. (Barth–Larsen theorem). *Let $X \subseteq \mathbf{P}^r$ be a non-singular subvariety of dimension n and codimension $e = r - n$.*

(i). *Restriction of cohomology classes determines an isomorphism*

$$H^i(\mathbf{P}^r; \mathbf{C}) \xrightarrow{\cong} H^i(X; \mathbf{C}) \quad \text{for } i \leq 2n - r = n - e.$$

(ii). *More generally, one has the vanishing of the relative homotopy groups*

$$\pi_i(\mathbf{P}^r, X) = 0 \quad \text{for } i \leq 2n - r + 1 = n - e + 1.$$

In particular, the maps

$$H^i(\mathbf{P}^r; \mathbf{Z}) \longrightarrow H^i(X; \mathbf{Z}) \ , \quad H_i(X; \mathbf{Z}) \longrightarrow H_i(\mathbf{P}^r; \mathbf{Z})$$

are isomorphisms for $i \leq 2n - r = n - e$.

We have separated the two statements for expository purposes: the second of course implies the first.

Statement (i) was originally proved by Barth [30] via delicate geometric and sheaf-theoretic arguments. In fact, he showed more generally that if $X, Y \subseteq \mathbf{P}^r$ are any two connected algebraic submanifolds having dimensions n and m respectively, and if X and Y meet properly, then

$$H^i\big(Y, X \cap Y; \mathbf{C}\big) \;=\; 0 \quad \text{for } i \;\leq\; \min\{n+m-r, 2n-r+1\};$$

Theorem 3.2.1 (i) is the case $Y = \mathbf{P}^r$. The extension (ii) to higher homotopy groups was established slightly later by Larsen [384] and Barth–Larsen [32], [31]. Hartshorne [279] found a very quick proof of (i) based on the hard Lefschetz theorem: we will present this shortly. In Section 3.5 we outline an argument from [211, §9] showing that one can deduce (ii) from a noncompact strengthening of the Lefschetz hyperplane theorem due to Goresky and MacPherson. Some other approaches and generalizations are indicated in Remarks 3.2.5 and 3.2.6.

Observe that Theorem 3.2.1 gives no information about subvarieties $X \subseteq \mathbf{P}^r$ having $\operatorname{codim} X \geq \dim X$. By contrast, as the codimension of X becomes small compared to its dimension, one gets stronger and stronger statements. For example:

Corollary 3.2.2. *Let $X \subseteq \mathbf{P}^r$ be a smooth subvariety of dimension n. If $2n - r \geq 1$, then X is simply connected.* □

Corollary 3.2.3. *Let $X \subseteq \mathbf{P}^r$ be a smooth subvariety of dimension n. If $2n - r \geq 2$, then restriction determines an isomorphism*

$$\operatorname{Pic}(\mathbf{P}^r) \xrightarrow{\cong} \operatorname{Pic}(X).$$

Some conjectures concerning subvarieties of small codimension in projective space are surveyed in the next subsection.

Proof of Corollary 3.2.3. The restrictions $H^i\big(\mathbf{P}^r, \mathbf{Z}\big) \longrightarrow H^i\big(X, \mathbf{Z}\big)$ are isomorphisms for $i \leq 2$ thanks to Theorem 3.2.1. It results from the Hodge decomposition that the maps

$$H^i\big(\mathbf{P}^r, \mathcal{O}_{\mathbf{P}^r}\big) \longrightarrow H^i\big(X, \mathcal{O}_X\big)$$

are likewise isomorphisms when $i \leq 2$. The corollary then follows by comparing the exponential sequences on $\mathbf{P}^r$ and on X. □

We now turn to Hartshorne's proof of Barth's theorem:

Proof of Theorem 3.2.1 (i). We begin with some notation. Write $j : X \hookrightarrow \mathbf{P} = \mathbf{P}^r$ for the inclusion, and $H^*(X)$, $H^*(\mathbf{P})$ for cohomology with complex coefficients. Denote by $\xi \in H^2(\mathbf{P})$ the hyperplane class, and by $\overline{\xi} = j^*(\xi) \in$

$H^2(X)$ its restriction to X. Assuming X has degree d, then the cohomology class η_X of X is given by d times the class of a codimension $e = r - n$ linear space: $\eta_X = d \cdot \xi^e$.

Hartshorne's idea is to study the Gysin map

$$j_* : H^i(X) \longrightarrow H^{i+2e}(\mathbf{P})$$

determined by pushforward. One has

$$j^* j_*(y) \;=\; y \cdot j^*(\eta_X) \;=\; y \cdot (d\overline{\xi}^e) \quad , \quad j_* j^*(x) \;=\; x \cdot \eta_X \;=\; x \cdot (d\xi^e).$$

Consider now the commutative diagram

$$\begin{array}{ccc} H^i(\mathbf{P}) & \xrightarrow{\quad j^* \quad} & H^i(X) \\ {\scriptstyle \cdot\, d\xi^e}\Big\downarrow & \swarrow{\scriptstyle j_*} & \Big\downarrow{\scriptstyle \cdot\, d\overline{\xi}^e} \\ H^{i+2e}(\mathbf{P}) & \xrightarrow[\quad j^* \quad]{} & H^{i+2e}(X), \end{array}$$

and assume that $i \leq n - e$. Then the hard Lefschetz theorem (Corollary 3.1.40) implies that the vertical homomorphism on the right is injective. Therefore the diagonal map j_* is likewise injective. On the other hand, the vertical map on the left is an isomorphism by inspection. Therefore j_* is onto, and hence an isomorphism. This implies that $j^* : H^i(\mathbf{P}) \longrightarrow H^i(X)$ is also an isomorphism when $i \leq n - e$, and the theorem is proved. □

Example 3.2.4. As Hartshorne [279, §2] notes, one can replace the ambient projective space $\mathbf{P}$ in this argument with other varieties having similar cohomological properties. For example, suppose that $V \subset \mathbf{P}^N$ is a smooth complete intersection of dimension r, and let $X \subseteq V$ be a non-singular subvariety of dimension n. The cohomology of V is controlled by the Lefschetz hyperplane theorem, and one finds that $H^i(V) \longrightarrow H^i(X)$ is an isomorphism for $i < 2n - r$ and injective when $i = 2n - r$. □

Remark 3.2.5. (Other approaches to Barth's theorem). Ogus [486] gave an algebraic proof of 3.2.1 (i) — allowing X to have local complete intersection singularities — via a study of the local cohomological dimension of $\mathbf{P}^r - X$. Some related work of Lyubeznik is summarized in Remark 3.5.14 below. Schneider and Zintl [525] found a proof using vanishing theorems for vector bundles. In a rather different spirit, Gromov [252, §1.1.2, Exercise A′] sketched a proof the Barth–Larsen theorem via Morse theory on a suitable path space. Schoen and Wolfson [526] used an approach along these lines to prove a more general statement, valid also on homogeneous varieties other than projective space. This Morse-theoretic approach has recently led to non-algebraic analogues of several of the results of this chapter: see Remarks 3.1.38, 3.2.7, and 3.3.12. □

Remark 3.2.6. (Extensions of the Barth–Larsen theorem). There has been considerable work on extending 3.2.1 to ambient varieties other than projective space. The situation that has received the most attention is that in which projective space $\mathbf{P}^r$ is replaced by a rational homogeneous variety $\mathbf{F} = G/P$. In this case the Barth–Larsen theorem was generalized by Sommese in the series of papers [549], [552], [553]. Given a subvariety $X \subseteq \mathbf{F}$, the conclusion is that one has a vanishing of the form

$$\pi_i\big(\mathbf{F}\,;\,X\big) \;=\; 0 \quad \text{for} \;\; i \leq \dim X - \operatorname{codim} X - \lambda,$$

where λ is an explicitly given "defect" that vanishes when $X = \mathbf{P}^n$. More recent results along the same lines appear in [487], [556], [526], [339]. Several authors have also focused on subvarieties $X \subseteq A$ of an abelian variety, notably Sommese [551] and more recently Debarre [109]. We refer to [211, §9] and [203, Chapter III], as well as Section 3.5.A, for further references and historical remarks.

In a different direction, an analogue of the Barth theorem for branched coverings $X \longrightarrow \mathbf{P}^n$ was given in [387]. In this result, which appears as Theorem 7.1.15 below, coverings of low degree emerge as analogues of projective subvarieties of small codimension. Some other related results appear later in 3.4.3 and 7.1.12. □

Remark 3.2.7. (Wilking's theorem). Wilking [606] established a result that, as pointed out in [182], can be seen as a differential-geometric analogue of the Barth–Larsen theorem. Specifically, let M be a compact r-dimensional Riemannian manifold of positive sectional curvature, and let $N \subseteq M$ be a totally geodesic compact submanifold of dimension n. Wilking proves that then $\pi_i(M, N) = 0$ for $i \leq 2n - r + 1$. Inspired by this, Fang, Mendonça, and Rong [182] extended many other connectedness statements from algebraic geometry to the setting of totally geodesic submanifolds of manifolds of positive curvature: see Remark 3.3.12. □

3.2.B Hartshorne's Conjectures

It was observed in the early 1970s that it is very hard to construct smooth subvarieties $X \subseteq \mathbf{P}^r$ with $\operatorname{codim} X \ll \dim X$ other than complete intersections.[3] Together with Barth's theorem, this led to the speculation that any smooth projective subvariety of sufficiently small codimension must in fact be a complete intersection of hypersurfaces. A precise conjecture was enunciated by Hartshorne in [279]:

Conjecture 3.2.8. (Hartshorne's conjecture on complete intersections). *Let $X \subseteq \mathbf{P}^r$ be a smooth irreducible variety of dimension n. If $3n > 2r$, then X is a complete intersection.*

[3] An illustration appears in Example 3.2.10.

As of this writing, Hartshorne's conjecture remains very much open already for subvarieties of codimension 2. In this case the conjecture asserts that a smooth subvariety $X \subseteq \mathbf{P}^{n+2}$ of dimension n is a complete intersection as soon as $n \geq 5$: in fact even when $n = 4$ no examples are known other than complete intersections.

Example 3.2.9. (Boundary examples). A couple of examples show that in any event 3.2.8 gives the best possible linear inequality. For instance, the Plücker embedding of the Grassmannian of lines in $\mathbf{P}^4$ is a six-dimensional variety $\mathbf{G} \subseteq \mathbf{P}^9$ that is not a complete intersection. Similarly there is a ten-dimensional spinor variety $\mathbf{S} \subseteq \mathbf{P}^{15}$ that likewise lies on the boundary of the conjecture. As Mori pointed out, non-homogeneous examples may then be constructed by taking pullbacks under branched coverings $\mathbf{P}^r \longrightarrow \mathbf{P}^r$. □

Hartshorne's conjecture has inspired a vast amount of activity: the papers [33], [510], [291], [4], [377] represent a very small sampling. Suffice it to say that if one imposes additional conditions on X — for example that it have very low degree compared to its dimension, or that it be defined by equations of very small degree — then one can show that it must indeed be a complete intersection. However, such hypotheses may be seen as changing the nature of the problem. Without any additional assumptions, the most striking progress to date is Zak's theorem on linear normality, to which we turn next.

Recall by way of background that if $X \subseteq \mathbf{P}^r$ is a complete intersection then X is *projectively normal*, meaning that the restriction maps

$$\rho_k : H^0\big(\mathbf{P}^r, \mathcal{O}_{\mathbf{P}^r}(k)\big) \longrightarrow H^0\big(X, \mathcal{O}_X(k)\big)$$

are surjective for all $k \geq 1$. In particular, ρ_1 is surjective: one says in this case that X is *linearly normal.* Based on some further classical examples, Hartshorne [279] was led to make a subsidiary conjecture:

> Let $X \subseteq \mathbf{P}^r$ be a smooth projective variety of dimension n. If $3n > 2(r-1)$, then X is linearly normal.

This was established by Zak [619] in 1979 as an outgrowth of his results on tangencies of algebraic varieties. A presentation of Zak's work, and a proof of his theorem on linear normality, appears in Section 3.4.B.

Example 3.2.10. (Codimension-2 degeneracy loci). The first non-trivial instance of Zak's theorem is that a smooth three-dimensional subvariety $X \subseteq \mathbf{P}^5$ is linearly normal. However, a construction of Peskine shows that X needn't be projectively normal. In fact, start with any vector bundle E of rank $e \geq 3$ on $\mathbf{P}^5$ having the property that $H^1\big(\mathbf{P}^5, E\big) \neq 0$: for example, one might take $E = \Omega^1_{\mathbf{P}^5}$. Fix next a large integer $k \gg 0$, with $k + c_1(E) > 0$, such that $E(k)$ is globally generated. Now choose $e-1$ general sections $s_1, \ldots, s_{e-1} \in H^0\big(\mathbf{P}^5, E(k)\big)$, and consider the degeneracy locus

$$X \ = \ \big\{ s_1 \wedge \ldots \wedge s_{e-1} \ = \ 0 \big\}$$

consisting of points in $\mathbf{P}^5$ at which the sections in question become linearly dependent (cf. Chapter 7.2). The resulting variety X is a smooth threefold, and the Eagon–Northcott complex (EN_0) described in Appendix B gives rise to a resolution

$$0 \longrightarrow \mathcal{O}_{\mathbf{P}^5}^{e-1} \longrightarrow E(k) \longrightarrow \mathcal{I}_X(a) \longrightarrow 0$$

with $a = ek + c_1(E)$, $\mathcal{I}_X \subseteq \mathcal{O}_{\mathbf{P}^5}$ being the ideal sheaf of X. In particular it follows that $H^1\big(\mathbf{P}^5, \mathcal{I}_X(a-k)\big) \neq 0$, and so ρ_{a-k} is not surjective. Note that if one tries to make the analogous construction on $\mathbf{P}^r$ when $r \geq 6$, then one expects the map $\mathcal{O}_{\mathbf{P}}^{e-1} \longrightarrow E(k)$ determined by the s_i to drop rank by two along a codimension 6 subset $Z \subseteq \mathbf{P}^r$, and X will be singular along Z. □

For subvarieties of codimension 2, Hartshorne's conjecture 3.2.8 is equivalent to some interesting questions about rank two vector bundles on projective space. Specifically, consider a smooth variety $X \subseteq \mathbf{P} = \mathbf{P}^{n+2}$ of dimension n. Assume that the canonical bundle of X is the restriction of a line bundle from the ambient projective space, i.e. assume that

$$\omega_X \;=\; \mathcal{O}_X(k) \quad \text{for} \quad \text{some } k \in \mathbf{Z}. \tag{*}$$

Then a construction of Serre (see [488, Chapter I, Section 5]) produces a rank two vector bundle E on $\mathbf{P}$, together with a section $s \in H^0\big(\mathbf{P}, E\big)$ having the property that $X = \mathrm{Zeroes}\big(s\big)$ is precisely the zero-locus of s. Moreover, X is a complete intersection if and only if E is the direct sum of two line bundles. On the other hand, it follows from Corollary 3.2.3 that the condition (*) is automatic if $n = \dim X \geq 4$. So in codimension two, 3.2.8 is equivalent to another conjecture from [279]:

Conjecture 3.2.11. (Hartshorne's splitting conjecture). *If $r \geq 7$, any rank two vector bundle E on $\mathbf{P}^r$ must split as a direct sum of line bundles.*

One expects that in fact any vector bundle E of rank $e \ll r$ splits, although there don't seem to be any specific guesses in the literature about where this splitting might start. This connection between bundles and subvarieties was one of the motivations for a great deal of activity in the late 1970s and early 1980s directed toward the geometry of bundles on projective space. The book [488] gives a nice overview of this work.

Remark 3.2.12. (Theorem of Evans and Griffith, I). Evans and Griffith [175] proved an interesting algebraic theorem implying (among other things) that if E is a vector bundle of rank e on $\mathbf{P}^r$ having the property that

$$H^i\big(\mathbf{P}^r, E(k)\big) \;=\; 0 \quad \text{for all} \quad k \in \mathbf{Z}$$

and every $0 < i < e$, then E splits as a sum of line bundles. By the discussion above, this shows that if $X \subseteq \mathbf{P}^{n+2}$ is a smooth variety of dimension $n \geq 4$, then X is a complete intersection if and only if X is projectively normal. Ein [140] found a quick proof of the Evans–Griffith criterion based on Castelnuovo–Mumford regularity and vanishing theorems for vector bundles. His argument is presented in Example 7.3.10. □

3.3 Connectedness Theorems

In this section we consider some connectedness theorems. A classical result of Bertini states that a general linear space section of an irreducible projective variety is irreducible provided that its expected dimension is positive. By specialization this implies the connectedness of an arbitrary linear section. Fulton and Hansen [209] realized that a variant of this statement — asserting the connectedness of the inverse image of the diagonal under maps to $\mathbf{P}^r \times \mathbf{P}^r$ — has many surprising consequences concerning subvarieties of and mappings to projective space.

We start with a review of the Bertini theorems, and then prove the Fulton–Hansen theorem. We also discuss without proof a local analogue of Bertini's theorem due to Grothendieck. Applications of the Fulton–Hansen theorem occupy Section 3.4.

3.3.A Bertini Theorems

In its simplest form, Bertini's theorem asserts the irreducibility of a general linear section of a variety mapping to projective space:

Theorem 3.3.1. (Bertini theorem for general linear sections). *Let X be an irreducible variety, and $f : X \longrightarrow \mathbf{P}^r$ a morphism. Fix an integer $d < \dim \overline{f(X)}$. If $L \subseteq \mathbf{P}^r$ is a general $(r-d)$-plane, then $f^{-1}(L)$ is irreducible.*

When we speak of a "general" plane, we mean of course that the statement is true for all planes L parameterized by a dense open subset of the Grassmannian of codimension d linear spaces in $\mathbf{P}^r$.

A detailed algebraic proof of this and related Bertini-type results appears in Jouanolou's notes [308, §I.6], and the discussion in [186, Chapter 3] is also very useful. At the expense of overkill, one could give a very quick proof using resolution of singularities and vanishing for big and nef line bundles (Theorem 4.3.1). The argument that follows — a variant of the presentation in [211, §1] — is close to the classical one.

Proof of Theorem 3.3.1. We will assume for simplicity of exposition that f is generically finite-to-one, although the proof goes through with only minor changes in general. Setting $n = \dim X = \dim \overline{f(X)}$, it suffices to prove the theorem in the borderline case $d = n-1$. To this end fix a general $(r-n)$-plane

$$\Lambda \subseteq \mathbf{P}^r.$$

We may assume that Λ is transverse to f in the sense that $f^{-1}(\Lambda) \subseteq X$ consists of a non-empty finite set of smooth points of X, each of which is reduced in its natural scheme structure. The set of all $(r-n+1)$-planes $L \subseteq \mathbf{P}^r$ passing through Λ is parameterized by a projective space $T = \mathbf{P}^{n-1}$: given $t \in T$ denote by $L_t \supseteq \Lambda$ the corresponding plane. Consider the set

$$V =_{\text{def}} \Big\{ (x,t) \mid f(x) \in L_t \Big\} \subseteq X \times T.$$

The issue is to establish the irreducibility of a general fibre of the second projection $p : V \longrightarrow T$.

To this end note that the first projection $V \longrightarrow X$ realizes V as the blowing up of X along $f^{-1}(\Lambda)$, and hence V is irreducible. Furthermore, if $O \in f^{-1}(\Lambda)$ is a fixed point, then the mapping

$$s : T \longrightarrow V \quad , \quad t \mapsto (O,t)$$

defines a section of p whose image lies in the smooth locus of V. Hence the following lemma applies to give the required irreducibility of fibres of p. □

Lemma 3.3.2. (Irreducibility of fibres). *Let $p : V \longrightarrow T$ be a dominating morphism of irreducible complex varieties. Assume that p admits a section $s : T \longrightarrow V$ whose image does not lie in the singular locus of V, i.e. with*

$$s(t) \in V - \operatorname{Sing}(V) \ \textit{ for a general point } \ t \in T.$$

Then the fibre $V_t = p^{-1}(t)$ is irreducible for general $t \in T$.

Intuitively, suppose that a general fibre V_t were reducible. Then one could define a monodromy action on its irreducible components, and since V itself is irreducible this action would necessarily be transitive. On the other hand the image of a suitable section will pick out one component in each fibre, leading to a contradiction. The quickest way to make this precise is to use considerations of smoothness to reduce to a connectedness statement.

Proof of Lemma 3.3.2. It is enough to prove the statement with V replaced by any non-empty Zariski open subset, since for dimensional reasons removing a proper closed subset can't affect the irreducibility of a general fibre. This being said, after shrinking T we can first remove the singular locus of V without changing the hypotheses. Then by the theorem on generic smoothness we can assume after further shrinking T that p is a smooth morphism. In this case it suffices to show that the fibres of p are connected. After shrinking T yet again we can suppose finally that p is topologically locally trivial ([586, Corollary 5.1]). But a locally trivial fibration between path-connected spaces that admits a section necessarily has connected fibres, and the lemma follows. □

Under suitable compactness hypotheses, Theorem 3.3.1 implies the connectedness of arbitrary linear sections:

Theorem 3.3.3. (Connectedness of arbitrary linear sections). *Let X be an irreducible variety, $f : X \longrightarrow \mathbf{P}^r$ a morphism, and $L \subseteq \mathbf{P}^r$ an arbitrary linear space of codimension $d < \dim \overline{f(X)}$. If X is complete then $f^{-1}(L)$ is connected. The same statement holds more generally if X is not complete provided that f is proper over a Zariski-open subset $V \subseteq \mathbf{P}^r$ and $L \subseteq V$.*

Proof. The following simple argument is due to Jouanolou [308]. Denote by $\mathbf{G}$ the Grassmannian parameterizing all codimension d linear spaces of $\mathbf{P}^r$, and let $W \subseteq \mathbf{G}$ be the open set consisting of linear spaces contained in V. We consider the correspondence

$$Z =_{\text{def}} \Big\{ \big(x, L'\big) \mid x \in f^{-1}(L') \Big\} \subseteq X \times W.$$

This is a Zariski-open subset of a Grassmannian bundle over X, and hence irreducible. The second projection $p_2 : Z \longrightarrow W$ is proper thanks to the fact that f is proper over V. Consider its Stein factorization

$$Z \xrightarrow{q} W' \xrightarrow{r} W.$$

By construction q has connected fibres and r is finite. Bertini's Theorem 3.3.1 implies that the general fibre of p_2 is irreducible, and therefore r is generically one-to-one. But r is a surjective branched covering of the normal variety W, and hence it must be everywhere one-to-one. The theorem follows. □

Remark 3.3.4. (Statements for fundamental groups). By applying similar arguments to possibly infinite-sheeted covering spaces, the preceding results can be generalized to statements involving the surjectivity of maps on fundamental groups. Assume that X is locally irreducible (in the classical topology). Then in the situation of Theorem 3.3.1, the induced homomorphism

$$\pi_1\big(f^{-1}(L)\big) \longrightarrow \pi_1\big(X\big)$$

is surjective. This implies in the setting of 3.3.3 that if U is any (classical) neighborhood of L in $\mathbf{P}^r$, then $\pi_1\big(f^{-1}(U)\big) \longrightarrow \pi_1\big(X\big)$ is likewise surjective. See [211, §1, §2] for details. These statements are due to Deligne. They are special cases of more general results of Goresky and MacPherson interpolating between Bertini's theorem and the Lefschetz theorems discussed in Section 3.1: see Theorem 3.5.10 below. □

Example 3.3.5. (Joining points by irreducible curves). Let X be an irreducible quasi-projective variety, and let $x_1, x_2 \in X$ be distinct points. Then x_1 and x_2 can be joined by an irreducible curve $\Gamma \subseteq X$. Moreover, there exists a smooth irreducible curve T and a map $T \longrightarrow X$ whose image contains the points in question. (Let $X' = \mathrm{Bl}_{x_1,x_2}(X)$ be the blowing up of X at the two points, with exceptional divisors E_1 and E_2. Fix a generically finite morphism $f : X' \longrightarrow \mathbf{P}^r$ that is finite along each E_i, and let $L \subseteq \mathbf{P}^r$ be a general linear space of codimension $= \dim X - 1$. By Bertini's theorem, $\Gamma' = f^{-1}(L)$ is an irreducible curve meeting E_1 and E_2. Then one can take $\Gamma \subseteq X$ to be the image of Γ', and T to be the normalization of Γ.) This argument appears in [447, p. 56], and is apparently due to Ramanujam (see [447, p. vii] and [508, p. 11]). □

3.3.B Theorem of Fulton and Hansen

We now come to the connectedness theorem of Fulton and Hansen. The essential interest of the statement lies in its applications, which are discussed in the next section.

Theorem 3.3.6. (Connectedness theorem of Fulton and Hansen). *Let X be an irreducible projective (or complete) variety, and*

$$f : X \longrightarrow \mathbf{P}^r \times \mathbf{P}^r$$

a morphism. Assume that $\dim f(X) > r$. *Then the inverse image $f^{-1}(\Delta) \subseteq X$ of the diagonal $\Delta \subseteq \mathbf{P}^r \times \mathbf{P}^r$ is connected.*

Proof. We reproduce the argument in [211], which uses an idea of Deligne's to pass from the diagonal $\Delta \subseteq \mathbf{P}^r \times \mathbf{P}^r$ to a linear embedding $L^r \subseteq \mathbf{P}^{2r+1}$. Let $[x] = [x_0, \ldots, x_r]$ and $[y] = [y_0, \ldots, y_r]$ be homogeneous coordinates on the the two factors of $\mathbf{P}^r \times \mathbf{P}^r$, and introduce the coordinates $[x, y] = [x_0, \ldots, x_r, y_0, \ldots, y_r]$ on $\mathbf{P}^{2r+1}$. Denote by V the complement in $\mathbf{P}^{2r+1}$ of the two linear spaces defined by $x_0 = \ldots = x_r = 0$ and $y_0 = \ldots = y_r = 0$. Then there is a natural morphism,

$$p : V \longrightarrow \mathbf{P}^r \times \mathbf{P}^r \quad , \quad [x, y] \mapsto \big([x] \, , \, [y] \big),$$

realizing V as the total space of a $\mathbf{C}^*$-bundle over $\mathbf{P}^r \times \mathbf{P}^r$. Let $L \subseteq V$ be the linear space of dimension r defined by the equations $x_i = y_i$ $(0 \leq i \leq r)$: L maps isomorphically to the diagonal $\Delta \subseteq \mathbf{P}^r \times \mathbf{P}^r$.

Given $f : X \longrightarrow \mathbf{P}^r \times \mathbf{P}^r$, denote by $X^* = X \times_{\mathbf{P}^r \times \mathbf{P}^r} V$ the pullback of f via p, and let $q : X^* \longrightarrow X$, $f^* : X^* \longrightarrow V$ be the projections. The situation is summarized in the following diagram:

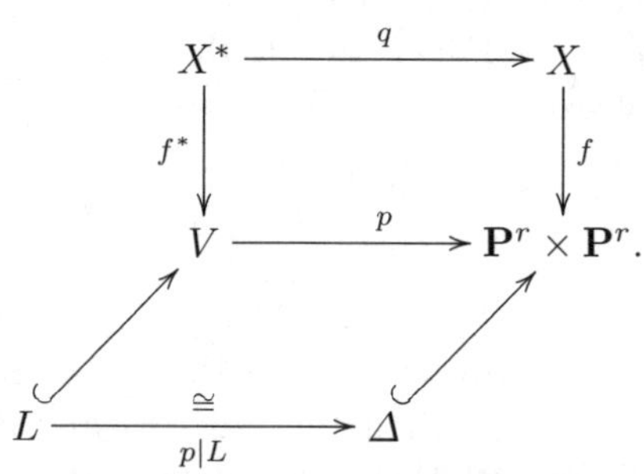

X^* is irreducible since X is, and f^* is proper over V since f is proper (X being complete). Viewing f^* as a map to $\mathbf{P}^{2r+1}$ via the inclusion $V \subseteq \mathbf{P}^{2r+1}$, one has

$$\dim \overline{f^*(X^*)} \ > \ r+1 \ = \ \operatorname{codim}_V L,$$

and therefore ${f^*}^{-1}(L)$ is connected thanks to 3.3.3. But q gives rise to an isomorphism

$$f^{*-1}(L) \;=\; f^{-1}(\Delta)$$

by virtue of the fact that L maps isomorphically to Δ, and the theorem is proved. □

Example 3.3.7. (Intersections). Let $X, Y \subseteq \mathbf{P}^r$ be irreducible subvarieties. If $\dim X + \dim Y > r$, then $X \cap Y$ is connected. □

Example 3.3.8. (Several factors). The Fulton–Hansen theorem extends to more than two factors: one considers a morphism

$$f : X \longrightarrow \mathbf{P}^r \times \ldots \times \mathbf{P}^r \quad (k\text{-fold product}),$$

and the small diagonal $\Delta \subseteq (\mathbf{P}^r)^k$. If X is irreducible and complete and $\dim f(X) > (k-1)r$, then $f^{-1}(\Delta)$ is connected. (Modify the construction appearing in the proof of 3.3.6.) □

Remark 3.3.9. As before there is a variant of Theorem 3.3.6 involving fundamental groups. Specifically, assume in the setting of the theorem that X is locally irreducible. Then

$$\pi_1\big(f^{-1}(\Delta)\big) \longrightarrow \pi_1\big(X\big)$$

is surjective. See [211, §3] for details and Section 3.5.B for more general statements. □

Example 3.3.10. (Irreducibility of inverse image of general graph). Let X be an irreducible quasi-projective variety, and let

$$f : X \longrightarrow \mathbf{P}^r \times \mathbf{P}^r$$

be a morphism such that $\dim \overline{f(X)} > r$. Given $g \in G =_{\text{def}} \mathrm{GL}(r+1)$, which we view in the evident way as an automorphism of $\mathbf{P}^r$, let $\Gamma_g \subseteq \mathbf{P}^r \times \mathbf{P}^r$ be its graph. Then $f^{-1}(\Gamma_g)$ is irreducible for general $g \in G$. (Returning to the proof of 3.3.6, let $L_g \subseteq V$ be the linear space defined by the equations

$$y_i \;=\; \sum a_{ij} x_j \;\; (0 \leq i \leq r),$$

(a_{ij}) being the matrix of g. Then L_g maps isomorphically to Γ_g, so it suffices to prove the irreducibility of $f^{*-1}(L_g)$. But all r-planes in a neighborhood of L are of the form L_g for some g, so this follows from Theorem 3.3.1.) This is a special case of a result of Debarre [111, Lemma 2.1 (2)]. □

Remark 3.3.11. (Weighted projective spaces). Bădescu [20, 1.7.14] shows that the Fulton–Hansen theorem remains true when $\mathbf{P}^r$ is replaced by a weighted projective space of dimension r. □

Remark 3.3.12. (Analogues for manifolds of positive curvature). Motivated in part by the theorem of Wilking (Remark 3.2.7), Fang, Mendonça, and Rong [182] established an analogue of the Fulton–Hansen theorem in which projective space is replaced by a closed smooth manifold of positive sectional curvature, and the morphism $f : X \longrightarrow \mathbf{P}^r \times \mathbf{P}^r$ is replaced by a minimally immersed closed submanifold of large asymptotic index. These authors also establish statements for higher homotopy in the spirit of Section 3.5.B. Among other things, this leads to uniform formulations of several classical results in the geometry of positive curvature. □

3.3.C Grothendieck's Connectedness Theorem

We conclude by briefly discussing a local result of Grothendieck from [258]. Grothendieck's theorem yields an alternative algebraic approach to theorems of Bertini type. It also gives a quantitative bound on the extent to which a linear section of an irreducible variety can fail to be irreducible. We do not give any proofs: the reader may consult [186, Chapter 3.1] for details and further information.

Grothendieck's idea is to introduce a numerical measure of how far a variety is from being irreducible. Because the main result is algebraic in nature, we will deal for the moment with fairly general schemes.

Definition 3.3.13. (Connectedness in dimension k). A Noetherian scheme X is *connected in dimension* k (or k*-connected*) if $\dim X > k$ and $X - T$ is connected for every closed subset $T \subseteq X$ of dimension $< k$. □

For the purposes of this discussion, the empty set $\varnothing$ is taken to have dimension -1.

Example 3.3.14. (i). X is connected if and only if X is 0-connected.
(ii). If X is irreducible of dimension n, then X is $(n-1)$-connected. □

Example 3.3.15. X is k-connected if and only if:

(i). Every irreducible component X' of X has dimension $> k$; and
(ii). Given any two irreducible components X', X'' of X, there is a sequence

$$X' = X_0 \ , \ X_1 \ , \ \ldots \ , \ X_{\ell-1} \ , \ X_\ell = X''$$

of irreducible components such that

$$\dim\big(X_{i-1} \cap X_i\big) \ \geq \ k$$

for $1 \leq i \leq \ell$. (See [186, Proposition 3.1.4].) □

Grothendieck's result [258, XIII.2.1] is a local analogue of Bertini's theorem:

Theorem 3.3.16. (Grothendieck connectedness theorem). *Let* $(A, \mathfrak{m})$ *be a complete local Noetherian ring and* $k \geq 1$ *a positive integer. If* $X =$ Spec(A) *is* k*-connected, and if* $f \in \mathfrak{m}$, *then* Spec(A/fA) *is* $(k-1)$*-connected.*

Observe that the spectrum of an n-dimensional Noetherian domain is $(n-1)$-connected. So by induction, 3.3.16 yields:

Corollary 3.3.17. *Let $(A, \mathfrak{m})$ be a complete local Noetherian domain of dimension n, and fix $d \leq n-1$ elements $f_1, \ldots, f_d \in \mathfrak{m}$. Then*

$$\operatorname{Spec}\big(A/(f_1, \ldots, f_d) \big)$$

is connected in dimension $n - d - 1$. □

We refer to [186, Chapter 3.1] for a nice account of the proof of Theorem 3.3.16. As explained there and in [258, Exposé XIII], one can recover and strengthen classical statements of Bertini type by normalizing and passing to affine cones. For example:

Corollary 3.3.18. *Let $X \subseteq \mathbf{P}^r$ be a closed algebraic subset that is connected in dimension k, and let $L \subseteq \mathbf{P}^r$ be a linear subspace of codimension d. Then $X \cap L$ is connected in dimension $k - d$.* □

Theorem 3.3.16 and related ideas also imply various interesting local statements, for instance Hartshorne's result that if $(A, \mathfrak{m})$ is a local ring with depth $A \geq 2$, then $\operatorname{Spec}(A) - \mathfrak{m}$ is connected. Again we refer to [186, Chapter 3] for more information.

Example 3.3.19. ([258, Example III.3.10]). Let $X \subseteq \mathbf{C}^6$ be the union of two 3-planes meeting transversally at the origin $0 \in \mathbf{C}^6$. Then X cannot be set-theoretically cut out near 0 by four or fewer equations. □

Example 3.3.20. (Subsets of normal varieties). Given a complex algebraic variety V with irreducible components $V_1, \ldots, V_t$, and an index $1 \leq i \leq t$, set

$$Z_V(V_i) \ = \ \big(\cup_{j \neq i} V_j \big) \cap V_i.$$

In other words, $Z_V(V_i)$ consists of all points of V_i that lie also in some other irreducible component of V. Now suppose that X is a normal variety of dimension n, and that $W \subseteq X$ is a closed subset that is locally cut out set-theoretically by $e \leq n-1$ equations. Fix an irreducible component W' of W. Then for any (closed) point $x \in Z_W(W')$,

$$\dim_x Z_W(W') \ \geq \ n - e - 1.$$

(It is sufficient to prove that $\operatorname{Spec}(\mathcal{O}_x W)$ is connected in dimension $n-e-1$, and this in turn follows from the $(n-e-1)$-connectedness of $\operatorname{Spec}(\widehat{\mathcal{O}_x W})$, where $\widehat{\mathcal{O}_x W}$ is the completion of $\mathcal{O}_x W$ at its maximal ideal.) □

3.4 Applications of the Fulton–Hansen Theorem

In this section we present some of the applications of the Fulton–Hansen theorem. Much of this material was surveyed in detail in the notes [211] of

Fulton and the author (from which we borrow liberally). So we limit ourselves here to highlights of the story.

3.4.A Singularities of Mappings

The first collection of results concerns singularities of mappings $f : X \longrightarrow \mathbf{P}^r$. The theme is that under mild conditions, singularities of f that one could expect on dimensional grounds will actually occur.

We start with some statements that appeared in the original paper [209] of Fulton and Hansen. Recall that a morphism $f : X \longrightarrow Y$ of varieties is *unramified* if the natural homomorphism $f^*\Omega^1_Y \longrightarrow \Omega^1_X$ is surjective. When X and Y are non-singular, this is equivalent to asking that the derivative $df_x : T_xX \longrightarrow T_{f(x)}Y$ be injective at every point, i.e. that f be an immersion in the sense of topologists.

Theorem 3.4.1. (Ramification). *Let X be a complete irreducible variety of dimension n, and $f : X \longrightarrow \mathbf{P}^r$ an unramified morphism. If $2n > r$, then f is a closed embedding.*

The dimensional hypothesis is of course necessary: the typical map from a smooth curve to the plane is an unramified morphism with finitely many double points.

Before giving the proof, we present some simple consequences.

Example 3.4.2. If $r \geq 3$, there are no irreducible singular hypersurfaces $X \subseteq \mathbf{P}^r$ having only normal crossing singularities.[4] (If X has only normal crossing singularities, then the normalization $X' \longrightarrow X$ is unramified.) □

Theorem 3.4.1 also shows that the algebraic analogue of 3.2.2 remains true for singular varieties.

Corollary 3.4.3. (Simple connectivity of singular projective varieties). *Let $X \subseteq \mathbf{P}^r$ be an irreducible subvariety of dimension n. If $2n > r$ then X is algebraically simply connected, i.e. X admits no non-trivial connected étale covers.*

Proof. Suppose that $p : Y \longrightarrow X$ is a connected étale cover, and let $Y' \subseteq Y$ be an irreducible component of Y. Then the composition

$$Y' \hookrightarrow Y \longrightarrow X \hookrightarrow \mathbf{P}^r$$

is unramified, and hence a closed embedding thanks to 3.4.1. This means that Y' maps isomorphically to X. Therefore p sections, and hence is trivial. □

[4] Recall that a variety X has only normal crossing singularities if every point has an analytic neighborhood in which X is locally isomorphic to a union of coordinate hyperplanes in affine space.

Remark 3.4.4. The analogue of 3.4.3 for topological fundamental groups is true: see [211, Corollary 5.3] for a proof following suggestions of Deligne. □

Fulton and Hansen deduce another striking consequence concerning secant and tangent varieties. Given a smooth irreducible subvariety $X \subseteq \mathbf{P}$ of dimension n embedded in a projective space $\mathbf{P}$, denote by

$$\mathrm{Sec}(X) \ , \ \mathrm{Tan}(X) \ \subseteq \ \mathbf{P}$$

(the closure of) the union of all secant and tangent lines to X in $\mathbf{P}$. Thus $\mathrm{Sec}(X) \supseteq \mathrm{Tan}(X)$, and these varieties have expected dimensions $2n+1$ and $2n$ respectively. Their actual dimensions may be smaller, but in this case the secant and tangent varieties must coincide:

Corollary 3.4.5. (Secant and tangent varieties). *With X as above, either*

$$\dim \mathrm{Sec}(X) = 2n+1 \quad \textit{and} \quad \dim \mathrm{Tan}(X) = 2n$$

or else $\mathrm{Sec}(X) = \mathrm{Tan}(X)$.

Proof. Assuming that $\mathrm{Tan}(X) \neq \mathrm{Sec}(X)$, it is enough to show that

$$\dim \mathrm{Tan}(X) \ = \ 2n.$$

Suppose to the contrary that $\dim \mathrm{Tan}(X) < 2n$. Then there exists a linear space $L \subseteq \mathbf{P}$ of codimension $\leq 2n$ that meets $\mathrm{Sec}(X)$ but not $\mathrm{Tan}(X)$. Linear projection from L then defines a morphism $X \longrightarrow \mathbf{P}^r$ with $r < 2n$ that is unramified but not an embedding. However Theorem 3.4.1 asserts that this is impossible. □

Remark 3.4.6. With a little more care about the definitions, one can allow X to be singular in 3.4.5: see [211, 5.4 and 5.5]. □

Turning to the proof of Theorem 3.4.1, we start with a heuristic explanation. Supposing that

$$f : X \longrightarrow \mathbf{P}^r$$

fails to be one-to-one, the issue is to produce a ramification point of f. To this end, one applies the connectedness theorem to the product

$$F = f \times f \ : \ X \times X \longrightarrow \mathbf{P}^r \times \mathbf{P}^r.$$

Observe that

$$F^{-1}\big(\Delta_{\mathbf{P}^r}\big) \ = \ \big\{ \, (x,y) \ \big| \ f(x) = f(y) \, \big\} \ \subseteq \ X \times X,$$

which by assumption contains points other than Δ_X. Moreover the dimensional hypotheses of the Fulton–Hansen theorem are satisfied, and so $F^{-1}(\Delta_{\mathbf{P}^r})$ is connected. Putting these facts together, we can find a sequence

$\{(x_k, y_k)\} \subseteq F^{-1}(\Delta_{\mathbf{P}^r})$ of double points of f, with $x_k \neq y_k$, having the property that

$$\lim_{k\to\infty}(x_k, y_k) = (x^*, x^*) \in \Delta_X, \quad \text{i.e.} \quad \lim_{k\to\infty} x_k = \lim_{k\to\infty} y_k = x^*$$

for some $x^* \in X$. Then one shows that f ramifies at x^*.

The formal proof is even quicker:

Proof of Theorem 3.4.1. Recall that a morphism $f : X \longrightarrow Y$ is unramified if and only if Δ_X is an open as well as a closed subscheme of

$$X \times_Y X = (f \times f)^{-1}(\Delta_Y)$$

(cf. [257, IV.17.4.2]). In particular, if f is unramified then Δ_X is a connected component of $X \times_Y X$. It follows from the Fulton–Hansen theorem that in the situation of 3.4.1, f is one-to-one. But a one-to-one proper unramified morphism is a closed embedding (cf. [280, III.7.3]). □

We discuss next a result of Gaffney and the author from [216] concerning the ramification of branched coverings of $\mathbf{P}^n$. Let $f : X \longrightarrow Y$ be a finite surjective mapping of irreducible varieties of dimension n, with Y smooth. Recall that for each $x \in X$ one can define the local degree $e_f(x)$ of f at x, which measures the number of sheets of the covering that come together at x. Specifically, set $y = f(x)$ and let $U = U(y)$ be a small ball (in the classical topology) about y. Then the inverse image of U splits as a disjoint union

$$f^{-1}\big(U(y)\big) = \coprod_{f(x')=y} V(x')$$

of connected neighborhoods of the inverse images of y.

Definition 3.4.7. (Local degree). The *local degree* $e_f(x)$ of f at $x \in X$ is the integer

$$e_f(x) = \deg\big(V(x) \longrightarrow U(y)\big). \quad \square$$

Thus $e_f(x)$ counts the number of preimages near x of a general point $y' \in Y$ near $y = f(x)$.

When X is also smooth, one can calculate $e_f(x)$ as the codimension in $\mathcal{O}_x X$ of the inverse image of the maximal ideal $\mathfrak{m}_y$ of $y = f(x)$:

$$e_f(x) = \dim_{\mathbf{C}} \frac{\mathcal{O}_x X}{\mathfrak{m}_y \cdot \mathcal{O}_x X}. \tag{3.10}$$

For fixed $y \in Y$ one has the basic formula

$$\sum_{f(x)=y} e_f(x) = \deg f. \tag{3.11}$$

The local degrees also satisfy a useful additivity property. Namely, consider two sequences $\{x_k\}$, $\{z_k\}$ of points in X, with $f(x_k) = f(z_k)$ for all k. Assume that $x_k \neq z_k$ but that

$$\lim x_k \;=\; \lim z_k \;=\; x^* \in X.$$

If e and f are integers with $e_f(x_k) \geq e$ and $e_f(z_k) \geq f$ for all k, then

$$e_f(x^*) \;\geq\; e + f. \tag{3.12}$$

Finally we will be concerned with the *higher ramification loci*

$$R_\ell(f) \;=\; \big\{ x \in X \;\big|\; e_f(x) > \ell \big\}.$$

Thus $R_1(f)$ is just the ramification locus of f, and in general the R_ℓ are algebraic subsets of X. We refer to [216] for further details on the definition and properties of the local degree.

The main theorem of [216] asserts that for branched coverings of projective space, these ramification loci are non-empty whenever it is dimensionally reasonable to expect them to be so.

Theorem 3.4.8. (Ramification of coverings of projective space). *Let X be an irreducible projective variety of dimension n and*

$$f : X \longrightarrow \mathbf{P}^n$$

a branched covering of degree d. Then there exists at least one point $x \in X$ at which

$$e_f(x) \;\geq\; \min\big\{d\, ,\, n+1\big\}.$$

More generally, $\operatorname{codim}_X R_\ell(f) \leq \ell$ *provided that* $\ell \leq \min\{d-1, n\}$.

The case $\ell = 1$ is just the assertion that $\mathbf{P}^n$ is (algebraically) simply connected.

As before, we start by presenting some examples and applications.

Example 3.4.9. If $f : X \longrightarrow \mathbf{P}^n$ is a branched covering of degree $d \leq n+1$, then there is a subvariety $T \subseteq \mathbf{P}^n$ of dimension $\geq n+1-d$ over which f has only one preimage. □

Corollary 3.4.10. (Simply connectedness of coverings of low degree). *Let X be an irreducible normal projective variety of dimension n that can be expressed as a branched covering $f : X \longrightarrow \mathbf{P}^n$ of degree $d \leq n$. Then X is algebraically simply connected.*

Proof. Suppose to the contrary that $p : Y \longrightarrow X$ is a connected étale covering of degree ≥ 2. Then Y is irreducible since X is normal, and we apply Theorem 3.4.8 to the composition

$$g = f \circ p \, : \, Y \longrightarrow \mathbf{P}^n.$$

The result in question guarantees the existence of a point $y \in Y$ at which at least $\min\{\deg g, n+1\}$ sheets of g come together. But

$$\min\left\{\deg\, g\,,\, n+1\right\} \;>\; d \;=\; \deg\, f$$

and so $e_g(y) \geq d+1$ for some $y \in Y$. On the other hand, $e_g\big(y\big) = e_f\big(p(y)\big)$ for every y since p is étale, and of course $e_f(x) \leq d$ for all $x \in X$: a contradiction. □

Example 3.4.11. The result quoted in Remark 3.3.9 implies that in the setting of 3.4.10, X is actually topologically simply connected. (In fact, let $f : X \longrightarrow \mathbf{P}^n$ be a normal covering of degree $d \leq n$. According to 3.4.9 there is a one-dimensional subset $T \subseteq \mathbf{P}^n$ over which f is one-to-one: let $T^* \longrightarrow T$ be its normalization. Then $T^* \times_{\mathbf{P}^n} X$ is homeomorphic to T^*. On the other hand, the homomorphism

$$\pi_1\big(T^* \times_{\mathbf{P}^n} X\big) \longrightarrow \pi_1\big(T^* \times X\big)$$

is surjective thanks to 3.3.9. It follows that $\pi_1(X) = \{1\}$.) □

Remark 3.4.12. Corollary 3.4.10 illustrates the principle that branched coverings of low degree share some of the properties of projective subvarieties of small codimension. In this spirit, a Barth-type theorem for coverings from [387] is presented in Section 7.1.C. □

Example 3.4.13. (A boundary example). Let $E \subseteq \mathbf{P}^n = \mathbf{P}$ be an elliptic normal curve of degree $n+1$. Denoting by $\mathbf{P}^* = \mathbf{P}^{n*}$ the dual projective space of hyperplanes in $\mathbf{P}$, consider the incidence correspondence

$$X \;=\; \Big\{\big(x\,,H\big) \mid x \in H\Big\} \;\subseteq\; E \times \mathbf{P}^*.$$

The second projection $f : X \longrightarrow \mathbf{P}^{n*}$ is a finite covering of degree $n+1$. On the other hand, the first projection realizes X as a projective bundle over E, and hence X is not simply connected. This shows that 3.4.10 is sharp. □

Proof of Theorem 3.4.8. We argue by induction on n, the statement being clear if $n = 1$. Assuming that $n \geq 2$, the inverse image $X' = f^{-1}(L)$ of a general hyperplane $L \subseteq \mathbf{P}^n$ is irreducible (Theorem 3.3.1), and so by induction the theorem is valid for the covering $f' : X' \longrightarrow L = \mathbf{P}^{n-1}$. If L is general then $e_{f'}(x) = e_f(x)$ for every $x \in X'$, and it follows that $\mathrm{codim}_X\big(R_\ell(f)\big) \leq \ell$ when $\ell \leq \min\{d-1, n-1\}$. It remains to show that R_n is non-empty if $d \geq n+1$.

To this end, pick an irreducible component S of R_{n-1} having dimension at least one, and apply the connectedness theorem to the mapping

$$F = f \times f|S : X \times S \longrightarrow \mathbf{P}^n \times \mathbf{P}^n.$$

Note that the diagonal $\Delta_S \subseteq S \times S$ embeds as an irreducible component of $F^{-1}(\Delta_{\mathbf{P}^N}) = X \times_{\mathbf{P}^n} S$. If $\Delta_S = F^{-1}(\Delta_{\mathbf{P}^N})$ then all the sheets of f come together along S, in which case $e_f(x) = d \geq n+1$ for every $x \in S$ thanks to (3.11). So we may suppose that $\Delta_S \subsetneqq F^{-1}(\Delta_{\mathbf{P}^n})$. Then Theorem 3.3.6 implies that there is a component $T \neq \Delta_S$ of $F^{-1}(\Delta_{\mathbf{P}^n})$ that meets Δ_S. So we may fix a sequence of points

$$\big(x_k, s_k\big) \in T \subseteq X \times S$$

with $x_k \neq s_k$ but $f(x_k) = f(s_k)$, converging to a point $(s^*, s^*) \in \Delta_S$. Thus $\lim x_k = \lim s_k = s^*$. Since $e_f(s_k) \geq n$ for all k, it follows from (3.12) that $e_f(s^*) \geq n+1$, as required. □

Remark 3.4.14. A variant is given in Example 6.3.56 below. Singularities of branched coverings of projective space also appear in the proof of Theorem 7.2.19, where they lead to lower bounds on the degrees of projective embeddings of varieties uniformized by bounded domains. □

Remark 3.4.15. (Codimensions of higher ramification loci). Grothendieck's connectedness theorem, in the form of Example 3.3.20, was used in [388] to prove an extension to higher ramification loci of the classical theorem on purity of the branch locus. Specifically, consider a branched covering $f : X \longrightarrow Y$ with X normal and Y non-singular. Then every irreducible component of the locus $R_\ell(f)$ has codimension $\leq \ell$ in X. Via a cone construction, this also leads to another proof of 3.4.8. □

Remark 3.4.16. (Weighted projective spaces). As we noted in Remark 3.3.11, Bădescu extended the Fulton–Hansen theorem to weighted projective spaces. He deduces that several of the results just presented extend to this weighted setting. We refer to Bădescu's forthcoming book [20] for an exposition of this and numerous related results. □

3.4.B Zak's Theorems

Zak ([619], [622]) found some interesting applications of the connectedness theorems to questions involving tangencies of projective varieties. Very surprisingly, he then used these results to prove Hartshorne's conjecture that smooth subvarieties of sufficiently small codimension in projective space are linearly normal. This subsection is devoted to a presentation of part of Zak's work.

We start with some notation and conventions. In the following we consider a smooth subvariety $X \subseteq \mathbf{P}^r$ of dimension n. Given a point $x \in X$, denote by $\mathbf{T}_x \subseteq \mathbf{P}^r$ the embedded tangent space to X at x: thus $\mathbf{T}_x$ is a projective space of dimension n passing through x. We say that X is tangent to a linear space $L \subseteq \mathbf{P}^r$ at $x \in X$ if $\mathbf{T}_x \subseteq L$. If $Y \subseteq \mathbf{P}^r$ is a subvariety and $y \in Y$ is a smooth point, we write $\mathbf{T}_y Y$ for the embedded tangent space to Y at y.

Zak's first basic result bounds the dimension of the locus along which X is tangent to a given linear space:

Theorem 3.4.17. (Zak's theorem on tangencies). *Let $X \subseteq \mathbf{P}^r$ be a smooth irreducible variety of dimension n, and assume that X is non-degenerate, i.e. not contained in any hyperplanes. Fix a linear space $L \subseteq \mathbf{P}^r$ of dimension k with $n \leq k \leq r-1$. Then the set*

$$\{ x \in X \mid \mathbf{T}_x \subseteq L \}$$

has dimension $\leq k - n$.

Zak's theorem has several surprising corollaries:

Corollary 3.4.18. (Finiteness of Gauss mappings). *In the setting of the theorem, let $\mathbf{G}$ denote the Grassmannian parameterizing linear subspaces $L \subseteq \mathbf{P}^r$ of dimension n, and consider the Gauss mapping*

$$\gamma : X \longrightarrow \mathbf{G} \quad , \quad x \mapsto \mathbf{T}_x.$$

Then γ is a finite morphism. □

Corollary 3.4.19. (Singularities of arbitrary hyperplane sections). *Still in the setting of Theorem 3.4.17, let $H \subseteq \mathbf{P}^r$ be any hyperplane. Then*

$$\dim \operatorname{Sing}\big(X \cap H\big) \ \leq \ r - 1 - n.$$

In particular, if $2n \geq r+1$ then $X \cap H$ is reduced, and if $2n \geq r+2$ then $X \cap H$ is normal.

Proof. In fact, $\operatorname{Sing}\big(X \cap H\big)$ is exactly the set of points at which X is tangent to H. □

Corollary 3.4.20. (Dimension of dual variety). *With $X \subseteq \mathbf{P}^r$ as above, let $X^* \subseteq \mathbf{P}^* = \mathbf{P}^{r*}$ denote the dual variety of X, i.e. the set of hyperplanes tangent to X at some point. Then* $\dim X^* \geq n$.

Proof. Consider the incidence correspondence

$$P \ = \ \big\{ (x, H) \mid \mathbf{T}_x \subseteq H \big\} \ \subseteq \ X \times \mathbf{P}^*.$$

The first projection realizes P as a $\mathbf{P}^{r-n-1}$-bundle over X, and hence $\dim P = r - 1$. By Zak's theorem the fibres of the second projection $P \longrightarrow X^*$ have dimension $\leq r - 1 - n$, and the assertion follows. □

Remark 3.4.21. (Tangents to complete intersections). It was observed in [211, Remark 7.5] that if X is a smooth non-degenerate complete intersection (of arbitrary codimension), then in fact a hyperplane can be tangent to X at only finitely many points: see Examples 6.3.6 and 6.3.8. □

Following Fulton's presentation in [203], we will deduce Zak's theorem from a generalization of Corollary 3.4.5. Specifically, with $X \subseteq \mathbf{P}^r$ as above, consider any closed irreducible subvariety $V \subseteq X$ of dimension v. Define

$$\mathrm{Sec}_V(X) =_{\mathrm{def}} \mathrm{closure}\left\{ p \in \mathbf{P}^r \;\middle|\; \begin{array}{c} p \in \text{ secant line } \overline{xv} \text{ joining} \\ x \in X\,,\ v \in V \end{array} \right\},$$

$$\mathrm{Tan}_V(X) =_{\mathrm{def}} \bigcup_{x \in V} \mathbf{T}_x,$$

where as above $\mathbf{T}_x$ denotes the embedded tangent space to X at x. These are closed irreducible subvarieties, with $\mathrm{Tan}_V(X) \subseteq \mathrm{Sec}_V(X)$, having dimensions at most $n + v + 1$ and $n + v$ respectively.

Proposition 3.4.22. *Either*

$$\dim \mathrm{Sec}_V(X) \;=\; n + v + 1 \quad \textit{and} \quad \dim \mathrm{Tan}_V(X) \;=\; n + v,$$

or else $\mathrm{Sec}_V(X) = \mathrm{Tan}_V(X)$.

Proof. Suppose to the contrary that $\dim \mathrm{Tan}_V(X) = t < n + v$ but that $\mathrm{Tan}_V(X) \neq \mathrm{Sec}_V(X)$. A generic linear space $L \subseteq \mathbf{P}^r$ of codimension $t + 1$ will then meet $\mathrm{Sec}_V(X)$ but not $\mathrm{Tan}_V(X)$. Linear projection from L defines a finite mapping $X \longrightarrow \mathbf{P}^t$, and the Fulton–Hansen theorem applies to the resulting morphism

$$V \times X \longrightarrow \mathbf{P}^t \times \mathbf{P}^t.$$

One deduces in the familiar fashion the existence of a sequence of distinct points $\big\{(v_k, x_k)\big\} \subseteq V \times X$ converging to $(v^*, v^*) \in \Delta_V$, having the property that the lines $\overline{v_k x_k}$ meet L. Then the limiting line

$$\ell^* \;=\; \lim_{k \to \infty} \overline{v_k x_k}$$

also meets L. On the other hand, ℓ^* — being a limit of secant lines — is a tangent line to X at v^*. But this is impossible since we assumed that L did not meet $\mathrm{Tan}_V(X)$. □

Zak's theorem follows at once:

Proof of Theorem 3.4.17. Suppose to the contrary that there is an irreducible set

$$V \;\subseteq\; \big\{ x \in X \;\big|\; \mathbf{T}_x \subseteq L \big\}$$

having dimension $\geq k + 1 - n$. Since evidently $\mathrm{Tan}_V(X) \subseteq L$, and $\dim L = k$, we find that $\mathrm{Tan}_V(X)$ has less than the expected dimension $n + \dim V \geq k + 1$. Therefore $\mathrm{Sec}_V(X) = \mathrm{Tan}_V(X)$ thanks to the preceding proposition. But then

$$X \;\subseteq\; \mathrm{Sec}_V(X) \;=\; \mathrm{Tan}_V(X) \;\subseteq\; L,$$

contradicting the non-degeneracy of X. □

Remark 3.4.23. (Join varieties). Flenner, O'Carroll, and Vogel [186] prove a statement that generalizes Proposition 3.4.22. Given irreducible subvarieties $X, Y \subseteq \mathbf{P}^r$ of dimensions n and m, denote by $X * Y \subseteq \mathbf{P}^r$ the *join* of X and Y, i.e. the closure of the union of all lines joining points of X with points of Y. Then define

$$\mathrm{LJoin}(X, Y) \ \subseteq \ \mathbf{P}^r$$

to be all points lying on a limit of lines $\overline{x_k y_k}$ as $x_k \in X$ and $y_k \in Y$ both approach a point $z \in X \cap Y$.[5] So for example if $Y = V$ in the setting of Proposition 3.4.22, then $X * Y = \mathrm{Sec}_V(X)$ and $\mathrm{LJoin}(X, Y) = \mathrm{Tan}_V(X)$. It is established in [186, Theorem 4.3.12] that either

$$\dim X * Y \ = \ n + m + 1 \quad \text{and} \quad \dim \mathrm{LJoin}(X, Y) \ = \ n + m,$$

or else $\mathrm{LJoin}(X, Y) = X * Y$. From the present point of view, the proof of 3.4.22 applies with little change. The cited authors also establish a local version of Zak's theorem on tangencies: see [186, Theorem 4.2.1]. □

Remark 3.4.24. (Degenerate secant and dual varieties). There is a substantial body of literature concerning varieties whose tangent, secant, or dual varieties exhibit exceptional behavior. The modern starting point is arguably the paper [249] of Griffiths and Harris, which relates the degeneracy of the varieties in question to geometric properties of the second fundamental form. This viewpoint has been systematically developed and applied by Landsberg: his notes [379] give a nice overview. In [142] and [141], Ein used vector bundle methods to study smooth projective varieties $X \subseteq \mathbf{P}^r$ whose dual varieties have smaller than expected dimensions. Among many other things, he recovers an unpublished result of Landman that if X^* has positive "defect"

$$\delta \ =_{\mathrm{def}} \ (r - 1) \ - \ \dim X^* \ > \ 0,$$

then $\delta \equiv \dim X \pmod 2$. The monograph [570] of Tevelev contains a great deal of information about dual varieties and their properties. □

Zak's most spectacular application of his theorem on tangencies was the proof of Hartshorne's conjecture on linear normality. Recall from Section 3.2.B the statement:

Theorem 3.4.25. (Zak's theorem on linear normality, I). *Let $X \subseteq \mathbf{P}^r$ be a smooth projective variety of dimension n. If $3n > 2(r - 1)$, then X is linearly normal, i.e. the natural mapping*

$$H^0\big(\mathbf{P}^r, \mathcal{O}_{\mathbf{P}^r}(1)\big) \longrightarrow H^0\big(X, \mathcal{O}_X(1)\big)$$

is surjective.

[5] Their notation "LJoin" derives from "limits of joins." See [186, 2.5 and Chapter 4] for a detailed account.

Observe that $X \subseteq \mathbf{P}^r$ *fails* to be linearly normal if and only if it is the linear projection of a non-degenerate embedding $X \subseteq \mathbf{P}^{r+1}$. Such a projection in turn exists if and only if the secant variety of X is a proper subvariety of $\mathbf{P}^{r+1}$. Therefore, setting $m = r + 1$, Theorem 3.4.25 is equivalent to

Theorem 3.4.26. (Zak's theorem on linear normality, II). *Let $X \subseteq \mathbf{P}^m$ be a smooth non-degenerate variety of dimension n. If $3n > 2(m-2)$ then*

$$\mathrm{Sec}(X) \;=\; \mathbf{P}^m.$$

We refer to [619] or [211] for an account of Zak's proof via Theorem 3.4.17. We will reproduce an alternative argument from [211] that deduces the result directly from the connectedness theorem for three factors.

Remark 3.4.27. (Classification of Severi varieties). Zak defines a *Severi variety* to be a smooth variety lying on the boundary of Theorem 3.4.26. In other words, a Severi variety is a smooth n-fold $X \subseteq \mathbf{P}^m$ with $3n = 2(m-2)$ having the property that $\mathrm{Sec}(X) \neq \mathbf{P}^m$. Zak [621] proved the remarkable theorem that there are only four such:

- The Veronese surface $V \subseteq \mathbf{P}^5$;
- The Segre variety $S = \mathbf{P}^2 \times \mathbf{P}^2 \subseteq \mathbf{P}^8$;
- The Grassmannian $G = \mathbf{G}(\mathbf{P}^1, \mathbf{P}^5) \subseteq \mathbf{P}^{14}$;
- The 16-dimensional E_6-variety $E \subseteq \mathbf{P}^{26}$, arising as the orbit of a highest weight vector for an irreducible representation of a simple group of type E_6.

The first three are classical,[6] and the fourth was pointed out to Zak by the author. A detailed account of Zak's argument appears in [395]. Several other approaches have been developed since [621] and [395]. Landsberg [378] gives a proof revolving around a careful analysis of the second fundamental form of X. More recently, Chaput [85] found a very quick argument that begins by showing that every Severi variety X is homogeneous: the key idea here is to exploit the fact that $\mathrm{Sec}(X)$ is a cubic hypersurface. The paper [19] of Atiyah and Berndt gives another view of these four fascinating varieties. □

Turning to the proof of Theorem 3.4.26, we start with some remarks on tri-secant varieties. Consider as above a smooth variety $X \subseteq \mathbf{P}^m$ of dimension n. Denote by $\mathrm{Trisec}(X) = \mathrm{Sec}^3(X)$ the variety swept out by all tri-secant two-planes to X:

$$\mathrm{Trisec}(X) \;=\; \mathrm{closure}\Big(\bigcup_{x,y,z \in X} \overline{xyz} \Big),$$

the union being taken over all triples of distinct non-collinear points in X. This is a closed irreducible subvariety of expected dimension $3n+2$ in $\mathbf{P}^m$.

[6] Hartshorne was led to the inequality in his conjecture on the basis of these three classical examples.

The first point is a classical lemma describing the tangent spaces to $\mathrm{Sec}(X)$ and $\mathrm{Trisec}(X)$:

Lemma 3.4.28. (Terracini's lemma). *Let $X \subseteq \mathbf{P}^m$ be a smooth irreducible variety.*

(i). *Let $x, y \in X$ be distinct points of X, and let p be a point on the line $\overline{xy}$. Then*

$$\mathrm{Span}\big(\mathbf{T}_x, \mathbf{T}_y\big) \subseteq \mathbf{T}_p \, \mathrm{Sec}(X), \tag{*}$$

and for general $x, y \in X$ and $p \in \overline{xy}$, equality holds in $(*)$.

(ii). *Similarly, let $x, y, z \in X$ be distinct non-collinear points of X, and let q be a point on the plane $\overline{xyz}$. Then*

$$\mathrm{Span}\big(\mathbf{T}_x, \mathbf{T}_y, \mathbf{T}_z\big) \subseteq \mathbf{T}_q \, \mathrm{Trisec}(X),$$

and again equality holds for general $x, y, z \in X$ and $q \in \overline{xyz}$.

Indication of proof of 3.4.28. It is enough to treat the affine situation in which $X \subseteq \mathbf{C}^m$, and we focus on (i). Fix open subsets $U, V \subseteq \mathbf{C}^n$ and local parameterizations

$$\phi : U \longrightarrow X \quad , \quad \psi : V \longrightarrow X$$

in neighborhoods of the given points. Then

$$f : U \times V \times \mathbf{C} \longrightarrow \mathrm{Sec}(X) \quad , \quad f \;=\; t\phi + (1-t)\psi$$

gives a local parameterization of Sec(X). The lemma follows by computing the derivative df, and using the fact that f is generically submersive. □

Corollary 3.4.29. *If $X \subseteq \mathbf{P}^m$ is a smooth projective variety that is not contained in any hyperplane, then $\mathrm{Sec}(X) = \mathbf{P}^m$ if and only if $\mathrm{Sec}(X) = \mathrm{Trisec}(X)$.*

Proof. Since in any event $\mathrm{Sec}(X) \subseteq \mathrm{Trisec}(X)$, it is enough to show that if $\mathrm{Sec}(X) \neq \mathbf{P}^m$, then $\mathrm{Sec}(X) \neq \mathrm{Trisec}(X)$. But this follows from an infinitesimal calculation using Terracini's lemma. Specifically, if $x, y \in X$ are general points, then $\mathbf{T}_x$ and $\mathbf{T}_y$ span a proper subspace $\Lambda \subseteq \mathbf{P}^m$ (namely $\Lambda = \mathbf{T}_p\mathrm{Sec}(X)$ for general $p \in \overline{xy}$). Since $X \subseteq \mathbf{P}^m$ is non-degenerate, we may choose a point $z \in X - \Lambda$, and then

$$\mathrm{Span}\big(\mathbf{T}_x, \mathbf{T}_y, \mathbf{T}_z\big) \supsetneqq \mathrm{Span}\big(\mathbf{T}_x, \mathbf{T}_y\big).$$

But by 3.4.28 (ii), this means that $\dim \mathrm{Trisec}(X) > \dim \mathrm{Sec}(X)$. □

Theorem 3.4.26 now follows from

Proposition 3.4.30. *Let $X \subseteq \mathbf{P}^m$ be a smooth projective variety of dimension n. If $3n > 2(m-2)$, then $\mathrm{Sec}(X) = \mathrm{Trisec}(X)$.*

Proof. We reproduce the presentation in [211, §7]. Suppose to the contrary that $\mathrm{Sec}(X) \neq \mathrm{Trisec}(X)$, so that we may choose distinct non-collinear points $x_0, y_0, z_0 \in X$ such that the trisecant plane $\overline{x_0 y_0 z_0}$ they span is not contained in $\mathrm{Sec}(X)$. A generic line $L \subseteq \overline{x_0 y_0 z_0}$ is then disjoint from X, and meets $\mathrm{Sec}(X)$ at only finitely many points. Fix such a line L, and consider the finite mapping $\pi : X \longrightarrow \mathbf{P}^{m-2}$ obtained by projection from L to a complementary $\mathbf{P}^{m-2}$.

Since $3n > 2(m-2)$ the connectedness theorem for three factors (Example 3.3.8) applies to the map

$$F = \pi \times \pi \times \pi : X \times X \times X \longrightarrow \mathbf{P}^{m-2} \times \mathbf{P}^{m-2} \times \mathbf{P}^{m-2}.$$

It implies that any irreducible component W of $F^{-1}(\Delta_{\mathbf{P}^{m-2}})$ sits in a sequence $W = W_0, W_1, \ldots, W_s$ of components such that W_i meets W_{i+1}, and W_s meets the union $D_X \subseteq X \times X \times X$ of the pairwise diagonals. We assume for simplicity that (x_0, y_0, z_0) lies on a component W that already meets D_X: we leave it to the reader to carry out at the end the slight additional argument needed to handle the general case. Given this assumption, we can find a family of triple points

$$\Big\{ (\, x_t \,,\, y_t \,,\, z_t \,) \Big\}_{t \in T} \ \subseteq \ X \times_{\mathbf{P}^{m-2}} X \times_{\mathbf{P}^{m-2}} X,$$

parameterized by a smooth irreducible curve T, containing (x_0, y_0, z_0), such that x_t, y_t, z_t are distinct for $t \in T - \{t^*\}$, while two or more members of the limiting triple

$$(x^*, y^*, z^*) \ =_{\text{def}} \ (x_{t^*}, y_{t^*}, z_{t^*})$$

coincide.

Now since x_0, y_0, z_0 are non-collinear, and since $L \cap \mathrm{Sec}(X)$ is finite, the points of intersection

$$a \ = \ \overline{x_t y_t} \cap L \ \ , \ \ b \ = \ \overline{x_t z_t} \cap L \ \ , \ \ c \ = \ \overline{y_t z_t} \cap L$$

are distinct and independent of t so long as $t \neq t^*$ (Figure 3.2). Hence if ℓ^*_{xy}, ℓ^*_{xz}, and ℓ^*_{yz} denote the limits of the corresponding secants as $t \to t^*$, then

$$\ell^*_{xy} \cap L \ = \ a \ \ , \ \ \ell^*_{xz} \cap L \ = \ b \ \ , \ \ \ell^*_{yz} \cap L \ = \ c.$$

In particular, these lines are distinct. But then all three of the limiting points x^*, y^*, z^* must coincide: for if e.g. $y^* = z^* \neq x^*$, then the secants $\overline{x_t y_t}$ and $\overline{x_t z_t}$ would degenerate as $t \to t^*$ to a common line. On the other hand, if $(x_t, y_t, z_t) \to (x^*, x^*, x^*)$ as $t \to t^*$, then $\ell^*_{xy}, \ell^*_{xz}, \ell^*_{yz}$ are tangent lines to X at x^*. In particular, the center of projection L meets the tangent space $\mathbf{T}_{x^*}$ in more than one point. But in this case $L \subseteq \mathbf{T}_{x^*} \subseteq \mathrm{Sec}(X)$, contradicting the choice of L. □

Remark 3.4.31. Faltings [177] and Peternell–Le Potier–Schneider [502] have given proofs of Zak's theorem having a more cohomological flavor. □

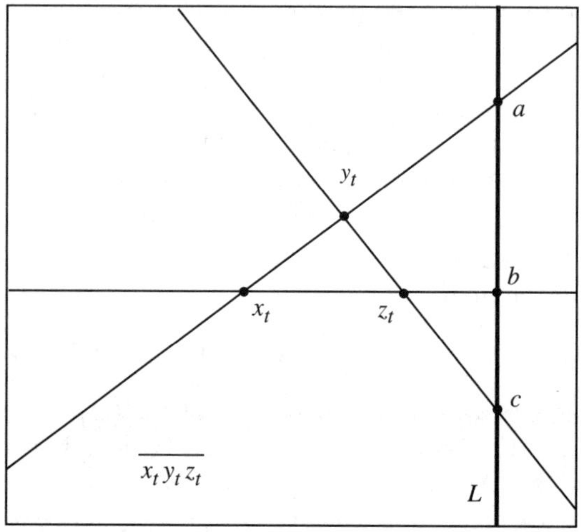

Figure 3.2. Proof of Proposition 3.4.30

Remark 3.4.32. (Superadditivity of join defects). An alternative approach to some of these questions grows out of work of Zak [620] on superadditivity of secant defects. Given irreducible varieties $X, Y \subseteq \mathbf{P}^m$, denote by $X * Y \subseteq \mathbf{P}^r$ the join variety of X and Y (Remark 3.4.23), so that for instance $X * X = \mathrm{Sec}(X)$. Assuming $X * Y \neq \mathbf{P}^m$, consider the *join defect*

$$\delta(X, Y) \;=\; \dim X + \dim Y + 1 - \dim X * Y.$$

Thus $\delta(X, Y)$ measures the amount by which $X * Y$ has less than the expected dimension. Write also

$$\mathrm{Sec}^k(X) \;=\; X * \mathrm{Sec}^{k-1}(X) \;=\; X * \ldots * X \quad (k \text{ times})$$

for the variety of k-secant $(k-1)$-planes to X. Zak proves that if $X \subseteq \mathbf{P}^m$ is a non-degenerate smooth subvariety with $\mathrm{Sec}^{k+1}(X) \neq \mathbf{P}^m$, then

$$\delta\big(\mathrm{Sec}^{k+1}(X)\,,\, X\big) \;\geq\; \delta\big(\mathrm{Sec}^k(X)\,,\, X\big) + \delta\big(\,X\,,\,X\,\big).$$

A more general result was obtained by Flenner and Vogel [187, Theorem 3.1] (see also [186, Chapter 4, §5]). Specifically, consider irreducible varieties $X, Y, Z \subseteq \mathbf{P}^m$ with $Z \subseteq Y_{\mathrm{reg}}$ non-degenerate. If $X * Y \neq \mathbf{P}^m$, then

$$\delta\big(\,X * Y\,,\,Z\,\big) \;\geq\; \delta\big(\,X\,,\,Z\,\big) + \delta\big(\,Y\,,\,Z\,\big), \tag{3.13}$$

$$\delta\big(\,X\,,\,Y\,\big) + \delta\big(\,Y\,,\,Z\,\big) \;\leq\; \dim Y. \tag{3.14}$$

Applying (3.14) with $X = Y = Z$, one finds that if $X \subseteq \mathbf{P}^m$ is a smooth non-degenerate variety whose secant variety $\mathrm{Sec}(X) = X * X$ is a proper subvariety of $\mathbf{P}^m$, then

$$2 \cdot \delta(X, X) \leq \dim X, \quad \text{i.e.} \quad 2 \cdot \dim \mathrm{Sec}(X) \geq 3 \dim X + 2.$$

The last statement is equivalent to the theorem on linear normality. □

3.4.C Zariski's Problem

One of the most striking applications of the connectedness theorem was discovered by Fulton [202], who used it to solve an old problem of Zariski concerning the fundamental group of the complement of a nodal curve in the plane. Zariski had proposed a proof to show that this fundamental group is abelian based on an assertion of Severi to the effect that the space of nodal plane curves of given degree and genus is connected. However, the proof of Severi's assertion was later found to have gaps. The question was studied from an algebraic point of view by Abhyanhkar [1] and Serre [531], who developed a line of attack that worked when every component of C is smooth. Fulton realized that one could use Theorem 3.3.6 to complete their argument in general.

Fulton's result is the following:

Theorem 3.4.33. (Coverings of the complement of a nodal curve). *Let $C \subseteq \mathbf{P}^2$ be a possibly reducible curve whose only singularities are simple nodal double points, and let $f : X \longrightarrow \mathbf{P}^2$ be a finite cover branched only over C. Then f is abelian, i.e. f is Galois and $\mathrm{Gal}(X/\mathbf{P}^2)$ is an abelian group.*

The theorem asserts that the algebraic fundamental group of $\mathbf{P}^2 - C$ is abelian. By extending Fulton's argument, Deligne [118] established that the topological fundamental group of $\mathbf{P}^2 - C$ is abelian. Slightly later, Harris [272] closed the circle by proving the assertion of Severi. We refer to [203, Chapter V, §4] for a historical overview, and an account of some subsequent developments. Here (following [211]) we run quickly through Fulton's argument.

We start with some general observations. Let Y be a smooth projective surface, and $C \subset Y$ a (possibly reducible) curve whose only singularities are simple double points. Consider a finite cover

$$f : X \longrightarrow Y$$

branched only over C. We assume that X is normal, and that f is Galois, with group $G = \mathrm{Gal}(X/Y)$. As usual, one says that f is *abelian* if G is. Given $x \in X$ we write $G_x \subseteq G$ for the subgroup of automorphisms fixing x.

The simple nature of C allows one to understand quite concretely the local structure of f. Specifically, fix points $y \in C$ and $x \in f^{-1}(y)$. There is a small classical neighborhood $U = U_y$ of y in Y having the property that

$$(U, U \cap C) \cong (B, B \cap \Gamma)$$

where B is a ball about the origin in $\mathbf{C}^2$, and Γ is the curve given in coordinates s, t by

$$\Gamma = \begin{cases} (s = 0) & \text{if } y \text{ is a smooth point of } C, \\ (st = 0) & \text{if } y \text{ is a node of } C. \end{cases}$$

Denote by $V = V_x$ the connected component of $f^{-1}(U)$ that contains x, and set

$$U^0 = U^0_y = U - (U \cap C) \ , \ V^0 = V^0_x = V - (V \cap f^{-1}(C)),$$
$$f_x = f \mid V : V \longrightarrow U \ , \ f^0 = f^0_x = f \mid V^0 : V^0 \longrightarrow U^0.$$

Thus f^0 is a connected covering space. Note that the subgroup G_x of automorphisms fixing x is naturally embedded in the group $\mathrm{Aut}(f^0)$ of deck transformations of f^0, which is in turn a quotient of $\pi_1(U^0)$.

Via the identification $U^0 = B - (B \cap \Gamma)$, the fundamental group and finite coverings of U^0 are readily described. Specifically, $\pi_1(U^0) = \mathbf{Z}$ when y is a smooth point of C, and $\pi_1(U^0) = \mathbf{Z} \oplus \mathbf{Z}$ when $y \in C$ is a node. Moreover, with respect to suitable analytic local coordinates z, w centered at x, f_x may be identified with the covering

$$(z\,,\,w) \mapsto (z^d\,,\,w) \tag{3.15}$$

when y is a smooth point of C. When $y \in C$ is a singular point, f_x is dominated by a covering

$$(z\,,\,w) \mapsto (z^d\,,\,w^e) \tag{3.16}$$

for some $d, e > 0$.[7]

Now let $D \subseteq X$ be an irreducible component of $f^{-1}(C)$. The *inertia subgroup*

$$I(D) \subseteq G = \mathrm{Gal}(X/Y)$$

of D is defined to be the set of all automorphisms fixing D pointwise:

$$I(D) = \big\{ \sigma \in G \bigm| \sigma | D = \mathrm{id}_D. \big\}.$$

Note that if $D, D' \subseteq f^{-1}(C)$ are two components having the same image, then the corresponding inertia groups $I(D), I(D')$ are conjugate in G.

We now use the local analysis to make a number of observations.

(i). $I(D)$ is a cyclic group.

In fact, $I(D)$ is determined by the geometry of f at a general point $x \in D$ where the normal form (3.15) applies.

---

[7] The covering (3.16) corresponds to the subgroup

$$d\mathbf{Z} \oplus e\mathbf{Z} \subseteq \mathbf{Z} \oplus \mathbf{Z} = \pi_1(U^0).$$

(ii). Suppose that two components $D, D' \subseteq f^{-1}(C)$ meet at a point $x \in X$. Then $I(D)$ and $I(D')$ commute.

Indeed, $I(D)$ and $I(D')$ are embedded in the stabilizer G_x of x, which in turn embeds in the abelian group $\mathrm{Aut}(f^0)$.

(iii). Let $N \subseteq \mathrm{Gal}(X/Y)$ be the (necessarily normal) subgroup generated by the inertia groups $I(D)$ as D ranges over all irreducible components of $f^{-1}(C)$. Then the corresponding covering $h : X/N \longrightarrow Y$ is étale.

In fact, all the inertia groups of h are trivial, so it is unramified in codimension one. Since X/N is normal and Y is smooth, the theorem on the purity of the branch locus implies that h is everywhere étale.

(iv). Let $E' \longrightarrow E$ be the normalization of an irreducible component E of C. Then $X \times_Y E'$ is locally unibranch, i.e. has only one branch at any point.

This can be checked using the local normal forms (3.15) and (3.16).

We summarize this discussion in a lemma.

Lemma 3.4.34. *Given $f : X \longrightarrow Y$ as above, assume that any two irreducible components of $f^{-1}(C)$ meet. If Y is algebraically simply connected, i.e. if it admits no non-trivial connected étale covers, then f is abelian.*

Proof. By (i), each of the inertia groups $I(D)$ is abelian. Thanks to (ii), the hypothesis on the components of $f^{-1}(C)$ implies that the group N that they generate is abelian. But it follows from (iii) and the simple connectivity of Y that $N = G$. □

Fulton's theorem now follows immediately:

Proof of Theorem 3.4.33. By passing to a Galois closure, there is no loss in generality in assuming that the given cover f is Galois. It is enough to prove that $f^{-1}(E)$ is irreducible for every irreducible component E of C. For since any two components of C meet, it will follow that any two components of $f^{-1}(C)$ likewise meet, and then Lemma 3.4.34 applies. Let $E' \longrightarrow E$ be the normalization of E. By (iv) above, $E' \times_{\mathbf{P}^2} X$ is unibranch, and by the Fulton–Hansen theorem it is connected. Therefore $E' \times_{\mathbf{P}^2} X$ is irreducible, and hence so too is its image $f^{-1}(E) = E \times_{\mathbf{P}^2} X$. □

Nori [481] developed a different approach to these questions that gives information for coverings of surfaces other than $\mathbf{P}^2$. We outline here the simplest of his results, namely:

Theorem 3.4.35. (Nori's theorem). *Consider a Galois covering $f : X \longrightarrow Y$ as above branched only over a nodal curve $C \subseteq Y$. Suppose that every irreducible component $E \subseteq C$ of the branch curve satisfies the inequality*

$$(E^2) \ > \ 2 \cdot r(E),$$

where $r(E)$ denotes the number of nodes on E. Then any two irreducible components of $f^{-1}(C)$ meet. Therefore the subgroup N generated by the inertia groups is abelian, and if Y is algebraically simply connected then f is also abelian.

This applies in particular when $Y = \mathbf{P}^2$ since for any irreducible nodal curve E of degree $e \geq 1$ the genus formula gives

$$e^2 - 2 \ > \ e(e-3) \ = \ 2p_g(E) - 2 + 2r(E) \ \geq \ 2r(E) - 2.$$

Sketch of Proof of Theorem 3.4.35. Let $E \subseteq X$ be an irreducible component of C, and let $D \subseteq Y$ be an irreducible component of $f^{-1}(E)$. Denote by $G_D \subseteq G = \mathrm{Gal}(X/Y)$ the stabilizer of D, consisting of automorphisms mapping D to itself. Let $V = Y/G_D$, so that f factors as the composition

$$X \xrightarrow{\ \lambda\ } V \xrightarrow{\ \phi\ } Y.$$

Set $F = \lambda(D) \subseteq V$. The first point to check is that ϕ is everywhere étale along F, and that F maps birationally to its image E in X. Given an arbitrary point $x \in D$, it is enough for this to establish the containment

$$G_x \ \subseteq \ G_D \tag{*}$$

of stabilizer subgroups: since $X/G_x \longrightarrow Y$ is étale over the image of x, (*) implies that ϕ is étale at $\lambda(x)$. As for (*), it can be verified using the local analysis of f presented above.

By what was just proven, V is smooth along F and in particular F is a Cartier divisor on V. We assert next that

$$(F^2) \ > \ 0. \tag{**}$$

This is verified by observing that the map $F \longrightarrow E$ is an isomorphism outside of a certain number — say s — pairs of smooth points on F that map to nodes of E. Then by comparing the normal bundles of F and E one finds that

$$(F^2)_V \ = \ (E^2)_Y \ - \ 2s,$$

where as usual, subscripts indicate on what surface the intersection products are computed. But of course $s \leq r(E)$, so the expression on the right is positive thanks to the numerical hypothesis of the theorem.

By construction, $\lambda^* F = aD$ for some $a > 0$. So it follows from (**) that $(D^2) > 0$ on X. But as this holds for any irreducible component $D \subseteq f^{-1}(C)$, the Hodge index theorem implies that

$$\left(D_1 \cdot D_2\right)^2 \ \geq \ (D_1^2) \cdot (D_2^2) \ > \ 0,$$

for any two components D_1 and D_2 of $f^{-1}(C)$. In particular, D_1 and D_2 meet. □

3.5 Variants and Extensions

In this final section we summarize some extensions and variants of the Fulton–Hansen theorem.

3.5.A Connectedness Theorems for Homogeneous Varieties

A first line of work has been to extend Theorem 3.3.6 to products of varieties other than projective space. Most of the existing results deal with homogeneous target spaces, and we survey some of these without proof.

Suppose to begin with that $\mathbf{F} = G/P$ is a rational homogeneous space, where G is a semi-simple linear algebraic group over $\mathbf{C}$ and $P \subseteq G$ is a parabolic subgroup. Denote by ℓ the minimum rank of the semi-simple factors of G. Faltings [176] proves:

Theorem 3.5.1. (Rational homogeneous spaces). *Let X be an irreducible complete variety and $f : X \longrightarrow \mathbf{F} \times \mathbf{F}$ a morphism. Assume that*

$$\operatorname{codim}_{\mathbf{F}\times\mathbf{F}} f(X) \ < \ \ell. \tag{*}$$

Then the inverse image $f^{-1}(\Delta)$ of the diagonal $\Delta \subseteq \mathbf{F} \times \mathbf{F}$ is connected. □

If $\mathbf{F} = \mathbf{P}^r$ — with $G = \mathrm{SL}_{r+1}(\mathbf{C})$ — then $\ell = r$ so one recovers the theorem of Fulton and Hansen. In general however $\dim \mathbf{F}$ is much larger than ℓ, in which case the dimensional hypotheses are stronger than in 3.3.6. Hansen [269] had previously established the theorem when $\mathbf{F}$ is a Grassmannian, and he gave examples moreover to show that the hypothesis (*) cannot be relaxed in this case.

On the other hand, Debarre [111] shows that one can weaken (*) under additional geometric hypotheses. Specifically, consider a product

$$\mathbf{P} \ = \ \mathbf{P}^{r_1} \times \ldots \times \mathbf{P}^{r_t}$$

of t projective spaces of various dimensions. For each non-empty subset $I \subseteq \{1,\ldots,t\}$ denote by $\mathbf{P}_I = \prod_{i\in I} \mathbf{P}^{r_i}$ the product of the corresponding factors, with $p_I : \mathbf{P} \longrightarrow \mathbf{P}_I$ the projection, and write $r_I = \sum r_i$ for the dimension of $\mathbf{P}_I$. Debarre shows that one can modify the proof of 3.3.6 to obtain:

Theorem 3.5.2. (Multi-projective spaces). *Let X be an irreducible complete variety and $f : X \longrightarrow \mathbf{P} \times \mathbf{P}$ a morphism. Writing $f_I : X \longrightarrow \mathbf{P}_I \times \mathbf{P}_I$ for the composition $(p_I \times p_I) \circ f$, assume that*

$$\dim \big(f_I \times f_I\big)\big(X\big) \ > \ r_I$$

for every $I \subseteq [1, r]$. Then the inverse image $f^{-1}(\Delta)$ of the diagonal $\Delta \subseteq \mathbf{P}\times\mathbf{P}$ is connected. □

As a corollary, Debarre is able to exhibit many subvarieties of $\mathbf{P}$ that satisfy the same sort of connectedness properties that Theorem 3.3.6 and Example 3.3.8 express for the small diagonal $\Delta \subseteq \mathbf{P}^r \times \ldots \times \mathbf{P}^r$. For instance ([111, Proposition 2.4]):

Corollary 3.5.3. *Let $Y \subseteq \mathbf{P}$ be an irreducible subvariety having the property that*

$$\dim p_i(Y) \;=\; \dim Y$$

for every $i \in [1, t]$. If $f : X \longrightarrow \mathbf{P}$ is a morphism from a complete irreducible variety X such that

$$\dim\, f(X) \;>\; \operatorname{codim} Y,$$

then $f^{-1}(Y)$ is connected. □

Example 3.5.4. (Connectedness and Künneth components). Suppose that $Y \subseteq \mathbf{P}^r \times \mathbf{P}^r$ is irreducible of dimension r. Then Y satisfies the hypothesis of the previous corollary if and only if each of the components of $[Y] \in H^{2r}\big(\mathbf{P}^r \times \mathbf{P}^r, \mathbf{Q}\big)$ under the Künneth decomposition

$$H^{2r}\big(\mathbf{P}^r \times \mathbf{P}^r, \mathbf{Q}\big) \;=\; \bigoplus_{j+k=r} H^{2j}\big(\mathbf{P}^r, \mathbf{Q}\big) \otimes H^{2k}\big(\mathbf{P}^r, \mathbf{Q}\big)$$

is non-zero. From this perspective, Theorem 3.3.6 reflects a numerical property of the diagonal. (Assume that each of the two projections $p_1, p_2 : Y \longrightarrow \mathbf{P}^r$ is surjective, and hence generically finite. Then the pullbacks of the hyperplane bundle give big and nef classes $h, \ell \in N^1(Y)$, and the Hodge index inequality 1.6.3 (ii) implies that $\big(h^j \cdot \ell^k\big) > 0$ whenever $j + k = r$. This in turn forces the non-vanishing of all the Künneth components of $[Y]$.) □

Remark 3.5.5. Debarre also proves analogous results for Grassmannians. □

In a quite different direction, Debarre [109], [110] has established analogues of the Fulton–Hansen theorem for abelian varieties. For ease of exposition we will state his result only in the case of simple abelian varieties.[8]

Theorem 3.5.6. (Simple abelian varieties). *Let A be a simple abelian variety, and let*

$$f : X \longrightarrow A \quad , \quad g : Y \longrightarrow A$$

be morphisms from irreducible normal projective varieties X and Y to A, with

$$\dim\, f(X) + \dim\, g(Y) \;>\; \dim\, A.$$

Then there is a finite étale covering $p : \tilde{A} \longrightarrow A$, together with factorizations

$$X \longrightarrow \tilde{A} \longrightarrow A \quad , \quad Y \longrightarrow \tilde{A} \longrightarrow A$$

of f and g, such that the fibre product $X \times_{\tilde{A}} Y$ is connected. □

[8] Recall that an abelian variety is *simple* if it contains no non-trivial proper abelian subvarieties.

Very roughly, the idea — which in a general way goes back to a proof of the Fulton–Hansen theorem given by Mumford [211, §3] — is to exploit the action of A by translation on the diagonal $\Delta \subseteq A \times A$. The statements in [109] and [110] do not assume that A is simple: instead Debarre imposes a non-degeneracy condition on f and g. See [112, Chapter VIII] for a nice account.

Debarre uses 3.5.6 to extend many of the applications of the Fulton–Hansen theorem to the setting of abelian varieties. We indicate two of these.

Example 3.5.7. (Unramified morphisms to abelian varieties). Let $f : X \longrightarrow A$ be an unramified morphism from a normal projective variety X to a simple abelian variety A. Assume that $2 \cdot \dim X > \dim A$. Then f factors as the composition

$$X \hookrightarrow \tilde{A} \longrightarrow A,$$

where $\tilde{A} \longrightarrow A$ is an isogeny of abelian varieties. (Argue as in the proof of 3.4.1: see [110, Theorem 5.1].) Note that compositions of the indicated type yield examples of unramified morphisms $X \longrightarrow A$ that are not one-to-one: this explains the presence of the covering p in Debarre's Theorem 3.5.6. □

Example 3.5.8. (Branched coverings of abelian varieties). Let A be a simple abelian variety of dimension n, and $f : X \longrightarrow A$ a branched covering of A by an irreducible normal projective variety X. Assume that f does not factor through any non-trivial isogenies $\tilde{A} \longrightarrow A$. Then there is at least one point $x \in X$ at which

$$e_f(x) \;\geq\; \min \big\{ \deg f \, , \, n + 1 \big\}.$$

(Argue as in 3.4.8, or see [110, Theorem 7.1].) Example 6.3.59 gives an additional positivity property of coverings of abelian varieties. □

3.5.B Higher Connectivity for Mappings to Projective Space

We noted in Remark 3.3.9 that the Fulton–Hansen theorem extends to a statement involving fundamental groups. Deligne conjectured a general result — since proven by Goresky–MacPherson [227, II.1.2, p. 153] — that implies statements for higher homotopy groups. Closely following [211, §9] we summarize some of these ideas here. While the proof of the theorem of Goresky–MacPherson lies far beyond the scope of this volume, we do reproduce from [211] the elementary formal arguments by which the result is applied in the present setting.

Convention 3.5.9. (Local complete intersection). Throughout this subsection, a *local complete intersection* is a connected but possibly reducible complex algebraic set that can locally be realized as a complete intersection in some smooth variety. □

The Goresky–MacPherson theorem can be seen as a non-compact analogue of the Lefschetz hyperplane theorem (Theorem 3.1.21) for varieties with at worst local complete intersection singularities:

Theorem 3.5.10. (Theorem of Goresky–MacPherson). *Let X be a local complete intersection of pure dimension n. Consider a finite-to-one morphism*

$$h : X \longrightarrow \mathbf{P}^r$$

of X to projective space, and let $L \subseteq \mathbf{P}^r$ be an arbitrary linear space of codimension d. Denote by L_ε an ε-neighborhood of L with respect to some real analytic Riemannian metric on $\mathbf{P}^r$. Then for sufficiently small $\varepsilon > 0$ one has

$$\pi_i\big(X \, , \, h^{-1}L_\varepsilon\big) \ = \ 0 \quad \text{for} \ \ i \leq n-d.$$

As indicated above, this — and more general statements also established in [227] — had been conjectured by Deligne.

Starting with a compact local complete intersection Z, together with a finite map $f : Z \longrightarrow \mathbf{P}^r \times \mathbf{P}^r$, Deligne showed that 3.5.10 easily leads to a statement for the higher homotopy groups of the pair $\big(Y, f^{-1}(\Delta_{\mathbf{P}^r})\big)$ ([211, Theorem 9.2]). However the non-vanishing of $\pi_2(\mathbf{P}^r)$ causes slight complications. We focus instead on a variant ([211, Theorem 9.6]) for mappings of the form $X \times Y \longrightarrow \mathbf{P}^r \times \mathbf{P}^r$ where the statements become a little cleaner.

We start with some notation. Denote by $\widehat{\mathbf{P}}^r$ the total space of the natural $\mathbf{C}^*$-bundle

$$\widehat{\mathbf{P}}^r \ = \ \mathbf{C}^{r+1} - \{0\} \longrightarrow \mathbf{P}^r.$$

If $f : X \longrightarrow \mathbf{P}^r$ is a morphism, we write

$$\widehat{X} \ = \ X \times_{\mathbf{P}^r} \widehat{\mathbf{P}}^r$$

for the pullback bundle, and $\widehat{f} : \widehat{X} \longrightarrow \widehat{\mathbf{P}}^r$ for the induced morphism. The essential technical result is the following consequence of the theorem of Goresky–MacPherson:

Lemma 3.5.11. *Let X and Y be compact local complete intersections of pure dimensions n and m respectively, and let*

$$f : X \longrightarrow \mathbf{P}^r \quad , \quad g : Y \longrightarrow \mathbf{P}^r$$

be finite (and hence proper) morphisms. Then

$$\pi_i\big(\widehat{X} \times \widehat{Y} \, , \, \widehat{X} \times_{\widehat{\mathbf{P}}^r} \widehat{Y}\big) \ = \ 0 \quad \text{for} \ \ i \leq n+m-r.$$

Sketch of Proof of Lemma 3.5.11. We return to the set-up of the proof of 3.3.6: let $V \subseteq \mathbf{P}^{2r+1}$ be the subset introduced there, and $L \subseteq V$ the linear r-plane mapping isomorphically to the diagonal $\Delta \subset \mathbf{P}^r \times \mathbf{P}^r$. Then there is a natural $\mathbf{C}^*$-bundle map $\widehat{\mathbf{P}}^r \times \widehat{\mathbf{P}}^r \longrightarrow V$, and the diagonal $\widehat{\mathbf{P}}^r = \widehat{\Delta} \subseteq \widehat{\mathbf{P}}^r \times \widehat{\mathbf{P}}^r$ is the inverse image of L. So starting with the finite map $F = f \times g : X \times X \longrightarrow \mathbf{P}^r \times \mathbf{P}^r$, one obtains the following commutative diagram of fibre squares with maps as indicated:

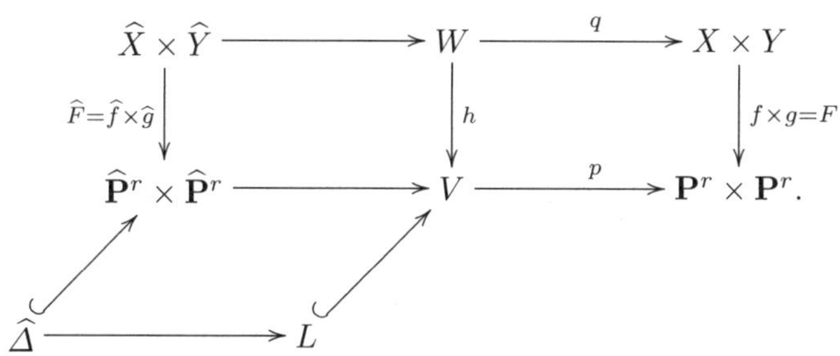

The horizontal maps are $\mathbf{C}^*$-bundles and the vertical morphisms are finite.

Theorem 3.5.10 implies that

$$\pi_i\big(W, h^{-1}L_\epsilon\big) = 0 \quad \text{for } \ i \le n + m - r,$$

but since h is proper over L the inclusion $\big(W, h^{-1}(L)\big) \subseteq \big(W, h^{-1}(L_\varepsilon)\big)$ induces an isomorphism on homotopy groups ([227, II.5.A]). On the other hand,

$$\pi_i\big(\widehat{X} \times \widehat{Y}\,,\, \widehat{F}^{-1}(\widehat{\Delta})\big) \;=\; \pi_i\big(W\,,\, h^{-1}(L)\big) \quad \text{for all } \ i$$

thanks to the fact that the pair appearing on the left is the pullback of the pair on the right under a bundle map. Observing that $\widehat{F}^{-1}(\widehat{\Delta}) = \widehat{X} \times_{\widehat{\mathbf{P}}^r} \widehat{Y}$, the Lemma follows. □

One now obtains a statement from [211] that unifies and extends several previous results.

Theorem 3.5.12. *Let X be a projective local complete intersection of pure dimension n, and $f : X \longrightarrow \mathbf{P}^r$ a finite mapping. Suppose that $Y \subseteq \mathbf{P}^r$ is a closed local complete intersection of pure codimension d. Then the induced homomorphism*

$$f_* \,:\, \pi_i\big(X\,,\, f^{-1}(Y)\big) \longrightarrow \pi_i\big(\mathbf{P}^r\,,\, Y\big) \tag{3.17}$$

is bijective when $i \le n - d$ and surjective when $i = n - d + 1$.

Before giving the proof, we consider some special cases. Suppose to begin with that $Y = L \subseteq \mathbf{P}^r$ is a linear space of codimension d. Since $\pi_i(\mathbf{P}^r, L) = 0$ for $i \le 2(r - d) + 1$, one finds that

$$\pi_i(X, f^{-1}L) \;=\; 0 \quad \text{for } \ i \le n - d.$$

So one recovers[9] the Lefschetz-type Theorem 3.5.10 of Goresky–MacPherson in the compact case. On the other hand, take $X = Y$, with f the inclusion: then the group on the left in (3.17) vanishes, and so one arrives at a homotopy-level Barth–Larsen theorem for local complete intersections.

[9] Not very surprisingly!

Corollary 3.5.13. *Let $X \subseteq \mathbf{P}^r$ be a closed local complete intersection of pure dimension n. Then*

$$\pi_i\big(\mathbf{P}^r\,,\,X\big) \;=\; 0 \quad \text{for } \; i \leq 2n - r + 1. \quad \square$$

Some other applications appear in [211, §9].

Sketch of Proof of Theorem 3.5.12. Keeping notation as in Lemma 3.5.11, it is equivalent to prove the statement after pulling back to the canonical $\mathbf{C}^*$-bundles over the spaces in question. Thus we need to establish that

$$\widehat{f}_* \,:\, \pi_i\big(\widehat{X}\,,\,\widehat{f}^{-1}(\widehat{Y})\big) \longrightarrow \pi_i\big(\widehat{\mathbf{P}}^r\,,\,\widehat{Y}\big) \tag{*}$$

is an isomorphism for $i \leq n - d$ and surjective when $i = n - d + 1$. Consider to this end the long exact sequences of the pairs in question:

$$\begin{array}{ccccccccc}
\cdots \longrightarrow & \pi_i\big(\widehat{X}\,,\,\widehat{f}^{-1}(\widehat{Y})\big) & \longrightarrow & \pi_{i-1}\big(\widehat{f}^{-1}(\widehat{Y})\big) & \xrightarrow{\;j_*\;} & \pi_{i-1}\big(\widehat{X}\big) & \longrightarrow \cdots \\
& \downarrow & & \big\downarrow{\scriptstyle \widehat{f}_*} & & \downarrow & \\
\cdots \longrightarrow & \pi_i\big(\widehat{\mathbf{P}}^r\,,\,\widehat{Y}\big) & \longrightarrow & \pi_{i-1}\big(\widehat{Y}\big) & \longrightarrow & \pi_{i-1}\big(\widehat{\mathbf{P}}^r\big) & \longrightarrow \cdots,
\end{array}$$

where j denotes the inclusion $\widehat{f}^{-1}(\widehat{Y}) \hookrightarrow \widehat{X}$. Identifying $\widehat{f}^{-1}(\widehat{Y})$ with $\widehat{X} \times_{\widehat{\mathbf{P}}^r} \widehat{Y}$ in the natural way, Lemma 3.5.11 asserts that

$$j_* \times \widehat{f}_* \,:\, \pi_{i-1}\big(\widehat{f}^{-1}(\widehat{Y})\big) \longrightarrow \pi_{i-1}\big(\widehat{X}\big) \,\times\, \pi_{i-1}\big(\widehat{Y}\big)$$

is bijective for $i \leq n - d$ and surjective if $i = n - d + 1$. Thus if $i \leq n - d$, then the top row in the diagram above forms a short exact sequence, and $\widehat{f}_*$ restricts to an isomorphism $\ker(j_*) \xrightarrow{\;\cong\;} \pi_{i-1}(\widehat{Y})$. Therefore the composition

$$\pi_i\big(\widehat{X}\,,\,\widehat{f}^{-1}(\widehat{Y})\big) \longrightarrow \pi_{i-1}\big(\widehat{f}^{-1}(\widehat{Y})\big) \longrightarrow \pi_{i-1}\big(\widehat{Y}\big)$$

is bijective when $i \leq n - d$; one sees similarly that it is surjective when $i = n - d + 1$. The required isomorphism (*) now follows from the observation that $\pi_{i-1}(\widehat{\mathbf{P}}^r\,,\,\widehat{Y}) \xrightarrow{\;\cong\;} \pi_{i-1}(\widehat{Y})$ when $i \leq n - d + 1$ since in fact $\pi_k(\widehat{\mathbf{P}}^r) = \pi_k(\mathbf{C}^{r+1} - \{0\}) = 0$ for $k \leq 2r$. $\square$

Remark 3.5.14. (Étale cohomological dimension). Lyubeznik [412, Theorem 10.5(iv)] has given an algebraic proof of Corollary 3.5.13, in which X is moreover allowed to be arbitrarily singular along a subset of dimension ≤ 1. His approach is to establish more generally upper bounds on the étale cohomological dimension of the complement $\mathbf{P}^r - X$. Lyubeznik also recovers and extends results of Peternell [500], [501] dealing with algebraic subsets $X \subseteq \mathbf{P}^r$ having arbitrary singularities: for example, that if every component of X has codimension $\leq c$, then $\pi_i(\mathbf{P}^r, X) = 0$ for $i \leq [\frac{r}{c}] - 1$. $\square$

Remark 3.5.15. (Mappings to homogeneous spaces). In the spirit of 3.5.1, Sommese and Van de Ven [556] proved a generalization of Theorem 3.5.12 for mappings $f : X \longrightarrow \mathbf{F}$ from a local complete intersection to a rational homogeneous space $\mathbf{F} = G/P$. Their result was in turn extended by Okonek [487], who gave a statement allowing for worse singularities. □

Notes

The material in Section 3.1 is quite classical. For a beautiful account of the Hodge-theoretic ideas leading to the hard Lefschetz theorem, we recommend Voisin's book [600], [599]. The presentation in Section 3.2 has been considerably influenced by Hartshorne's articles [279] and [278], while 3.3.A and 3.3.B follow [211]. Large parts of Section 3.4 are likewise adapted from [211], although Fulton's survey [203] has also been helpful. Section 3.5 very closely follows Section 9 of [211]. The argument in 3.1.9 was shown to us by Kollár.

Special cases of the Fulton–Hansen connectedness theorem (Theorem 3.3.6) originally appeared in Barth's paper [28]. However, Barth seems not to have been aware of the many applications of the result, and his paper was overlooked. Barth's argument was rediscovered by Fulton and Hansen in [209]. As indicated in the text, the simpler argument appearing here is due to Deligne.

4

Vanishing Theorems

This chapter is devoted to the basic vanishing theorems for integral divisors. The prototype is Kodaira's result that if A is an ample divisor on a smooth complex projective variety X, then $\mathcal{O}_X(K_X + A)$ has vanishing higher cohomology. An important extension, due to Kawamata and Viehweg, asserts that the statement remains true assuming only that A is nef and big. These and related vanishing theorems have a vast number of applications to questions central to the focus of this book.

We start in Section 4.1 with some preparatory material, notably involving covering constructions of various sorts. Proofs of the classical theorems of Kodaira and Nakano, following the very elementary approach of Ramanujam [509], appear in the second section. In Section 4.3 we present the extension to nef and big divisors, and give some applications. We conclude in Section 4.4 with an account of one of the generic vanishing theorems of Green and the author from [242].

Recent years have seen an explosive growth in the technology surrounding vanishing theorems and the conceptual framework in which to view them. Several excellent expositions have already appeared — notably [174], [362], [364] — explaining these sophisticated new ideas. By contrast, the present discussion is decidedly lowbrow. Our main thrust leans more towards applications of vanishing theorems than to the internal working of the results themselves. Therefore we have endeavored to keep the exposition in this chapter as down to earth as possible.[1] In this spirit we also separate the theorems for integral divisors from their **Q**-divisor analogues (which appear in Chapter 9).

As always, we work with algebraic varieties defined over the complex numbers. Unless otherwise stated, all varieties are assumed to be non-singular.

[1] However, many of the ideas that play an important role in recent work appear here in embryonic form. Parts of the present discussion might therefore be useful to the novice preparing to study the more advanced presentations just cited.

4.1 Preliminaries

This section is devoted to some preliminary definitions and results. After a few remarks about resolutions of singularities, we discuss some covering constructions that will be useful on several occasions.

4.1.A Normal Crossing Divisors and Resolution of Singularities

We give some definitions and state some results concerning simple normal crossing divisors and resolution of singularities.

Let X be a non-singular variety of dimension n.

Definition 4.1.1. (Normal crossing divisor). An effective divisor $D = \sum D_i$ on X has *simple normal crossings* (and D is a *simple normal crossing* or *SNC divisor*) if D is reduced, each component D_i is smooth, and D is defined in a neighborhood of any point by an equation in local analytic coordinates of the type

$$z_1 \cdot \ldots \cdot z_k = 0$$

for some $k \leq n$. A divisor $E = \sum a_i D_i$ has *simple normal crossing support* if the underlying reduced divisor $\sum D_i$ has simple normal crossings. □

In other words, intersections among the components of D should occur "as transversely as possible," i.e. the singularities of D should locally look no worse that those of a union of coordinate hyperplanes. See Figure 4.1.

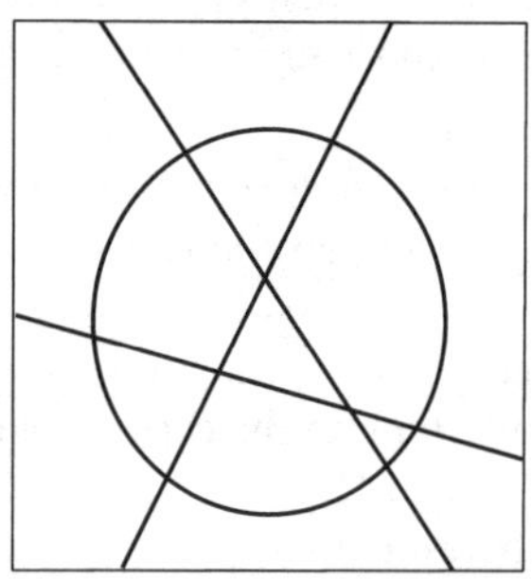

Figure 4.1. A normal crossing divisor

Example 4.1.2. (Hyperplane arrangements). Let $D = \sum H_i$ be the reduced divisor on $\mathbf{P}^n$ whose components are a collection of hyperplanes $H_i \subset \mathbf{P}^n$, corresponding to a finite set of points $S \subset \mathbf{P}^{n*}$ in the dual projective space. Then D is an SNC divisor if and only if the points of S are in linearly general position, i.e. for $r < n$ no $r+2$ lie on an r-dimensional linear subspace of $\mathbf{P}^{n*}$. □

The normal crossing condition derives its importance from the fact that it singles out a class of divisors whose singularities one can understand completely. Confronted with an arbitrary divisor, the first step in many situations is to perform some blowings-up to bring it into normal crossing form. Hironaka's theorem on resolution of singularities guarantees that this is possible:

Theorem 4.1.3. (Resolution of singularities). *Let X be an irreducible complex algebraic variety (possibly singular), and let $D \subseteq X$ be an effective Cartier divisor on X.*

(i). *There is a projective birational morphism*

$$\mu : X' \longrightarrow X,$$

where X' is non-singular and μ has divisorial exceptional locus $\mathrm{except}(\mu)$, *such that*

$$\mu^* D + \mathrm{except}(\mu)$$

is a divisor with SNC support.

(ii). *One can construct X' via a sequence of blowings-up along smooth centers supported in the singular loci of D and X. In particular, one can assume that μ is an isomorphism over*

$$X \; - \; \big(\mathrm{Sing}(X) \cup \mathrm{Sing}(D)\big).$$

One calls X' an *embedded* or *log resolution* of D, or of the pair (X, D). Recall that the exceptional locus of μ is the set of points at which μ fails to be biregular. When X itself is non-singular — which is the main case with which we shall be concerned — this is the same as the locus of points that lie on a positive dimensional fibre of μ.

Theorem 4.1.3 was of course first established in Hironaka's great paper [288]. Bierstone and Milman [56], and Encinas–Villamayor [166], [167], have recently given streamlined arguments. Building on ideas of De Jong from [108], there are now quite quick proofs of statement (i) due to Abramovich–De Jong and Bogomolov–Pantev [2], [64]. Paranjape gives a particularly simple and clean account of the argument in [494].[2]

Once we are beyond the classical results of Kodaira and Nakano, our work with vanishing theorems makes repeated and essential use of the existence of resolutions. As in [368] the (now more elementary) first assertion of Theorem

[2] The published versions of the three papers just cited explicitly mention only the possibility of putting $\mu^*(D)$ into normal crossing form. However, Paranjape has posted a revised version of his paper in which the stronger statement involving $\mathrm{except}(\mu)$ is worked out: see `math.AG/9806084` in the math arXiv. In this paper the exceptional locus of μ is understood to be the μ-saturation of the set of points where μ fails to be biregular. When X is normal this coincides with the definition stated above.

4.1.3 is often sufficient, but on several occasions it will be convenient to invoke the stronger result (ii).[3]

One advantage of knowing something about the construction of a resolution is that it allows one easily to control some of the properties of the resulting map:

Corollary 4.1.4. *Assume that X is smooth, and that $\mu : X' \longrightarrow X$ is an embedded resolution of an effective divisor $D \subseteq X$ constructed as in 4.1.3 (ii) by a sequence of blowings-up along smooth centers.*

(i). *Denoting by K_X and $K_{X'}$ the canonical divisors of X and X' respectively, one has*

$$\mu_* \mathcal{O}_{X'}(K_{X'}) = \mathcal{O}_X(K_X) \quad \textit{and} \quad R^j \mu_* \mathcal{O}_{X'}(K_{X'}) = 0 \quad \text{for } j > 0.$$

(ii). *Assume that X is projective and let H be an ample divisor on X. Then for suitable integers $p \gg 0$ and $b_j \geq 0$,*

$$\mu^*(pH) - \sum b_j E_j$$

is an ample divisor on X', the sum being taken over the exceptional divisors of μ.

Proof. Working inductively, it is enough to prove the statements for a single blowing-up of a smooth variety along a smooth center. But in this case they are elementary (see [280, Theorem II.8.24]). □

Remark 4.1.5. Statement (ii) of the corollary holds in fact for any birational morphism between smooth projective varieties: see [368, Lemma 2.62]. The first equality in (i) is also an elementary general fact (cf. Lemma 9.2.19) while the second is a special case of the Grauert–Riemenschneider vanishing theorem (Theorem 4.3.9).

4.1.B Covering Lemmas

On several occasions we will need to be able to extract "roots" of divisors or line bundles. We discuss here some covering constructions that allow one to do this.

Cyclic coverings. We begin with a local description of the m-fold cyclic covering branched along a given divisor on a variety. Suppose then that X is an affine variety, and $s \in \mathbf{C}[X]$ is a non-zero regular function. We wish to define a variety Y on which the mth root $\sqrt[m]{s}$ of s makes sense. To this

[3] Where appropriate we will indicate rearrangements to the arguments that would allow one to use only the weaker Hironaka theorem (i).

end, start with the product $X \times \mathbf{A}^1$ of X and the affine line. Taking t for the coordinate on $\mathbf{A}^1$, consider the subvariety $Y \subseteq X \times \mathbf{A}^1$ defined by the equation $t^m - s = 0$:

$$\begin{array}{ccccc} \{ t^m - s = 0 \} = Y & & \subseteq & & X \times \mathbf{A}^1 \\ & \searrow^{\pi} & & \swarrow_{\mathrm{pr}_1} & \\ & & X. & & \end{array}$$

The natural mapping $\pi : Y \longrightarrow X$ is a cyclic covering branched along the divisor $D = \mathrm{div}(s)$. Setting $s' = t \mid Y \in \mathbf{C}[Y]$, one has the equality

$$(s')^m = \pi^* s$$

of functions on Y, so we have indeed extracted the desired root of s.

This local construction globalizes:

Proposition 4.1.6. (Cyclic coverings). *Let X be a variety, and L a line bundle on X. Suppose given a positive integer $m \geq 1$ plus a non-zero section*

$$s \in \Gamma\big(X, L^{\otimes m}\big)$$

defining a divisor $D \subseteq X$. Then there exists a finite flat covering

$$\pi : Y \longrightarrow X,$$

where Y is a scheme having the property that the pullback $L' = \pi^* L$ of L carries a section

$$s' \in \Gamma\big(Y, L'\big) \quad \textit{with} \quad \big(s'\big)^m = \pi^* s.$$

The divisor $D' = \mathrm{div}(s')$ maps isomorphically to D. Moreover if X and D are non-singular, then so too are Y and D'.

As above, one should of course think of s' as giving the mth root $\sqrt[m]{s}$ of s.

Proof. One can prove this by taking an affine open covering $\{U_i\}$ of X which locally trivializes L, and carrying out the preceding construction over each U_i: the fact that s is a section of the mth power of a line bundle allows one to glue together the resulting local coverings. However, it is perhaps cleaner to note that one can directly globalize the local discussion.

Specifically, let $\mathbf{L}$ be the total space of the line bundle L with $p : \mathbf{L} \longrightarrow X$ the bundle projection. In other words, $\mathbf{L} = \mathrm{Spec}_{\mathcal{O}_X} \mathrm{Sym}(L^*)$.[4] Then there is a "tautological" section

$$T \in \Gamma\big(\mathbf{L}, p^* L\big).$$

[4] Note that this is what is called $\mathbb{V}(L^*)$ in [280].

In fact, a section of p^*L is specified geometrically by giving for each point $a \in \mathbf{L}$ a vector in the fibre of p over $x = p(a)$. But a *is* such a vector, and we set $T(a) = a$. More formally, T is determined by a homomorphism

$$\mathcal{O}_{\mathbf{L}} \longrightarrow p^*L$$

of $\mathcal{O}_{\mathbf{L}}$-modules, or equivalently a mapping

$$\mathrm{Sym}_{\mathcal{O}_X}(L^*) \longrightarrow L \otimes \mathrm{Sym}_{\mathcal{O}_X}(L^*) \tag{*}$$

of quasi-coherent sheaves on X. The term on the left in (*) is naturally a summand of that on the right, and the map in question is the canonical inclusion. One should view T as a "global fibre coordinate" in $\mathbf{L}$: for instance $\{T = 0\}$ defines the zero-section of $\mathbf{L}$.

Having said this, the proof of the Proposition is immediate. One simply takes $Y \subseteq \mathbf{L}$ to be the divisor of the section

$$T^m - p^*s \in \Gamma\big(\mathbf{L}, p^*L^{\otimes m}\big),$$

and $s' = T \mid Y$. The assertions of the proposition are then clear from the local construction. □

Remark 4.1.7. It follows from the construction that there is a canonical isomorphism

$$\pi_*\mathcal{O}_Y = \mathcal{O}_X \oplus L^{\otimes -1} \oplus \ldots \oplus L^{\otimes(1-m)}. \quad \square$$

Remark 4.1.8. It will be useful to note that normal crossing divisors pull back nicely under the covering just constructed. Specifically, assume in the situation of 4.1.6 that X and D are smooth, and suppose that $\sum D_i$ is a reduced effective divisor on X such that $D + \sum D_i$ has simple normal crossings. Put $D_i' = \pi^* D_i$. Then it follows from the local discussion that $D' + \sum D_i'$ is a simple normal crossing divisor on Y. □

Remark 4.1.9. This cyclic covering construction is all that is needed for the classical Kodaira vanishing theorem in the next section. □

Bloch–Gieseker and Kawamata coverings. To begin, we complement the previous result with a rather general construction allowing one to take roots of arbitrary bundles. It has its genesis in the paper [60] of Bloch and Gieseker, with further refinements by Kollár and Mori [368, Proposition 2.67]:

Theorem 4.1.10. (Bloch–Gieseker coverings). *Let X be an irreducible quasi-projective variety of dimension n, let L be a line bundle on X, and fix any positive integer m. Then there exists a reduced and irreducible variety Y, a finite flat surjective mapping*

$$f : Y \longrightarrow X,$$

together with a line bundle N on Y such that

$$f^*L \;=\; N^{\otimes m}.$$

If X is non-singular then we can take Y to be so as well. In this case, given a simple normal crossing divisor

$$D \;=\; \sum D_i$$

on X, we can arrange that its pullback f^*D be a simple normal crossing divisor on Y. Finally, if $\dim X \geq 2$ then we can further ensure that the preimage $D_i' = f^*D_i$ of each irreducible component of D is irreducible.

Proof. Suppose first that the result is known for the pullback of the hyperplane bundle under a quasi-finite mapping $X \longrightarrow \mathbf{P}^r$. We can write an arbitrary bundle L as the difference $L = A \otimes B^*$ of two such bundles, e.g. by asking that A and B be very ample.[5] By taking roots first of A and then of the pullback of B, we arrive at the required statement for L.

Thus we may assume that there is a finite-to-one mapping $\phi : X \longrightarrow \mathbf{P}^r$ such that $L = \phi^*\mathcal{O}_{\mathbf{P}}(1)$. Fix next a branched covering

$$\nu : \mathbf{P}^r \longrightarrow \mathbf{P}^r \;\; \text{with} \;\; \nu^*\mathcal{O}_{\mathbf{P}^r}(1) \;=\; \mathcal{O}_{\mathbf{P}^r}(m),$$

e.g. the "Fermat covering" $\nu(\,[\,T_0, \ldots, T_r\,]\,) \;=\; [\,T_0^m, \ldots, T_r^m\,]$. Given

$$g \in G =_{\text{def}} \mathrm{GL}(r+1),$$

acting on $\mathbf{P}^r$ in the natural way, denote by $\nu_g : \mathbf{P}^r \longrightarrow \mathbf{P}^r$ the composition $\nu_g = g \circ \nu$. Define

$$Y = Y_g = X \times_{\mathbf{P}^r} \mathbf{P}^r$$

to be the fibre product of X and $\mathbf{P}^r$ with respect to ϕ and ν_g, so that one has the Cartesian square

$$\begin{array}{ccc} Y = Y_g & \xrightarrow{\;\overline{\phi}\;} & \mathbf{P}^r \\ {\scriptstyle f=f_g}\big\downarrow & & \big\downarrow{\scriptstyle \nu_g} \\ X & \xrightarrow[\;\phi\;]{} & \mathbf{P}^r. \end{array}$$

The indicated map $f : Y_g \longrightarrow X$ is finite and flat since ν_g is, and by construction $f^*L = \overline{\phi}^*\big(\mathcal{O}_{\mathbf{P}^r}(1)^{\otimes m}\big)$. We will show that if $g \in G$ is sufficiently general, then $f_g : Y_g \longrightarrow X$ has all the stated properties.

[5] Recall that a line bundle on a quasi-projective variety X is very ample if it is the restriction of the hyperplane line bundle under an embedding of X as a locally closed subset of projective space.

Note first that Y_g is the inverse image of the graph $\Gamma_{g^{-1}} \subseteq \mathbf{P}^r \times \mathbf{P}^r$ of g^{-1} under the map

$$\phi \times \nu : X \times \mathbf{P}^r \longrightarrow \mathbf{P}^r \times \mathbf{P}^r.$$

So the irreducibility of Y_g for general g follows from Example 3.3.10. For reducedness, we can certainly assume that Y_g is generically reduced. However, if $f : V \longrightarrow W$ is a flat mapping of schemes with W integral and V generically reduced, then in fact V must be everywhere reduced.[6] Thus we can take Y_g to be globally reduced. Given a reduced irreducible Cartier divisor $E \subseteq X$, the same argument applied to E in place of X shows that one can arrange for f^*E to be reduced and irreducible provided that $\dim X \geq 2$ (so that $\dim E \geq 1$).

Kleiman's theorem on the transversality of the generic translate ([343], [280, III.10.8]) immediately implies that one can take Y to be non-singular if X is so, but it will be convenient to recall the argument. Consider the mapping

$$m : G \times \mathbf{P}^r \longrightarrow \mathbf{P}^r \quad , \quad (g, y) \mapsto g \cdot \nu(y),$$

and form the fibre product $W = X \times_{\mathbf{P}^r} (G \times \mathbf{P}^r)$. Thus we have a commutative diagram:

$$\begin{array}{ccccc} & & b & & \\ W & \longrightarrow & G \times \mathbf{P}^r & \xrightarrow{\mathrm{pr}_1} & G \\ {\scriptstyle n}\downarrow & & \downarrow{\scriptstyle m} & & \\ X & \xrightarrow[\phi]{} & \mathbf{P}^r. & & \end{array}$$

As in [280, III.10.8], m is a smooth morphism, and hence n is smooth as well. Therefore if X is non-singular, then so too is W. On the other hand, the fibre of b over $g \in G$ is just Y_g, i.e. $b^{-1}(g) = Y_g$. The non-singularity of Y_g for general g then follows from the theorem on generic smoothness. Similarly, if $E \subseteq X$ is a smooth divisor, then for general g its preimage in $Y = Y_g$ will likewise be non-singular.

Finally, assume X non-singular, and suppose that we are given a normal crossing divisor $D = \sum D_i$ on X. We have already seen that we can arrange that each $D_i' = f^*D_i$ be reduced and irreducible if $\dim X \geq 2$. It remains to show that we can take f^*D to be a simple normal crossing divisor. To this end, note that each component D_i is smooth, and any collection of the D_i meet transversely or not at all. Referring to the diagram above, the pre-images

$$F_i \; = \; n^{-1}(D_i) \; \subseteq \; W$$

6 This is verified by the following algebraic argument, shown to us by Mike Roth. Let A be a domain, and $A \hookrightarrow B$ a flat extension of commutative rings. Assume that there exists an element $0 \neq \delta \in A$ such that B_δ is reduced. Then B is reduced. For if $\phi \in B$ is nilpotent, then ϕ must die in B_δ, i.e. $\delta^m \phi = 0$ for some $m > 0$. But multiplication by δ^m gives an injective map $A \xrightarrow{\cdot \delta^m} A$, and by flatness it must remain injective as a mapping $B \xrightarrow{\cdot \delta^m} B$. So $\phi = 0$.

are likewise smooth subvarieties, any collection of which meet transversely or not at all. The following lemma shows that if $g \in G$ is sufficiently general, then the same statement holds for the fibres of F_i over g. But these are the components of f^*D, and this implies that f^*D has simple normal crossings. □

Lemma 4.1.11. *Let $b : W \longrightarrow G$ be a morphism of non-singular varieties, and suppose given a finite collection of smooth subvarieties $V_i \subseteq W$ meeting transversely. Then for general $g \in G$, the fibres $(V_i)_g$ of the V_i over g are smooth and meet transversely (or not at all) in W_g.*

Proof. This follows by applying the theorem on generic smoothness to the V_i and all of their intersections. □

By combining these two constructions, we show that given an SNC divisor $D = \sum D_i$ one can pass to a covering on which each of the components D_i becomes divisible by any preassigned integer $m_i > 0$:

Proposition 4.1.12. (Kawamata coverings). *Let X be a non-singular quasi-projective variety, let $D = \sum_{i=1}^{t} D_i$ be a simple normal crossing divisor on X, and fix positive integers $m_1, \ldots, m_t > 0$. Then there exists a smooth variety Y and a finite flat covering $h : Y \longrightarrow X$ with the property that*

$$h^*D_i \;=\; m_i D_i'$$

for some smooth divisor D_i' on Y, where $D' = \sum D_i'$ has simple normal crossings.

Proof. By induction on the number of components, it suffices to produce a covering $h : Y \longrightarrow X$ where

$$h^*D_1 = m_1 D_1' \quad , \quad h^*D_i = D_i' \quad \text{for} \quad 2 \leq i \leq t,$$

and $\sum D_i'$ has simple normal crossings. To this end, first take a Bloch–Gieseker covering $f : X_1 \longrightarrow X$ to arrange that

$$D_1^+ \;=_{\text{def}}\; f^*D_1 \;\equiv_{\text{lin}}\; m_1 A_1^+$$

for some (possibly ineffective) divisor A_1^+ on X_1, while making sure that f^*D still has simple normal crossings. Then take an m_1-fold cyclic covering $p : Y \longrightarrow X_1$ branched along D_1^+, so that $p^*D_1^+ \;=\; m_1 D_1'$ for a smooth divisor D_1'. Set $h = p \circ f : Y \longrightarrow X$, and for $i \geq 2$ write $D_i' = h^*D_i$. It follows from Remark 4.1.8 that $\sum D_i'$ still has normal crossings, and we have the required covering. □

Remark 4.1.13. (Kawamata's construction). Kawamata [315, Theorem 17] originally proved 4.1.12 by an explicit geometric construction involving a Kummer covering. □

Injectivity. In the pages that follow we will use these covering constructions to establish vanishing theorems. Starting with a line bundle L on X for which one wants to prove a vanishing, the general strategy is to pull L back to a covering $f : Y \longrightarrow X$ on which the problem simplifies in some way. The next lemma shows that it will be enough to get the vanishing on Y:

Lemma 4.1.14. (Injectivity lemma). *Let $f : Y \longrightarrow X$ be a finite surjective morphism of irreducible complex projective varieties, with X normal, and let E be a vector bundle on X. Then the natural homomorphism*

$$H^j\big(X, E\big) \longrightarrow H^j\big(Y, f^*E\big)$$

induced by f is injective. In particular, if

$$H^j\big(Y, f^*E\big) \ = \ 0 \quad \text{for } \textit{some } \ j \geq 0,$$

then $H^j\big(X, E\big) = 0$.

Proof. By passing to its normalization, we can assume without loss of generality that Y is also normal. Note next that

$$\begin{aligned} H^j\big(Y, f^*E\big) \ &= \ H^j\big(X, f_*f^*E\big) \\ &= \ H^j\big(X, E \otimes f_*\mathcal{O}_Y\big) \end{aligned}$$

thanks to the projection formula and the finiteness of f. On the other hand, since X and Y are normal there is a trace map

$$\mathrm{Tr}_{Y/X} \ : \ f_*\mathcal{O}_Y \longrightarrow \mathcal{O}_X,$$

which gives rise to a splitting of the natural inclusion $\mathcal{O}_X \longrightarrow f_*\mathcal{O}_Y$. Therefore the map $E \longrightarrow E \otimes f_*\mathcal{O}_Y$ likewise splits, and so $H^j\big(X, E\big)$ embeds as a summand of

$$H^j\big(X, E \otimes f_*\mathcal{O}_Y\big) = H^j\big(Y, f^*E\big).$$

The stated injectivity follows. □

4.2 Kodaira and Nakano Vanishing Theorems

In this section we prove the classical vanishing theorems of Kodaira and Nakano. We follow the down-to-earth approach of Ramanujam [509], who uses the Hodge decomposition to deduce the results directly from the Lefschetz hyperplane theorem. Here we require only the elementary cyclic covering construction 4.1.6 together with Lemma 4.1.14.

Theorem 4.2.1. (Kodaira vanishing theorem). *Let X be a smooth irreducible complex projective variety of dimension n, and let A be an ample divisor on X. Then*

$$H^i\big(X, \mathcal{O}_X(K_X + A)\big) \;=\; 0 \quad \text{for} \;\; i > 0.$$

Equivalently,

$$H^j\big(X, \mathcal{O}_X(-A)\big) \;=\; 0 \quad \text{for} \;\; j < n = \dim X.$$

The equivalence of the two vanishings comes from Serre duality.

The crucial input to the proof is a holomorphic manifestation of the Lefschetz hyperplane theorem. The statement already appeared in Example 3.1.24, but we repeat it here for the reader's convenience.

Lemma 4.2.2. *Let X be a smooth complex projective variety of dimension n, and let $D \subseteq X$ be a smooth effective ample divisor. Denote by*

$$r_{p,q} \,:\, H^q\big(X, \Omega^p_X\big) \longrightarrow H^q\big(D, \Omega^p_D\big)$$

the natural maps determined by restriction. Then

$$r_{p,q} \;\; \text{is} \;\; \begin{cases} \text{bijective} & \text{for} \;\; p+q \le n-2, \\ \text{injective} & \text{for} \;\; p+q = n-1. \end{cases}$$

Proof. The Lefschetz hyperplane theorem implies that the restriction maps

$$r_j \,:\, H^j\big(X, \mathbf{C}\big) \longrightarrow H^j\big(D, \mathbf{C}\big)$$

are isomorphisms when $j \le n-2$ and injective when $j = n-1$. On the other hand, the Hodge theorem and the Dolbeaut isomorphisms give functorial decompositions

$$H^j\big(X, \mathbf{C}\big) \;=\; \bigoplus_{p+q=j} H^q\big(X, \Omega^p_X\big) \quad , \quad H^j\big(D, \mathbf{C}\big) \;=\; \bigoplus_{p+q=j} H^q\big(D, \Omega^p_D\big).$$

Under these identifications the restriction r_j on complex cohomology is the direct sum $\oplus_{p+q=j}\, r_{p,q}$ of the corresponding homomorphisms $r_{p,q}$, and the lemma follows. □

Proof of Theorem 4.2.1. Since A is ample, there exists a smooth divisor

$$D \;\in\; |mA|$$

for $m \gg 0$. Let $p : Y \longrightarrow X$ be the m-fold cyclic covering branched along D constructed in Proposition 4.1.6, and set $A' = p^*A$. By construction there is a smooth ample divisor $D' \in |A'|$, and 4.1.14 shows that it suffices to prove that

$$H^j\big(Y, \mathcal{O}_Y(-D')\big) \;=\; 0 \quad \text{for} \;\; j < n.$$

Therefore we are reduced to establishing the theorem under the additional hypothesis that the given ample divisor is effective and non-singular.

Suppose then that $D \subseteq X$ is a smooth ample divisor. The case $p = 0$ and $q = j$ of Lemma 4.2.2 asserts that the restriction map

$$H^j\big(X, \mathcal{O}_X\big) \longrightarrow H^j\big(D, \mathcal{O}_D\big)$$

is an isomorphism when $j \leq n-2$ and injective when $j = n-1$. But then the exact sequence

$$0 \longrightarrow \mathcal{O}_X(-D) \longrightarrow \mathcal{O}_X \longrightarrow \mathcal{O}_D \longrightarrow 0$$

shows that $H^j\big(X, \mathcal{O}_X(-D)\big) = 0$ when $j \leq n-1$, as required. □

Observe that the proof just completed uses only part of the information provided by Lemma 4.2.2. The remaining instances lead to Nakano's generalization of Kodaira vanishing:

Theorem 4.2.3. (Nakano vanishing theorem). *Let X be a smooth complex projective variety of dimension n, and let A be an ample divisor on X. Then*

$$H^q\big(X, \Omega^p_X \otimes \mathcal{O}_X(A)\big) \;=\; 0 \quad \text{for } \; p+q > n. \tag{4.1}$$

Equivalently,

$$H^s\big(X, \Omega^r_X \otimes \mathcal{O}_X(-A)\big) \;=\; 0 \quad \text{for } \; r+s < n. \tag{4.2}$$

Theorem 4.2.1 appears as the case $p = n$ of the first statement, and the case $r = 0$ of the second. The equivalence of the two vanishings again follows from Serre duality.

As before, the strategy is to use cyclic coverings to reduce to the case in which A is represented by a smooth effective divisor D. For this it will be convenient to make some preliminary remarks about bundles of forms with logarithmic poles.

Suppose then that X is a smooth variety, and $D \subset X$ is a non-singular divisor. As customary, we denote by $\Omega^1_X\big(\log D\big)$ the sheaf of one-forms on X with logarithmic poles along D: if $z_1, \ldots, z_n$ are local analytic coordinates on X, with $D = \big(z_n = 0\big)$, then $\Omega^1_X\big(\log D\big)$ is locally generated by $dz_1, \ldots, dz_{n-1}, \frac{dz_n}{z_n}$. One puts

$$\Omega^p_X\big(\log D\big) \;=\; \Lambda^p\Big(\Omega^1_X\big(\log D\big)\Big).$$

The next lemma summarizes the properties we require.

Lemma 4.2.4. *Assume as above that $D \subseteq X$ is a smooth divisor.*

(i). *Viewing Ω^p_D as a sheaf on X via extension by zero, one has exact sequences*

$$0 \longrightarrow \Omega^p_X \longrightarrow \Omega^p_X\big(\log D\big) \xrightarrow{\;res\;} \Omega^{p-1}_D \longrightarrow 0, \tag{4.3a}$$

$$0 \longrightarrow \Omega^p_X\big(\log D\big) \otimes \mathcal{O}_X(-D) \longrightarrow \Omega^p_X \longrightarrow \Omega^p_D \longrightarrow 0. \tag{4.3b}$$

(ii). *Let $\pi : Y \longrightarrow X$ be the m-fold cyclic covering branched along D constructed in 4.1.6, and let $D' \subseteq Y$ be the divisor mapping isomorphically to D, so that $\pi^* D = mD'$. Then*

$$\pi^* \big(\Omega^p_X(\log D) \big) \;=\; \Omega^p_Y(\log D').$$

The homomorphism $\Omega^p_X(\log D) \longrightarrow \Omega^{p-1}_D$ appearing on the right in (4.3a) is the residue map (cf. [174, §2]), while the map on the right in (4.3b) is given by restriction of forms.

Indication of Proof of Lemma 4.2.4. The two sequences in (i) are special cases of [174, 2.3]; to give the flavor, we prove (4.3b) when $p = 1$. Choose local analytic coordinates $z_1, \ldots, z_n$ such that $D =_{\text{locally}} \{ z_n = 0 \}$. Then the kernel of $\Omega^1_X \longrightarrow \Omega^1_D$ is locally generated by $z_n\, dz_1, \ldots, z_n\, dz_{n-1}, dz_n$. But these are exactly the local generators of the subsheaf $\Omega^p_X(\log D)\big(- D\big) \subseteq \Omega^1_X(\log D)$. For (ii) it is enough to prove the stated isomorphism when $p = 1$, and this likewise follows immediately from a calculation in local coordinates. □

Proof of Theorem 4.2.3. We will establish (4.2). Since A is ample, there exists for $m \gg 0$ a smooth divisor $D \in |mA|$. We assert first that it suffices to prove the vanishing

$$H^s\Big(X \,,\, \Omega^r_X(\log D) \otimes \mathcal{O}_X(-A) \Big) \;=\; 0 \quad \text{for} \;\; r + s < n. \qquad (*)$$

In fact, we can assume by induction on $\dim X$ that we already know Nakano vanishing for D, so that

$$H^s\big(D, \Omega^{r-1}_D \otimes \mathcal{O}_X(-A)\big) \;=\; 0 \quad \text{for} \;\; (r-1) + s < (n-1).$$

Using the exact sequence (4.3a), (*) then gives (4.2) for X.

For (*) form the m-fold cyclic covering $\pi : Y \longrightarrow X$ branched along D (Proposition 4.1.6). According to Lemma 4.1.14 it suffices to prove that

$$H^s\Big(Y \,,\, \pi^*\big(\Omega^r_X(\log D) \otimes \mathcal{O}_X(-A) \big) \Big) \;=\; 0 \quad \text{for} \;\; r + s < n.$$

Now $\pi^*\big(\Omega^r_X(\log D)\big) \;=\; \Omega^r_Y\big(\log D'\big)$ thanks to 4.2.4, and $\pi^* A$ is represented by the smooth divisor $D' \subseteq Y$. So the question is reduced to the vanishing

$$H^s\Big(Y \,,\, \Omega^r_Y(\log D') \otimes \mathcal{O}_Y(-D') \Big) \;=\; 0 \quad \text{for} \;\; r + s < n.$$

But this follows from Lemma 4.2.2 by using the exact sequence (4.3b). □

Remark 4.2.5. (Bogomolov's vanishing theorem). Let D be an SNC divisor on a smooth projective variety X, and let L be any divisor on X. Bogomolov [62] proves that

$$H^0\big(X, \Omega^p_X(\log D) \otimes \mathcal{O}_X(-L)\big) \ = \ 0 \quad \text{for} \quad p < \kappa(L).$$

This is one of the principal inputs to Viehweg's proof in [589] of Kawamata–Viehweg vanishing (Theorem 9.1.18). The paper [589] also contains a proof of Bogomolov's result. See also [174, Corollary 6.9]. □

4.3 Vanishing for Big and Nef Line Bundles

This section is devoted to the basic vanishing theorem, due to Kawamata and Viehweg, for big and nef divisors. The result is stated and proved in the first subsection, while the second is devoted to some applications.

4.3.A Statement and Proof of the Theorem

During the years following the appearance of [353], mathematicians came to realize that one could weaken the amplitude hypotheses appearing in Kodaira vanishing. For example, Mumford [446] proved that at least parts of the statement of 4.2.1 remain true assuming only that A is semiample and big. In the case of surfaces even stronger results were obtained by Ramanujam [509].

A clean and in some respects definitive statement was discovered by Kawamata [316] and Viehweg [589] in the early 1980s. They showed that Kodaira vanishing holds for any nef and big line bundle:

Theorem 4.3.1. (Vanishing for nef and big divisors). *Let X be a smooth complex projective variety of dimension n, and D a nef and big divisor on X. Then*

$$H^i\big(X, \mathcal{O}_X(K_X + D)\big) \ = \ 0 \quad \text{for} \quad i > 0. \tag{4.4}$$

Equivalently,

$$H^j\big(X, \mathcal{O}_X(-D)\big) \ = \ 0 \quad \text{for} \quad j < n = \dim X. \tag{4.5}$$

Remark 4.3.2. Recall from 2.2.16 that a nef divisor D is big if and only if its top self-intersection is strictly positive: $\big(D^n\big) > 0$. However, we do not use this fact in the proof of 4.3.1. □

Remark 4.3.3. (Kawamata–Viehweg vanishing for Q-divisors). The papers [316], [589] of Kawamata and Viehweg also contain the fundamental vanishing theorem for **Q**-divisors that we discuss in Section 9.1. We could — and eventually will — prove this more general result with little extra effort. However applications of the theorem on **Q**-divisors have a rather different flavor than the consequences flowing directly from 4.3.1. Therefore, we have felt it best to treat the more general statement separately. By the same token, we try to reserve the title Kawamata–Viehweg vanishing for the more general Theorem 9.1.18, even though 4.3.1 is also due to these authors. □

Example 4.3.4. The analogue of Nakano vanishing can fail for big and nef divisors. For instance, let $X = \mathrm{Bl}_P(\mathbf{P}^3)$ be the blowing-up of $\mathbf{P}^3$ at a point, and let $H \cong \mathbf{P}^2$ be the pullback of a general hyperplane to X. Then $\dim H^1\big(X, \Omega^1_X\big) = 2$ while $\dim H^1\big(H, \Omega^1_H\big) = 1$. It follows using the exact sequences in 4.2.4 that

$$H^1\big(X, \Omega^1_X \otimes \mathcal{O}_X(-H)\big) \neq 0$$

even though H is nef and big. The point here is that while the groups $H^{0,k}(X)$ are birational invariants of a smooth variety, the spaces $H^{p,q}(X)$ $(p, q \geq 1)$ are not. So one cannot expect Nakano-type statements to hold under essentially birational hypotheses of positivity. □

We now turn to the proof of Theorem 4.3.1. Following Kawamata's original article [316], the plan is to reduce 4.3.1 to a statement showing that one can weaken positivity assumptions in the presence of strong geometric hypotheses:

Lemma 4.3.5. (Norimatsu's lemma). *Let X be a smooth projective variety, A an ample divisor on X, and E an SNC divisor on X. Then*

$$H^i\big(X, \mathcal{O}_X(K_X + A + E)\big) \;=\; 0 \quad \text{for} \quad i > 0,$$

or equivalently,

$$H^j\big(X, \mathcal{O}_X(-A - E)\big) \;=\; 0 \quad \text{for} \quad j < \dim X.$$

Proof. Write $E = \sum_{i=1}^{t} E_i$. One argues by induction on the number t of components of E, the case $t = 0$ being Kodaira vanishing. Assuming the result known for normal crossing divisors with $\leq k$ components, consider the natural exact sequence.

$$0 \longrightarrow \mathcal{O}_X\Big(-A - \sum_{i=1}^{k+1} E_i\Big) \longrightarrow \mathcal{O}_X\Big(-A - \sum_{i=1}^{k} E_i\Big) \longrightarrow \mathcal{O}_{E_{k+1}}\Big(-A - \sum_{i=1}^{k} E_i\Big) \longrightarrow 0.$$

The induction hypothesis, applied on E_{k+1} and X, gives

$$H^j\Big(E_{k+1}, \mathcal{O}_{E_{k+1}}\big(-A - \sum_{i=1}^{k} E_i\big)\Big) \;=\; 0 \quad \text{for} \quad j < (n-1),$$
$$H^j\Big(X, \mathcal{O}_X\big(-A - \sum_{i=1}^{k} E_i\big)\Big) \;=\; 0 \quad \text{for} \quad j < n.$$

This yields the analogous vanishing for the term on the left, and the lemma follows. □

Proof of Theorem 4.3.1. Since D is big, for $m \gg 0$ we can write

$$mD \equiv_{\text{lin}} H + N \tag{4.6}$$

where H is an ample divisor and N an effective divisor on X. We now proceed in two steps.

Step 1. We argue first that it suffices to prove the theorem under the additional assumption that the divisor N in (4.6) has SNC support.

To this end, we will show to begin with that after passing to a birational modification $\mu : X^* \longrightarrow X$ we can arrange for $D^* = \mu^* D$ — which of course remains nef and big — to admit an expression of the form

$$m' D^* \equiv_{\text{lin}} H^* + N^*, \tag{4.7}$$

where H^* is ample and N^* is effective with SNC support. Then we will argue that knowing the theorem on X^* implies the statement on X.

Turning to details, start with an arbitrarily singular divisor N in (4.6) and construct an embedded resolution

$$\mu : X^* \longrightarrow X$$

of N via a sequence of blowings-up along smooth centers (Theorem 4.1.3 (ii)).[7] Thus $\mu^* N$ has simple normal crossing support, and moreover:

$$\mu_* \mathcal{O}_{X^*}(K_{X^*}) = \mathcal{O}_X(K_X) \quad , \quad R^q \mu_* \mathcal{O}_{X^*}(K_{X^*}) = 0 \ \text{ for } q > 0 \tag{4.8}$$

by virtue of 4.1.4 (i). The plan of course is to pull back (4.6), but we have to deal with the fact that $\mu^* H$ is no longer ample. Write

$$\mu^* N = \sum a_j F_j,$$

where the divisors F_j appearing here include all μ-exceptional divisors, and each $a_j \geq 0$. Thanks to 4.1.4 (ii), if $p \gg 0$ then there exist $b_j \geq 0$ such that

$$\mu^*(pH) - \sum b_j F_j$$

is an ample divisor on X^*. Then

$$\mu^*\Big(pmD \Big) \equiv_{\text{lin}} \Big(\mu^*(pH) - \sum b_j F_j\Big) + \Big(\sum (pa_j + b_j) F_j \Big)$$

is the sum of an ample and an effective divisor with SNC support. Thus we have constructed the promised decomposition (4.7).

Now suppose we know that the theorem holds for $\mu^* D$ on X^*, i.e. assume that

$$H^j(X^*, \mathcal{O}_{X^*}(K_{X^*} + \mu^* D)) = 0 \quad \text{for } j > 0.$$

[7] With a little extra work, one can make do with Theorem 4.1.3 (i): see Remark 4.3.6.

Then it follows from (4.8) and the projection formula that

$$H^j\big(X, \mathcal{O}_X(K_X + D)\big) = 0 \quad \text{for } j > 0,$$

so we get the desired vanishing on X. This completes Step 1: we can assume in (4.6) that N has SNC support.

Step 2. We now prove (4.4) assuming that

$$mD \;\equiv_{\text{lin}}\; H + N,$$

where H is ample and N has SNC support.

Write $N = \sum_{i=1}^{t} e_i\, E_i$, with $e_i > 0$, and set

$$e^* \;=\; e_1 \cdot \ldots \cdot e_t \quad , \quad e_i^* = \frac{e^*}{e_i}.$$

We use Proposition 4.1.12 to construct a covering $h : Y \longrightarrow X$ of X by a smooth variety Y, with

$$h^*\, E_i \;=\; m\, e_i^*\, E_i' \qquad \text{and} \qquad E' \;=_{\text{def}} \sum E_i' \text{ an SNC divisor on } Y.$$

Put $D' = h^*D$ and $H' = h^*H$, so that

$$mD' \;\equiv_{\text{lin}}\; H' \;+\; me^*\, E'. \tag{4.9}$$

Now D' is nef since D is, and therefore $H' + m(e^* - 1)D'$ — being the sum of an ample and nef divisor — is ample. So it follows from (4.9) that

$$me^*\big(D' \,-\, E'\big) \;\equiv_{\text{lin}}\; H' + m(e^* - 1)\, D'$$

is ample. Hence $D' \equiv_{\text{lin}} A' + E'$ for some ample divisor A' on Y, and therefore

$$H^j\big(Y, \mathcal{O}_Y(-D')\big) \;=\; H^j\big(Y, \mathcal{O}_Y(-A' - E')\big) \;=\; 0 \quad \text{for } j < n$$

thanks to Norimatsu's lemma. The desired vanishing (4.5) then follows from Lemma 4.1.14. □

Remark 4.3.6. We indicate how to modify the argument so as to use only the weak Hironaka Theorem 4.1.3 (i). Keeping the notation of the proof just completed, we used 4.1.3 (ii) only in order to be able to invoke Corollary 4.1.4. As noted in Remark 4.1.5, statement (ii) of that corollary as well as the first equality in statement (i) hold quite generally. So it is enough to show that the vanishing of the higher direct images in (4.8) holds for any embedded resolution $\mu : X^* \longrightarrow X$ of N. This is a special case of Grauert–Riemenschneider vanishing (Theorem 4.3.9), which one can prove directly in the case at hand. Specifically, observe that if we go through the argument with D replaced by $D+rH$ for any integer $r > 0$ then we still get the decomposition

(4.7) with H^* replaced by $H^* + \mu^*(mrH)$. Therefore we can assume in Step 1 that we know the vanishing

$$H^j\big(X^*, \mathcal{O}_{X^*}(K_{X^*} + \mu^*(D + rH))\big) = 0 \quad \text{for } \; j > 0 \text{ and all } \; r > 0.$$

But by Lemma 4.3.10 below, this implies the vanishing of the higher direct images in (4.8). □

Example 4.3.7. (An extension of Theorem 4.3.1). Let D be a nef divisor on a smooth projective n-fold X. Assume that $\big(D^{n-k} \cdot H^k\big) > 0$ for some natural number k and ample divisor H. Then

$$H^i\big(X, \mathcal{O}_X(K_X + D)\big) = 0 \quad \text{for } \; i > k.$$

This statement is again due to Kawamata and Viehweg, and contains Theorem 4.3.1 as the case $k = 0$. (One can suppose that H is a very ample smooth divisor, and that the result is known for all varieties of dimension $< n$. Then consider the exact sequence

$$0 \longrightarrow \mathcal{O}_X(K_X + D) \longrightarrow \mathcal{O}_X(K_X + D + H) \longrightarrow \mathcal{O}_H(K_H + D) \longrightarrow 0.$$

The induction hypothesis gives vanishings for the term on the right, whereas all the higher cohomology of the term in the middle vanishes thanks to Kodaira vanishing.) □

Remark 4.3.8. (Kollár's theorems). In his fundamental papers [356] and [357], Kollár proved several theorems applicable to line bundles that may not be big. We state here two of these from [356].

> **Theorem. (Kollár's injectivity theorem).** *Let X be a smooth projective variety and L a semiample divisor on X. Given $k \geq 1$, fix any divisor $D \in |kL|$. Then the homomorphisms*
>
> $$H^i\big(X, \mathcal{O}_X(K_X + mL)\big) \xrightarrow{\cdot D} H^i\big(X, \mathcal{O}_X(K_X + (m+k)L)\big)$$
>
> *determined in the natural way by D are injective for all $i \geq 0$ and $m > 0$.*

(Recall from Definition 2.1.26 that a divisor L is semiample if pL is free for some $p > 0$.) Observe that this implies Kodaira vanishing, since if L is ample then

$$H^i\big(X, \mathcal{O}_X(K_X + (m+k)L)\big) \;=\; 0$$

when $i > 0$ and $k \gg 0$. The second theorem shows that the direct images $R^j f_* \mathcal{O}_X(K_X)$ have exceptionally good cohomological properties: in particular, they satisfy the same sort of vanishings as canonical bundles themselves.

Theorem. (Kollár vanishing theorem). *Let $f : X \longrightarrow Y$ be a surjective morphism of projective varieties, with X smooth. Then:*

(i). $R^j f_* \mathcal{O}_X(K_X)$ *is torsion-free for all j.*

(ii). $R^j f_* \mathcal{O}_X(K_X) = 0$ for $j > \dim X - \dim Y$.

(iii). *For any ample divisor A on Y and every $j \in \mathbf{N}$,*

$$H^i\Big(Y\,,\, R^j f_* \mathcal{O}_X(K_X) \otimes \mathcal{O}_Y(A)\Big) \;=\; 0 \quad \text{for } \; i > 0.$$

An application of this vanishing appears in Section 6.3.E. Many others, as well as further results, can be found in [356], [357]. □

4.3.B Some Applications

In this subsection we present some applications of the vanishing theorem for big and nef divisors.

Theorem of Grauert and Riemenschneider. We start with the Grauert-Riemenschneider vanishing theorem [234], which one can view as a local version of 4.3.1.

Theorem 4.3.9. (Grauert–Riemenschneider vanishing). *Let $f : X \longrightarrow Y$ be a generically finite and surjective projective morphism of varieties, with X smooth. Then*

$$R^i f_* \mathcal{O}_X(K_X) \;=\; 0 \quad \text{for } \; i > 0.$$

Our proof of Theorem 4.3.9 — whose strategy goes back at least to Mumford [452, pp. 258–259] — depends on the possibility of using a global vanishing theorem to kill the higher direct images of a sheaf. The next statement summarizes the principle at work here.

Lemma 4.3.10. *Let $f : V \longrightarrow W$ be a morphism of irreducible projective varieties, and let A be an ample divisor on W. Suppose that $\mathcal{F}$ is a coherent sheaf on V with the property that*

$$H^j\Big(V\,,\, \mathcal{F} \otimes \mathcal{O}_V(f^* mA)\Big) = 0 \quad \textit{for all } \; j > 0 \textit{ and all } m \gg 0.$$

Then $R^j f_* \mathcal{F} = 0$ for every $j > 0$.

Proof of Lemma 4.3.10. Choose m sufficiently large so that

$$H^i\big(W, R^j f_* \mathcal{F} \otimes \mathcal{O}_W(mA)\big) = 0$$

for all $i > 0$ and $j \geq 0$, and so that in addition the sheaves $R^j f_* \mathcal{F} \otimes \mathcal{O}_W(mA)$, if non-zero, are globally generated. Then the Leray spectral sequence degenerates and shows that

$$H^j\Big(V\,,\,\mathcal{F}\otimes\mathcal{O}_V(f^*mA)\Big) \;=\; H^0\Big(W\,,\,R^jf_*\mathcal{F}\otimes\mathcal{O}_W(mA)\Big).$$

We have arranged that the group on the right is non-zero if $R^jf_*\mathcal{F}\neq 0$ and $m\gg 0$ is sufficiently positive. But if $j>0$ this would violate the hypothesis, and the l emma follows. □

Proof of Theorem 4.3.9. Since the theorem is local on Y, we can assume to begin with that Y is quasi-projective. The idea then is to "compactify" f in order to reduce to the situation in which both Y and X are projective. Specifically, we assert that one can construct a fibre square

$$\begin{array}{ccc} X & \hookrightarrow & \overline{X} \\ {\scriptstyle f}\big\downarrow & & \big\downarrow{\scriptstyle \bar{f}} \\ Y & \hookrightarrow & \overline{Y} \end{array}$$

where $\overline{X},\overline{Y}$ are projective, $\overline{X}$ is smooth, and $\bar{f}$ is generically finite. In fact, let $\overline{Y}$ be any projective closure of Y and take $\overline{X}_0$ to be any projective variety sitting in a Cartesian diagram as above, so that $\overline{X}_0\times_{\overline{Y}}Y=X$. Then take $\overline{X}\longrightarrow\overline{X}_0$ to be any resolution of singularities that is an isomorphism over X, as in Theorem 4.1.3 (ii). Since

$$R^i\bar{f}_*\mathcal{O}_{\overline{X}}(K_{\overline{X}})\mid Y \;=\; R^if_*\mathcal{O}_X(K_X),$$

it is enough to prove 4.3.9 for $\bar{f}$. Therefore we are reduced to proving the theorem under the additional assumption that both X and Y are projective.

Now let A be an ample divisor on Y. Then $f^*\big(mA\big)$ is a nef divisor on X, and it is big thanks to the fact that $\bar{f}$ is generically finite. Therefore

$$H^i\Big(\overline{X}\,,\,\mathcal{O}_{\overline{X}}\big(K_{\overline{X}}+f^*(mA)\big)\Big) \;=\; 0 \quad \text{for } i>0 \text{ and all } m\geq 1$$

by virtue of 4.3.1. The theorem now follows from the previous lemma. □

Remark 4.3.11. Once again, with a little more work it would be possible to invoke only the weak Hironaka Theorem 4.1.3 (i): see the proof of Theorem 9.4.1. □

Example 4.3.12. (Grauert–Riemenschneider canonical sheaf). Let X be an irreducible projective variety, and $\mu : X'\longrightarrow X$ any resolution of singularities. Define $\mathcal{K}_X$ to be the coherent $\mathcal{O}_X$-module

$$\mathcal{K}_X \;=\; \mu_*\mathcal{O}_{X'}(K_{X'}).$$

Then $\mathcal{K}_X$ is independent of the choice of resolution. Moreover, if D is any big and nef Cartier divisor on X, then

$$H^i\big(X,\mathcal{K}_X\otimes\mathcal{O}_X(D)\big) \;=\; 0 \quad \text{for } i>0. \qquad (*)$$

(For the independence statement, use the facts that any two resolutions can be dominated by a third, and that if $\nu : X'' \longrightarrow X'$ is a birational morphism of smooth varieties, then

$$\nu_*\big(\mathcal{O}_{X''}(K_{X''})\big) \;=\; \mathcal{O}_{X'}(K_{X'}).$$

For (*), use 4.3.9 to reduce to the vanishing $H^i\big(X', \mathcal{O}_{X'}(K_{X'} + \mu^* D)\big) = 0$ for $i > 0$.) □

Example 4.3.13. (Vanishing on varieties with rational singularities). Recall that a normal variety V has *rational singularities* if there exists a resolution $\mu : V' \longrightarrow V$ with

$$R^i\mu_*\mathcal{O}_{V'} \;=\; 0 \quad \text{for} \quad i > 0. \tag{*}$$

(It is equivalent to ask that this property hold for every resolution.) If X is a projective variety having only rational singularities, and if D is a big and nef divisor on X, then

$$H^j\big(X, \mathcal{O}_X(-D)\big) \;=\; 0 \quad \text{for} \quad j < \dim X.$$

(The hypothesis (*) reduces the question to the vanishing of the cohomology group $H^j\big(X', \mathcal{O}_{X'}(-f^*D)\big)$.) □

Remark 4.3.14. (Fujita's vanishing theorem). We recall that some applications of these results were presented in Section 1.4.D. □

A vanishing theorem for smooth subvarieties of projective space. As another application, we prove a vanishing theorem for the ideal sheaf of a smooth subvariety of projective space. It is a special case of the results of [55]. We refer to Section 1.8.C — where the result in question was stated as Theorem 1.8.40 — for an overview of the interest in vanishings of this type.

Theorem 4.3.15. *Let $X \subseteq \mathbf{P}^r$ be a smooth subvariety of dimension n and codimension $e = r - n$. Assume that X is scheme-theoretically cut out by hypersurfaces of degree d in the sense that $\mathcal{I}_X(d)$ is globally generated, $\mathcal{I}_X$ being the ideal sheaf of X. Then*

$$H^i\big(\mathbf{P}^r, \mathcal{I}_X(k)\big) \;=\; 0 \quad \text{for} \quad i > 0 \text{ and } k \;\geq\; e \cdot d - r.$$

We start with a useful lemma that allows one to realize ideals of smooth subvarieties via a blow-up.

Lemma 4.3.16. *Let V be an algebraic variety, and $W \subseteq V$ a non-singular subvariety contained in the smooth locus of V, with ideal sheaf $\mathcal{I} = \mathcal{I}_W \subseteq \mathcal{O}_V$. Consider the blowing-up*

$$\mu \,:\, V' \,=\, Bl_W(V) \longrightarrow V$$

of V along W, with exceptional divisor $E \subseteq V'$. Then for every integer $a \geq 0$,

$$\mu_*\mathcal{O}_{V'}(-aE) \;=\; \mathcal{I}^a \quad \textit{and} \quad R^j\mu_*\mathcal{O}_{V'}(-aE) \;=\; 0 \textit{ when } j>0.$$

In particular,

$$H^i\big(V', \mathcal{O}_{V'}(\mu^*L - aE)\big) \;=\; H^i\big(V, \mathcal{O}_V(L)\otimes\mathcal{I}^a\big)$$

for every integer i and every divisor L on V.

Indication of Proof. One argues inductively on a using the sequences

$$0 \longrightarrow \mathcal{O}_{V'}(-(a+1)E) \longrightarrow \mathcal{O}_{V'}(-aE) \longrightarrow \mathcal{O}_E(-aE) \longrightarrow 0,$$

noting that $\mu_*\mathcal{O}_E(-aE) = S^aN^*$, where $N^* = \mathcal{I}/\mathcal{I}^2$ is the conormal bundle to W in V. (Compare [280, Proposition V.3.4].) □

Proof of Theorem 4.3.15. Let $\mu : Y = \mathrm{Bl}_X(\mathbf{P}^r) \longrightarrow \mathbf{P}^r$ be the blowing-up of X, with exceptional divisor E, and denote by H the pullback of the hyperplane class. Thus Y is a smooth projective variety, and

$$\begin{aligned} K_Y &\equiv_{\text{lin}} \mu^*(K_{\mathbf{P}^r}) + (e-1)E \\ &\equiv_{\text{lin}} -(r+1)H + (e-1)E. \end{aligned}$$

Since $\mathcal{I}_X(d)$ is globally generated, its inverse image $dH - E$ is free and hence nef. Therefore

$$D_\ell =_{\text{def}} e\cdot\big(dH - E\big) + \ell H$$

is a big and nef divisor on Y provided that $\ell \geq 1$. Vanishing then gives

$$\begin{aligned} H^i\Big(Y\,,\,\mathcal{O}_Y(K_Y + D_\ell)\Big) &= H^i\Big(Y\,,\,\mathcal{O}_Y\big((ed+\ell-r-1)H - E\big)\Big) \\ &= 0 \quad \text{for} \quad i>0,\ \ell>0. \end{aligned}$$

The Theorem then follows from the case $a = 1$ of 4.3.16. □

Remark 4.3.17. As Kleiman observes, it is enough in 4.3.16 to assume that $W \subseteq V$ is a local complete intersection contained in the smooth locus of V. However, if $X \subseteq \mathbf{P}^r$ is singular, then the blow-up $Y = \mathrm{Bl}_X(\mathbf{P}^r)$ will generally be so as well, in which case vanishing does not directly apply. Thus Theorem 4.3.15 doesn't appear to extend formally to the setting of a local complete intersection $X \subseteq \mathbf{P}^r$. □

Example 4.3.18. (A vanishing for powers). In the situation of the theorem, given a positive integer $a > 0$ one has more generally the vanishing

$$H^i\big(\mathbf{P}^r, \mathcal{I}_X^a(k)\big) \;=\; 0 \quad \text{for} \quad i>0 \text{ and } k \;\geq\; d\cdot(e+a-1) - r.$$

(Argue as above, using Lemma 4.3.16.) □

Example 4.3.19. (Normal generation of "hyperadjoint" linear series). Let X be a non-singular projective variety of dimension n, and let B be a *very ample* divisor on X. If $k \geq n+1$ then for every $m \geq 1$ the maps

$$S^m H^0\big(X, \mathcal{O}_X(K_X + kB)\big) \longrightarrow H^0\big(X, \mathcal{O}_X(m(K_X + kB))\big)$$

are surjective. In other words, the linear series $|K_X + kB|$ defines a projectively normal embedding of X for every $k \geq n+1$.[8] This result appears in [55] and [5]. It is generalized to higher syzygies in [153]: see Theorem 1.8.60. (For simplicity we focus on the crucial case $m = 2$. Consider the product $X \times X$ with projections $\mathrm{pr}_1, \mathrm{pr}_2 : X \times X \longrightarrow X$: write

$$L_k \;=\; \mathrm{pr}_1^* \mathcal{O}_X(K_X + kB) \otimes \mathrm{pr}_2^* \mathcal{O}_X(K_K + kB).$$

It is enough to show that

$$H^1\big(X \times X, L_k \otimes \mathcal{I}_\Delta\big) = 0 \;\text{ when }\; k \geq n+1,$$

$\Delta \subseteq X \times X$ being the diagonal, and $\mathcal{I}_\Delta$ its ideal sheaf. For this, note first that

$$\mathcal{O}_{X \times X}(\mathrm{pr}_1^* B + \mathrm{pr}_2^* B) \otimes \mathcal{I}_\Delta$$

is globally generated: since B is very ample the assertion reduces to the case $X = \mathbf{P}^n$, where it can be verified explicitly. Then blow up Δ and argue as in the proof of 4.3.15.) □

4.4 Generic Vanishing Theorem

We discuss here a theorem of Green and the author from [242] in which one weakens the positivity hypotheses even further but in return settles for generic statements.

Let X be a smooth projective variety of dimension n, and denote by $\mathrm{Pic}^0(X)$ the identity component of the Picard variety of X. One thinks of $\mathrm{Pic}^0(X)$ as parameterizing all topologically trivial holomorphic or algebraic line bundles on X. We seek conditions under which a *generic* line bundle $P \in \mathrm{Pic}^0(X)$ satisfies a Kodaira-type vanishing

$$H^j\big(X, P\big) \;=\; 0 \quad \text{for } \; j \;<\; n \;=\; \dim X.$$

However, one cannot expect that this holds for every variety X. An elementary example illustrates the sort of phenomena that have to be taken into account.

---

8 As noted following Theorem 1.8.60, if $X = \mathbf{P}^n$ and B is the hyperplane divisor then $K_X + (n+1)B$ is trivial and so does not define an embedding of X. However, in other cases the bundle in question is actually very ample.

Example 4.4.1. Let C be a smooth projective curve of genus ≥ 2, Y a simply connected variety of dimension $n-1$, and set $X = C \times Y$. Then $\mathrm{Pic}^0(Y) = 0$, so $\mathrm{Pic}^0(X) = \mathrm{Pic}^0(C)$, i.e. every topologically trivial bundle on X is of the form $\mathrm{pr}_1^* \eta$ for some $\eta \in \mathrm{Pic}^0(C)$. But then the Künneth formula shows that

$$H^1\big(X, P\big) \neq 0$$

for every $P \in \mathrm{Pic}^0(X)$. □

What's happening in this example is that we've exhibited an n-dimensional variety that "looks one-dimensional" from the point of view of Pic^0. This shows that the actual dimension of a variety is not the relevant invariant for the problem at hand. Instead, we focus on its Albanese dimension.

Recall that the Albanese variety of a smooth projective variety X is the abelian variety

$$\mathrm{Alb}(X) = \frac{H^0\big(X, \Omega^1_X\big)^*}{H_1\big(X, \mathbf{Z}\big)}.$$

Integration of one-forms determines a holomorphic mapping

$$\mathrm{alb}_X : X \longrightarrow \mathrm{Alb}(X),$$

called the Albanese mapping of X. It is characterized by the property that any morphism $X \longrightarrow T$ from X to a complex torus or abelian variety factors through alb_X.

Definition 4.4.2. (Albanese dimension). The *Albanese dimension* of X, written $\mathrm{alb.dim}(X)$, is the dimension of the image $\mathrm{alb}_X(X)$ of X under alb_X. One says that X has *maximal Albanese dimension* if

$$\mathrm{alb.dim}(X) = \dim X,$$

or equivalently if $\mathrm{alb}_X : X \longrightarrow \mathrm{Alb}(X)$ is generically finite over its image. □

One then has:

Theorem 4.4.3. (Generic vanishing theorem). *Let X be a smooth projective variety of dimension n. If X has maximal Albanese dimension, then*

$$H^j\big(X, P\big) = 0 \quad \text{for} \quad j < n \quad \textit{and generic} \quad P \in \mathrm{Pic}^0(X). \tag{4.10}$$

In other words, there is a non-empty Zariski-open subset of $\mathrm{Pic}^0(X)$ parameterizing line bundles satisfying the vanishing (4.10).

Remark 4.4.4. The proof will show that if $a : X \longrightarrow T$ is a generically finite mapping from X to some complex torus or abelian variety, then (4.10) holds for the pullback to X of a generic element of $\mathrm{Pic}^0(T)$.

Corollary 4.4.5. *Assume that X has Albanese dimension k. Then*

$$H^j\big(X,P\big) \;=\; 0$$

for $j < k$ and generic $P \in \mathrm{Pic}^0(X)$.

Proof. We may suppose that $k < n = \dim X$. In this case, if $H \subseteq X$ is a generic hyperplane section of X then $\mathrm{alb.\,dim}(H) = \mathrm{alb.\,dim}(X)$. By induction on dimension (and Remark 4.4.4) we may assume that $H^j\big(H, P \mid H\big) = 0$ for $j < k$ and generic $P \in \mathrm{Pic}^0(X)$. The assertion then follows from the sequence $0 \longrightarrow P \otimes \mathcal{O}_X(-H) \longrightarrow P \longrightarrow P \mid H \longrightarrow 0$, since the term on the left is governed by Kodaira vanishing. □

The essential idea for the proof of Theorem 4.4.3 is to study the deformation theory of the groups $H^j\big(X,P\big)$ as P varies over $\mathrm{Pic}^0(X)$. Roughly speaking, one wants to argue that if $j < \mathrm{alb.\,dim}\, X$ then one can "deform away" any non-zero cohomology class. We will explain here the outline of how one carries this out, referring to [242] for full details.

The starting point is Grothendieck's variational theory of cohomology. Suppose that $f : Y \longrightarrow S$ is a flat projective morphism, and L is a line bundle on Y. Given $t \in S$ write Y_t, L_t for fibres over t. Then Grothendieck shows that there is a bounded complex $E^\bullet$ of vector bundles, defined locally on S, that computes the cohomology of L. More precisely, assume that S is affine, with coordinate ring $A = \mathbf{C}[S]$. Then for any A-module M — viewed as a sheaf on S — one can consider the A-modules

$$H^j\big(Y, L \otimes_A M\big) \;=\; R^j f_*\big(L \otimes_A M\big).$$

Grothendieck proves that there is a complex $E^\bullet$:

$$0 \longrightarrow E^0 \xrightarrow{\;d_0\;} E^1 \xrightarrow{\;d_1\;} \;\cdots\; \xrightarrow{\;d_{m-1}\;} E^m \longrightarrow 0 \tag{4.11}$$

of vector bundles on S with the property that

$$H^j\big(E^\bullet \otimes_A M\big) \;=\; R^j f_*\big(L \otimes_A M\big) \tag{4.12}$$

as δ-functors of M. We refer to [447, Chapter II.5] or [280, III.12] for excellent presentations. As a special case, take $M = \mathbf{C}(t)$ to be the residue field at a point $t \in S$: one gets an isomorphism

$$H^j\big(Y_t, L_t\big) \;=\; H^j\big(E^\bullet(t)\big),$$

where $E^\bullet(t) = E^\bullet \otimes \mathbf{C}(t)$ is the complex of vector spaces obtained from $E^\bullet$ by evaluation at t.

Assume now that S is non-singular, and fix a point $0 \in S$ plus a non-zero tangent vector $v \in T_0 S$. These data determine a *derivative complex* $D_v\big(E^\bullet, 0\big)$:

$$\cdots \longrightarrow H^{j-1}\big(E^\bullet(0)\big) \xrightarrow{D_v(d_{j-1})} H^j\big(E^\bullet(0)\big) \xrightarrow{D_v(d_j)} H^{j+1}\big(E^\bullet(0)\big) \longrightarrow \cdots \tag{4.13}$$

of vector spaces at 0. The quickest construction is to recall that $v \in T_0S$ corresponds to an embedding $D \subseteq S$ of the dual numbers into S. So one gets a short exact sequence

$$0 \longrightarrow \mathbf{C}(0) \longrightarrow \mathcal{O}_D \longrightarrow \mathbf{C}(0) \longrightarrow 0$$

of $\mathcal{O}_S$-modules. Tensoring by $E^\bullet$ yields a short exact sequence of complexes, which in turn gives rise to connecting homomorphisms

$$\begin{array}{ccc} H^j\big(\, E^\bullet \otimes \mathbf{C}(0)\,\big) & \xrightarrow{D_v(d_j)} & H^{j+1}\big(\, E^\bullet \otimes \mathbf{C}(0)\,\big) \\ \Vert & & \Vert \\ H^j\big(E^\bullet(0)\big) & & H^{j+1}\big(E^\bullet(0)\big) \end{array} \tag{4.14}$$

One checks that $D_v(d_{j+1}) \circ D_v(d_j) = 0$, and (4.13) is established.[9]

The philosophy is that the derivative complex $D_v\big(E^\bullet, 0\big)$ controls the deformation theory of the groups $H^j\big(E^\bullet(t)\big)$. A general statement along these lines appears in [242, Theorem 1.6]. Here we need only a special case:

Proposition 4.4.6. *Always supposing that S is smooth, assume that each of the cohomology sheaves $H^j\big(E^\bullet\big)$ is locally free (or zero) in a neighborhood of $0 \in S$. Then for every $v \in T_0S$ the derivative complex $D_v\big(E^\bullet, 0\big)$ is trivial, i.e. the maps*

$$D_v(d_j) \;:\; H^j\big(E^\bullet(0)\big) \longrightarrow H^{j+1}\big(E^\bullet(0)\big)$$

are zero.

We remark that the hypothesis on the cohomology sheaves is equivalent to asking that the dimensions $h^j\big(E^\bullet(t)\big)$ be constant in a neighborhood of $0 \in S$.

Proof of Proposition 4.4.6. It follows from [280, III.12.5 and III.12.6] that under the stated hypothesis the functor

---

9 For a more concrete construction, start by shrinking S so that each of the bundles E^j is trivial, say

$$E^j \;=\; V^j \otimes_{\mathbf{C}} \mathcal{O}_S$$

where V^j is a complex vector space of dimension p_j. Then the differentials $d_j : E^j \longrightarrow E^{j+1}$ are given by $p_j \times p_{j+1}$ matrices of functions on S. Differentiating the relation $d_{j+1} \circ d_j = 0$ yields

$$D_v(d_{j+1}) \circ d_j \;+\; d_{j+1} \circ D_v(d_j) \;=\; 0, \tag{*}$$

which shows that $D_v(d_j)$ maps cycles to cycles and boundaries to boundaries. Hence $D_v(d_j)$ induces a homomorphism $H^j\big(E^\bullet(0)\big) \longrightarrow H^{j+1}\big(E^\bullet(0)\big)$, and differentiating (*) shows that the composition of two of these maps is zero.

$$M \mapsto H^j(E^\bullet \otimes_A M)$$

is exact for every j. Therefore the connecting homomorphism in (4.14) vanishes. (See also [174, Lemma 12.6].) □

We now return to topologically trivial line bundles on a smooth projective variety X, and apply this discussion to the Poincaré bundle $\mathcal{P}$ on the product $X \times \mathrm{Pic}^0(X)$. Given $\xi \in \mathrm{Pic}^0(X)$, denote by P_ξ the corresponding topologically trivial line bundle on X, so that $P_\xi = \mathcal{P} \mid (X \times \{\xi\})$. Fix now a point $\xi \in \mathrm{Pic}^0(X)$ and a tangent vector

$$v \in T_\xi \mathrm{Pic}^0(X) = H^1(X, \mathcal{O}_X).$$

Then the associated derivative complex $D_v(X, P_\xi)$ takes the form

$$\ldots \longrightarrow H^{j-1}(X, P_\xi) \longrightarrow H^j(X, P_\xi) \longrightarrow H^{j+1}(X, P_\xi) \longrightarrow \ldots . \qquad (4.15)$$

It is verified in [242, Lemma 2.3] that the differentials appearing here are determined in the expected fashion:

Lemma 4.4.7. *The maps in the derivative complex* (4.15) *are given by cup product with the class* $v \in H^1(X, \mathcal{O}_X)$. □

Corollary 4.4.8. *At a general point* $\xi \in \mathrm{Pic}^0(X)$, *the homomorphisms*

$$H^j(X, P_\xi) \otimes H^1(X, \mathcal{O}_X) \longrightarrow H^{j+1}(X, P_\xi)$$

determined by cup product vanish for every j.

Proof. Like any finite collection of coherent sheaves on $\mathrm{Pic}^0(X)$, the direct images $R^j p_{2,*}\mathcal{P}$ of the Poincaré line bundle under the projection

$$p_2 : X \times \mathrm{Pic}^0(X) \longrightarrow X$$

are locally free (but possibly zero) in a neighborhood of a general point $\xi \in \mathrm{Pic}^0(X)$. The assertion then follows from 4.4.6 and 4.4.7. □

Remark 4.4.9. (Arbitrary line bundles). So far we have not used the hypothesis that the line bundle P_ξ is topologically trivial. The analogue of 4.4.8 remains valid for the bundles parameterized by a general point $\xi \in \mathrm{Pic}^\lambda(X)$ of the component of $\mathrm{Pic}(X)$ corresponding to any algebraic equivalence class λ. □

The next step is to reinterpret 4.4.8 Hodge-theoretically. Specifically, any topologically trivial line bundle $P \in \mathrm{Pic}^0(X)$ comes from a unitary representation $\pi_1(X) \longrightarrow U(1)$. The Hodge decomposition (and Dolbeaut theorem) for the corresponding twisted coefficient system gives rise to conjugate linear isomorphisms

$$H^j(X, P) \cong H^0(X, \Omega^j_X \otimes P^*).$$

Moreover, these isomorphisms are compatible with the Hodge conjugate isomorphism

$$H^1(X, \mathcal{O}_X) \cong H^0(X, \Omega^1_X)$$

in the sense that cup product on cohomology goes over to wedge product of forms. Therefore Corollary 4.4.8 is equivalent to a statement on holomorphic forms:

Corollary 4.4.10. *At a general point $\xi \in \mathrm{Pic}^0(X)$, the map*

$$H^0(X, \Omega^j_X \otimes P^*_\xi) \otimes H^0(X, \Omega^1_X) \longrightarrow H^0(X, \Omega^{j+1}_X \otimes P^*_\xi) \tag{4.16}$$

determined by wedge product vanishes for every j. □

A pointwise computation now shows that if X has maximal Albanese dimension and $H^0(X, \Omega^j_X \otimes P^*) \neq 0$, then the vanishing of the maps in (4.16) is possible only when $j = n$.

Proof of Theorem 4.4.3. Suppose that $0 \neq s \in \Gamma(X, \Omega^j_X \otimes P^*_\xi)$ is a non-zero section for general $\xi \in \mathrm{Pic}^0(X)$. Then

$$s \wedge \omega \ = \ 0 \quad \text{for all } \ \omega \in \Gamma(X, \Omega^1_X) \tag{*}$$

thanks to 4.4.10. Now fix a general point $x \in X$. We may suppose to begin with that $s(x) \neq 0$. Moreover, the assumption that X has maximal Albanese dimension is equivalent to the condition that global holomorphic one-forms span the cotangent space to X at a general point.[10] So we may choose $\omega_1, \ldots, \omega_n \in \Gamma(X, \Omega^1_X)$ in such a way that

$$\omega_1(x), \ldots, \omega_n(x) \ \in \ T^*_x X$$

form a basis of $T^*_x X$. Upon fixing a local trivialization of P^*_ξ near x, we can view $s(x)$ as an element of $\Lambda^j \, T^*_x X$, and

$$s(x) \wedge \omega_i(x) \ = \ 0 \ \in \ \Lambda^{j+1} \, T^*_x X$$

for every i by virtue of (*). But then the following elementary lemma implies that $j = n$, as required. □

Lemma 4.4.11. *Let V be an n-dimensional vector space and $u \in \Lambda^j V$ a non-zero vector. If $j < n$, then there exists a vector $w \in V$ such that $u \wedge w \neq 0$.* □

---

[10] Recall that the homomorphism

$$H^0(X, \Omega^1_X) \otimes_{\mathbf{C}} \mathcal{O}_X \longrightarrow \Omega^1_X$$

determined by evaluation is identified with the coderivative of the Albanese mapping alb_X.

Remark 4.4.12. (Further developments). These ideas are developed further in the papers [242], [45], [244], [46], [536], [13], [77], [95]. For instance, consider the cohomology support loci $V_i = \{\xi \in \mathrm{Pic}^0(X) \mid H^i(X, P_\xi) \neq 0\}$. It is established in [244] and [536] that each V_i is a torsion translate of a subtorus of $\mathrm{Pic}^0(X)$. Applications to the birational geometry of irregular varieties appear in [154], [262], [88], [89]; see Theorem 10.1.8 for one such. Hacon [263] has recently found a new approach to the generic vanishing theorems. □

Notes

The construction of cyclic coverings (Proposition 4.1.6), as well as its use in the present context, goes back to Mumford [446, pp. 97–98]. His later paper [452] contains several interesting proofs and applications of vanishing theorems.

As indicated in the text, Theorem 4.1.10 is an elaboration of constructions appearing in [60] and [368]. Our presentation of Theorem 4.2.3 draws on notes [146] from Ein's lectures at Catania, and our account of Theorem 4.3.1 follows the exposition in [533].

5

Local Positivity

In this chapter we discuss local positivity. The theory originates with Demailly's idea for quantifying how much of the positivity of an ample line bundle can be localized at a given point of a variety. The picture turns out to be considerably richer and more structured than one might expect at first glance, although the existing numerical results are (presumably!) not optimal.

In the first section we define the Seshadri constants measuring local positivity, and give their formal properties. We prove in Section 5.2 the existence of universal lower bounds on these invariants that hold at a very general point of any variety of fixed dimension. Section 5.3 focuses on abelian varieties, where Seshadri constants are related to metric invariants introduced by Buser and Sarnak. Finally, in Section 5.4 we discuss a variant involving local positivity along an arbitrary ideal sheaf: here the complexity of the ideal and its powers comes into play.

A word about notation: it seems most natural for these questions to work with line bundles, but it is also traditional to use additive notation. In keeping with the convention outlined at the end of Section 1.1.A, we will therefore use symbols suggestive of bundles, but will speak of divisors.

5.1 Seshadri Constants: Definition and Formal Properties

This section is devoted to the formal properties of Seshadri constants. We start with the basic definition.

Definition 5.1.1. (Seshadri constants). Let X be an irreducible projective variety and L a nef divisor on X. Fix a point $x \in X$, and let

$$\mu : X' = \mathrm{Bl}_x(X) \longrightarrow X$$

be the blowing up of x, with exceptional divisor $E \subseteq X'$. The *Seshadri constant*

$$\varepsilon(X, L; x) = \varepsilon(L; x)$$

of L at x is defined to be the non-negative real number

$$\varepsilon(L; x) = \max\{\varepsilon \geq 0 \mid \mu^*L - \varepsilon \cdot E \text{ is nef }\}. \quad \square$$

By the blowing up of X at a (possibly singular) point, we mean of course the blowing up of X along the maximal ideal sheaf $\mathfrak{m}_x \subseteq \mathcal{O}_X$.

Remark 5.1.2. The intuition to keep in mind is that $\varepsilon(L; x)$ measures how much of the positivity of L can be concentrated at x. Theorem 5.1.17 gives a concrete illustration of this idea. □

Example 5.1.3. (Numerical nature of Seshadri constants). It follows from the definition that $\varepsilon(L; x)$ depends only on the numerical equivalence class of L. □

Example 5.1.4. (Homogeneity). Seshadri constants satisfy the homogeneity property

$$\varepsilon(mL; x) = m \cdot \varepsilon(L; x)$$

for every $m \in \mathbf{N}$ and every $x \in X$. □

The name of these invariants stems from their connection to Seshadri's criterion for amplitude (Theorem 1.4.13):

Proposition 5.1.5. *In the situation of Definition 5.1.1 one has*

$$\varepsilon(L; x) = \inf_{x \in C \subseteq X} \left\{ \frac{(L \cdot C)}{\operatorname{mult}_x C} \right\},$$

the infimum being taken over all reduced irreducible curves $C \subseteq X$ passing though x.

Proof. As in the proof of Seshadri's criterion, to establish the nefness of $\mu^*L - \varepsilon E$ on $X' = \operatorname{Bl}_x(X)$, it is enough to intersect with the proper transforms of curves on X. So let $C \subseteq X$ be an integral curve passing through x, and denote by $C' \subseteq X'$ its proper transform. Then

$$(\mu^*L - \varepsilon E) \text{ is nef} \iff \big((\mu^*L - \varepsilon E) \cdot C'\big) \geq 0$$

for every such C. On the other hand,

$$(C' \cdot E) = \operatorname{mult}_x(C)$$

(see Lemma 5.1.10), and the assertion follows. □

Example 5.1.6. (Very ample divisors). If L is *very* ample, then $\varepsilon(L; x) \geq 1$ for every $x \in X$. (If $C \subseteq \mathbf{P}$ is a curve sitting in some projective space, then projection from a point $x \in C$ shows that $\operatorname{mult}_x(C) \leq \deg C$.) See Example 5.1.18 for a generalization. □

Example 5.1.7. (Projective spaces and Grassmannians). Let L be the hyperplane divisor on $\mathbf{P}^n$. Then $\varepsilon\big(\mathbf{P}^n, L; x\big) = 1$ for every $x \in \mathbf{P}^n$. Similarly, if $\mathbf{G}$ is a Grassmannian and L is the divisor defining the Plücker embedding, then $\varepsilon\big(\mathbf{G}, L; x\big) = 1$ for all x. (The upper bound $\varepsilon \leq 1$ follows from the presence of a line passing through every point.) □

Example 5.1.8. (Principally polarized abelian surfaces). Let $A = \operatorname{Pic}^1(\Gamma)$ be the Jacobian of a smooth curve Γ of genus two, and embed Γ in A in the canonical way. Then $\varepsilon\big(A, \Gamma; x\big)$ is independent of $x \in A$ since A is homogeneous, and we write simply $\varepsilon\big(A, \Gamma\big)$. Consider the homomorphism

$$m_2 \,:\, A \longrightarrow A \quad , \quad x \mapsto 2x$$

given by multiplication by two. The six Weierstrass points of Γ all have the same image under m_2, and so

$$C \;=_{\text{def}}\; m_2(\Gamma)$$

has a point of multiplicity 6. One computes that $C \equiv_{\text{num}} 4 \cdot \Gamma$, and therefore

$$\varepsilon\big(A, \Gamma\big) \;\leq\; \frac{\big(4\Gamma \cdot \Gamma\big)}{6} \;=\; \frac{4}{3}. \tag{*}$$

This example is due to Steffens [559], who shows that in fact equality holds in (*) at least when $\rho(A) = 1$ (compare Remark 5.3.13). It is also a special case of results of Bauer–Szemberg and Bauer concerning the Seshadri constants of an abelian surface with an arbitrary polarization: see Section 5.3.C. □

Singular subvarieties of higher dimension also lead to upper bounds on Seshadri constants.

Proposition 5.1.9. *Let $V \subseteq X$ be an irreducible subvariety of dimension ≥ 1 passing through x. Then*

$$\varepsilon\big(X, L; x\big) \;\leq\; \left(\frac{\big(L^{\dim V} \cdot V\big)}{\operatorname{mult}_x V} \right)^{\frac{1}{\dim V}}. \tag{5.1}$$

Moreover, equality holds for some V (possibly $V = X$).

In particular, if $n = \dim X$ then one has the basic bound

$$\varepsilon\big(L; x\big) \;\leq\; \sqrt[n]{\frac{\big(L^n\big)}{\operatorname{mult}_x X}}\,. \tag{5.2}$$

For Proposition 5.1.9 we start by recording a useful characterization of the multiplicity of a variety at a point. It is a special case of [208, p. 79].

Lemma 5.1.10. *Let V be a variety and $x \in V$ a fixed point. Denote by $\mu : V' \longrightarrow V$ the blowing-up of V at x, with exceptional divisor $E \subseteq V'$. Then*

$$(-1)^{(1+\dim V)} \cdot \big(E^{\dim V}\big) \;=\; \mathrm{mult}_x(V). \quad \square$$

Proof of Proposition 5.1.9. Let $V' \subseteq X' = \mathrm{Bl}_x(X)$ be the proper transform of V. Then $V' = \mathrm{Bl}_x(V)$, with exceptional divisor $E \,|\, V'$. If $\mu^*L - \varepsilon E$ is nef, then

$$\Big(\big(\mu^*L - \varepsilon E\big)^{\dim V} \cdot V'\Big) \;\geq 0$$

by virtue of Kleiman's theorem (Theorem 1.4.9), and the inequality (5.1) follows from 5.1.10. If $\mu^*L - \varepsilon E$ is nef but not ample, then the theorem of Campana and Peternell (2.3.18) implies that the divisor in question must have zero intersection number with some subvariety $V' \subseteq X'$. In this case equality holds in (5.1) for $V = \mu(V')$. $\square$

Example 5.1.11. (Variation of Seshadri constants in families). Let $p : X \longrightarrow T$ be a surjective proper morphism, L a divisor on X, and $x : T \longrightarrow X$ a section of p. Assume that p is smooth along the image of x. Write X_t, L_t for the indicated fibres over $t \in T$, and $x_t \in X_t$ for the image of x at t. Then $\varepsilon\big(X_t, L_t; x_t\big)$ is constant off the union of countably many proper subvarieties of T. If $t^* \in T$ is a very general point, then $\varepsilon\big(X_t, L_t; x_t\big) \leq \varepsilon\big(X_{t^*}, L_{t^*}; x_{t^*}\big)$ for every $t \in T$, i.e. Seshadri constants can only "jump down." (Blow up along the image of x, and apply 1.4.14.) $\square$

Remark 5.1.12. (Computations). It is very difficult to compute $\varepsilon\big(L; x\big)$ except in the simplest situations. In view of 5.1.5 and 5.1.9, upper bounds can be established by exhibiting suitable singular curves or subvarieties. However, lower bounds are generally much harder to come by: in effect one has to prove the non-existence of singular curves or subvarieties of arbitrarily high degree. $\square$

Remark 5.1.13. (Irrational Seshadri constants). As an illustration of the preceding remark, we note that there is no instance in which $\varepsilon\big(X, L; x\big)$ has been shown to be an irrational number. However it seems rather unlikely that Seshadri constants should always be rational. $\square$

Remark 5.1.14. (Nagata's conjecture). These questions are closely related to a celebrated conjecture of Nagata concerning the blow-up

$$\mu : P \;=\; \mathrm{Bl}_r(\mathbf{P}^2) \longrightarrow \mathbf{P}^2$$

of the projective plane along a finite set $S \subseteq \mathbf{P}^2$ consisting of r very general points. Denote by H the pullback to P of a line, and let $E \subseteq P$ be the exceptional divisor of μ (so that E is a sum of r disjoint (-1)-curves). Nagata [463] conjectured in effect that the $\mathbf{R}$-divisor

$$H \;-\; \sqrt{\tfrac{1}{r}} \cdot E$$

is nef on P provided that $r \geq 9$. Note that if $\varepsilon > \frac{1}{\sqrt{r}}$ then $\big((H - \varepsilon E)^2\big) < 0$, so the class in question cannot be nef in this case. Therefore the assertion of Nagata's conjecture is that the Seshadri-like constant associated to a very general set of points is maximal. The statement is elementary — and was established by Nagata — when r is a perfect square.[1] However, in spite of a great deal of effort in recent years, the conjecture seems not to be known for any other values of $r > 9$. See however [613] and [57] for some partial results. A symplectic analogue, discussed in Remark 5.1.23, is solved in [58]. □

At least when L is ample, Demailly [123] has given a nice interpretation of Seshadri constants in terms of jets. We start with a definition:

Definition 5.1.15. (Separation of jets). Let B be a line bundle on a projective variety X, and fix a smooth point $x \in X$ with maximal ideal $\mathfrak{m}_x \subseteq \mathcal{O}_X$. One says that $|B|$ *separates s-jets at x* if the natural map

$$H^0\big(X, B\big) \longrightarrow H^0\big(X, B \otimes \mathcal{O}_X/\mathfrak{m}_x^{s+1}\big) =_{\text{def}} J_x^s\big(B\big)$$

taking sections of B to their s-jets is surjective. □

Thus $|B|$ separates 0-jets if and only if B is free at x, and in this case $|B|$ separates 1-jets if and only if the derivative $d_x\phi_{|B|}$ at x of the rational map $\phi_{|B|} : X \dashrightarrow \mathbf{P}H^0\big(X, B\big)$ is injective. We also define an invariant measuring how many jets the linear series in question separates:

Definition 5.1.16. With X and B as in 5.1.15, denote by $s\big(B; x\big)$ the largest natural number s such that $|B|$ separates s-jets at x. (If x is a base point of $|B|$, put $s\big(B; x\big) = -1$.) □

Of course one makes analogous definitions starting from a divisor L or from a possibly incomplete linear series $|V| \subseteq |B|$.

Now suppose that L is ample. Then one expects $|kL|$ to separate more and more jets as k grows. Demailly's observation is that $\varepsilon\big(L; x\big)$ controls the rate of growth of $s\big(kL; x\big)$ as a function of k.

Theorem 5.1.17. (Growth of jet separation). *Let L be an ample divisor on an irreducible projective variety X, and let $x \in X$ be a smooth point. Then*

$$\varepsilon\big(X, L; x\big) \;=\; \lim_{k \to \infty} \frac{s\big(kL; x\big)}{k}\,.$$

Proof. Write $\varepsilon = \varepsilon\big(L; x\big)$ and $s_k = s\big(kL; x\big)$. We will prove from left to right the inequalities

$$\varepsilon \;\geq\; \limsup \frac{s_k}{k} \;\geq\; \liminf \frac{s_k}{k} \;\geq\; \varepsilon.$$

[1] If S consists of the d^2 points of intersection of two plane curves of degree d, then $dH - E$ is globally generated and hence nef. It follows from 1.4.14 or 5.1.11 that this class is also nef when S is any sufficiently general set of d^2 points.

To this end we first show that $\varepsilon \geq \frac{s_k}{k}$ for every k. In fact, fix any reduced and irreducible curve $C \ni x$. Since $|kL|$ separates s_k-jets at x, we can find a divisor $F_k \in |kL|$ with $\text{mult}_x(F_k) \geq s_k$ and $C \not\subseteq F_k$. Then

$$\begin{aligned} k\,(L \cdot C) &= (F_k \cdot C) \\ &\geq \text{mult}_x(F_k) \cdot \text{mult}_x(C) \\ &\geq s_k \cdot \text{mult}_x(C). \end{aligned}$$

Using 5.1.5, this implies that $\varepsilon \geq \frac{s_k}{k}$, as asserted.

It remains to prove that $\liminf \frac{s_k}{k} \geq \varepsilon$. To this end, fix integers $p_0, q_0 \gg 0$ with $0 < \varepsilon - \frac{p_0}{q_0} \ll 1$. As in 5.1.1 write $\mu : X' \longrightarrow X$ for the blowing-up of $x \in X$, with exceptional divisor E. Then $\mu^*(q_0 L) - p_0 E$ is ample. So by Fujita's vanishing theorem 1.4.35 there is a natural number m_0 such that

$$H^1\Big(X' \,,\, \mathcal{O}_{X'}\big(m(\mu^*(q_0L) - p_0E) + P\big)\Big) \;=\; 0 \qquad (*)$$

for every $m \geq m_0$ and every nef line bundle P. Now given any integer $k > m_0 q_0$, write $k = mq_0 + q_1$ with $0 \leq q_1 < q_0$. Applying (*) with $P = \mu^*(q_1 L)$ one finds that

$$H^1\Big(X' \,,\, \mathcal{O}_{X'}\big(\mu^*(kL) - mp_0E \big)\Big) \;=\; 0. \qquad (**)$$

The H^1 appearing in (**) is isomorphic to the group $H^1\big(X, \mathcal{O}_X(kL) \otimes \mathfrak{m}_x^{mp_0}\big)$, whose vanishing implies that $|kL|$ separates $(mp_0 - 1)$-jets (Lemma 4.3.16). Therefore

$$\frac{s_k}{k} \;\geq\; \frac{mp_0 - 1}{k} \;\geq\; \frac{mp_0 - 1}{(m+1)q_0} \;=\; \left(\frac{m}{m+1}\right)\frac{p_0}{q_0} - \frac{1}{(m+1)q_0}.$$

By first choosing $\frac{p_0}{q_0}$ to closely approximate ε and then taking k — and hence also m — to be sufficiently large, we can arrange that the last expression on the right is arbitrarily close to ε. Therefore $\liminf \frac{s_k}{k} \geq \varepsilon$, and we are done. □

Example 5.1.18. (Ample and free divisors). Assume that L is ample and $|L|$ is free. Then $\varepsilon\big(L; x\big) \geq 1$ for every $x \in X$. (Given any irreducible curve $C \ni x$, there is a divisor $F_x \in |L|$ passing through x that does not contain C.) This generalizes Example 5.1.6. □

On smooth varieties, Kodaira-type vanishings lead to more precise results for adjoint bundles.

Proposition 5.1.19. (Seshadri constants and adjoint bundles). *Let X be a smooth projective variety of dimension n, and let L be a big and nef divisor on X.*

(i). *If $\varepsilon\big(L; x\big) > n + s$ at some point $x \in X$, then $K_X + L$ separates s-jets at x.*

(ii). *If* $\varepsilon(L;x) > 2n$ *at some point* $x \in X$*, then the rational mapping*

$$\phi = \phi_{|K_X+L|} \; : \; X \dashrightarrow \mathbf{P}$$

defined by $K_X + L$ *is birational onto its image.*

(iii). *If* $\varepsilon(L;x) > 2n$ *for every* $x \in X$*, then* $K_X + L$ *is very ample.*

Indication of Proof of Proposition 5.1.19. As usual, let $\mu : X' \longrightarrow X$ be the blowing up of $x \in X$, with exceptional divisor E. Then the canonical bundle on X' is given by $K_{X'} = \mu^* K_X + (n-1)E$, and so

$$K_{X'} + \mu^* L - (n+s)E = \mu^*\big(K_X + L\big) - (s+1)E.$$

Now if $\varepsilon(L;x) > n+s$, then $\mu^* L - (n+s)E$ is nef and big, and hence

$$H^1\Big(X', \mathcal{O}_{X'}\big(\mu^*(K_X+L) - (s+1)E\big)\Big) = H^1\Big(X, \mathcal{O}_X\big(K_X+L\big) \otimes \mathfrak{m}_x^{s+1}\Big) = 0$$

thanks to Theorem 4.3.1 (and Lemma 4.3.16). Therefore $K_X + L$ separates s-jets at x, and this proves (i). For (ii), note first that if the stated inequality holds at any one point x, then it holds for a very general point of X (Example 5.1.11). Now let

$$\nu : \mathrm{Bl}_{\{x,y\}}(X) \longrightarrow X$$

be the blowing up of X at very general points $x, y \in X$, with exceptional divisors E_x and E_y lying over x and y respectively. Then each of the $\mathbf{Q}$-divisors

$$\nu^*\big(\tfrac{1}{2}L\big) - nE_x \quad , \quad \nu^*\big(\tfrac{1}{2}L\big) - nE_y$$

is nef and big, and hence so too is their sum $\nu^* L - n\big(E_x + E_y\big)$. Arguing as above we find that

$$H^1\big(X, \mathcal{O}_X(K_X + L) \otimes \mathfrak{m}_x \otimes \mathfrak{m}_y\big) = 0,$$

and hence that $K_X + L$ separates x and y. In other words, $\phi_{|K_X+L|}$ is one-to-one off a countable union of proper subvarieties of X, and statement (ii) follows. Assertion (iii) is similar. (See [149] for details.) □

Remark 5.1.20. Suppose that L is a big and nef divisor on a smooth projective n-fold X with the property that

$$\varepsilon\big(L;x\big) > \frac{n+s}{a}$$

for some $a \in \mathbf{N}$ and some $x \in X$. Then it follows from 5.1.19 and the homogeneity properties of Seshadri constants that $K_X + aL$ separates s-jets at x. Statements (ii) and (iii) of 5.1.19 admit similar generalizations. □

Finally, we mention without proof an interesting connection with symplectic geometry, due to McDuff and Polterovich. Let X be a smooth projective variety of dimension n and L an ample line bundle on X. Choose a Kähler form ω_L representing $c_1(L) \in H^2\big(X, \mathbf{C}\big)$. Then $\big(X, \omega_L\big)$ is a symplectic manifold, and it follows from Moser's lemma [423, Theorem 3.17] that up to symplectomorphism it depends only on $c_1(L)$, not the particular choice of ω_L.

Consider now the open ball

$$B(\lambda) \subseteq \mathbf{C}^n = \mathbf{R}^{2n} \tag{5.3}$$

of radius λ, i.e. $B(\lambda) = \big\{ z \in \mathbf{C}^n \mid \|z\| < \lambda \big\}$. One views $B(\lambda)$ as a symplectic manifold by equipping it with the standard symplectic form

$$\omega_{\text{std}} = \sum dx_\alpha \wedge dy_\alpha \tag{5.4}$$

on $\mathbf{C}^n$, $z_\alpha = x_\alpha + \sqrt{-1}y_\alpha$ being the usual coordinates on $\mathbf{C}^n$. In his seminal paper [251], Gromov discovered that it is very interesting to ask how large a ball can be embedded symplectically into a given symplectic manifold. Specifically, following [423, §12.1], one considers the Gromov width of such a manifold:

Definition 5.1.21. (Gromov width). The *Gromov width* $w_G\big(M, \omega\big)$ of a symplectic manifold $\big(M, \omega\big)$ is the supremum of all $\lambda > 0$ for which there exists a $\mathcal{C}^\infty$ embedding

$$j : B(\lambda) \hookrightarrow M \quad \text{with } j^*\omega = \omega_{\text{std}}. \quad \square$$

Returning to the symplectic manifold arising from an ample line bundle L on a smooth projective variety X, McDuff and Polterovich [422, Corollary 2.1.D] prove in effect:

Theorem 5.1.22. (Seshadri constants and Gromov width). *Set*

$$\varepsilon\big(X, L\big) = \max_{x \in X} \varepsilon\big(X, L; x\big).$$

Then the Gromov width of $\big(X, \omega_L\big)$ *satisfies the inequality*

$$w_G\big(X, \omega_L\big) \geq \sqrt{\frac{\varepsilon\big(X, L\big)}{\pi}}.$$

In other words, a lower bound on the holomorphic invariant $\varepsilon\big(L; x\big)$ yields a lower bound on the Gromov width. The reader may consult [422] for the proof, which revolves around the ideas of symplectic blowing up and down. We will draw on an elementary partial converse in Section 5.3 below.

Remark 5.1.23. (Symplectic packings). These matters are also closely connected with the symplectic packing problem. Given a symplectic manifold

(M, ω) and a natural number $r > 0$ one asks whether M can be filled by r disjoint symplectic balls of equal radius.[2] This is already very interesting for $(M, \omega) = (B, \omega_{\text{std}})$ where $B = B(1)$ is the ball of radius 1. Drawing on the analogue of 5.1.22 for several points, McDuff and Polterovich showed in [422] that Nagata's conjecture (Example 5.1.14) would imply that the four-dimensional ball B admits full packings by $r \geq 9$ balls.[3]

Shortly thereafter, Biran [58] completely solved the symplectic packing problem in dimension four using Seiberg–Witten technology. In fact, Biran proves that given any symplectic 4-manifold (M, ω), there is a constant $r_0 = r_0(M, \omega)$ such that M admits full packings by any number $r \geq r_0$ balls. Moreover, he gives a geometric interpretation of $r_0(M, \omega)$: if k_0 is a positive integer such that the cohomology class $k_0[\omega]$ is represented by a symplectically embedded surface of genus ≥ 1, then one can take $r_0 = k_0^2 \int_M \omega \wedge \omega$. We refer to [59] for a nice survey of the symplectic packing problem and its relation to algebraic geometry. □

Remark 5.1.24. (A generalization of Nagata's conjecture). The theorem of Biran discussed in the previous remark suggests a natural extension of Nagata's conjecture (Remark 5.1.14):

> **Conjecture.** *Let X be a smooth projective surface and L an ample divisor on X. Choose an integer k_0 such that there exists a smooth curve of genus ≥ 1 in the linear system $|k_0 L|$, and set $r_0 = k_0^2 \cdot (L^2)$. Consider the blowing-up of*
>
> $$\mu : \mathrm{Bl}_r(X) \longrightarrow X$$
>
> *of X at r very general points, and denote by E and H respectively the exceptional divisor on $\mathrm{Bl}_r(X)$ and the pullback of L. Then the* **R**-*divisor*
>
> $$H - \sqrt{\frac{(L^2)}{r}} \cdot E$$
>
> *is nef provided that $r \geq r_0$.*

Note that when $X = \mathbf{P}^2$ this exactly reduces to the statement of 5.1.14. □

[2] More precisely, the problem asks whether one can find symplectic embeddings of r disjoint balls whose volume comes arbitrarily close to the volume of (M, ω).

[3] The connection stems from the fact that the unit ball $B(1) \subseteq \mathbf{C}^n$ is symplectomorphic to $\mathbf{P}^n - \mathbf{P}^{n-1}$ with the Fubini–Study form, $\mathbf{P}^{n-1} \subseteq \mathbf{P}^n$ being a hyperplane.

5.2 Lower Bounds

The results of the previous section point to the geometric interest in lower bounds on Seshadri constants. In this section we discuss a theorem of Ein, Küchle, and the author [149] giving such inequalities (which however are not expected to be optimal). Roughly speaking, the picture is that there are universal bounds that hold at a very general point of every variety of given dimension, while at special points Seshadri constants can become arbitrarily small.

The exposition proceeds in three parts. We start by discussing the background and statements of the results. We next present some preparatory material on multiplicities of points on divisors: we have chosen to go through this in some detail in the thought that it might find applications in other contexts as well. Finally, we sketch the proof of the existence of universal generic lower bounds on local positivity.

5.2.A Background and Statements

The first thing to note is that there do not exist uniform lower bounds on Seshadri constants holding at every point of every variety.

Example 5.2.1. (Miranda's example). Fix any $\delta > 0$. We produce a smooth surface X, a point $x \in X$, and an ample divisor L on X such that

$$\varepsilon\big(X, L; x\big) \ < \ \delta.$$

To this end, start with a curve $\Gamma \subseteq \mathbf{P}^2$ of degree d having an m-fold point at $p \in \Gamma$, with $m > \frac{1}{\delta}$. Choose next a second curve $\Gamma' \subseteq \mathbf{P}^2$ of degree d that meets Γ transversely. By taking $d \gg 0$ and Γ' sufficiently general, we can suppose that all the curves in the pencil spanned by Γ and Γ' are irreducible. Now let

$$\mu : X = \mathrm{Bl}_{\Gamma \cap \Gamma'}(\mathbf{P}^2) \longrightarrow \mathbf{P}^2$$

be the surface obtained by blowing up the base points of that pencil. Thus X admits a mapping

$$\pi : X \longrightarrow \mathbf{P}^1,$$

and Γ, Γ' are isomorphic to curves $C\,,\, C' \subseteq X$ appearing as fibres of π. In particular, C has multiplicity m at a point $x \in C$. The d^2 points of intersection $\Gamma \cap \Gamma'$ in $\mathbf{P}^2$ determine sections of π: fix one of them, say $S \subseteq X$. The situation is illustrated in Figure 5.1.

Now take $a \in \mathbf{N}$, and put

$$L \ = \ L_a \ = \ aC + S.$$

We assert that L_a is ample provided that $a \geq 2$. In fact, the three intersection numbers

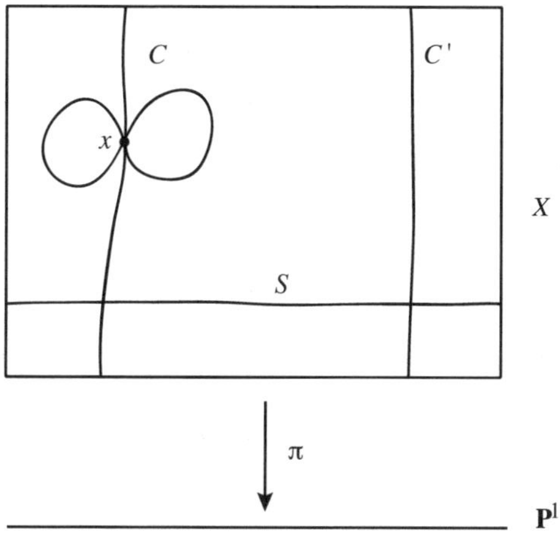

Figure 5.1. Miranda's example

$$(L \cdot L) = 2a - 1 \ , \ (L \cdot S) = a - 1 \ , \ (L \cdot C) = 1$$

are strictly positive. Moreover if $T \subseteq X$ is any curve dominating $\mathbf{P}^1$, then $(T \cdot C) > 0$, from which it follows that $(L \cdot T) > 0$, and by construction any curve contained in a fibre is numerically equivalent to a positive multiple of C. So the stated amplitude follows from Nakai's criterion. But then we find

$$\varepsilon(L; x) \leq \frac{(L \cdot C)}{\text{mult}_x(C)} = \frac{1}{m},$$

as required. Note that by varying a we can make $(L \cdot L)$ as large as we like. □

Example 5.2.2. (Higher-dimensional examples). As Viehweg remarked, examples in dimension two quickly lead to higher-dimensional ones as well. Specifically, let (X, L) be the output from 5.2.1 and for $n \geq 3$ set

$$Y = X \times \mathbf{P}^{n-2} \quad \text{and} \quad N = \text{pr}_1^* L + \text{pr}_2^* H,$$

H being the hyperplane divisor on $\mathbf{P}^{n-2}$. By taking curves in $X \times \{z\}$, one sees that

$$\varepsilon(N; (x, z)) \leq \varepsilon(L; x) \quad \text{for all} \quad z \in \mathbf{P}^{n-2}.$$

In particular, $\varepsilon(N; y)$ can be arbitrarily small in codimension two. □

Note that in Miranda's example, the Seshadri constant is small only at a very special point. Somewhat surprisingly, Ein and the author showed by a simple argument in [152] that this is a general fact on surfaces.

Proposition 5.2.3. (Seshadri constants on surfaces). *Let X be a smooth projective surface, and L an ample divisor on X. Then*

$$\varepsilon(L; x) \geq 1$$

for all except perhaps countably many points $x \in X$.

An outline of the proof appears at the end of this subsection.

Proposition 5.2.3 suggested the following

Conjecture 5.2.4. Let X be any irreducible projective variety, and L any nef and big divisor on X. Then

$$\varepsilon(X, L; x) \geq 1$$

for all $x \in X$ outside the union of countably many proper subvarieties of X.

The existence of such a bound would be quite striking. It would mean in effect that at a very general point, any ample line bundle has at least the same local positivity as the hyperplane bundle on projective space.

The basic known result, from [149], proves the existence of universal generic bounds in a fixed dimension. However, the actual statement is (presumably) suboptimal by a factor of $n = \dim X$:

Theorem 5.2.5. (Universal generic bounds). *Let X be an irreducible projective variety of dimension n, and L a big and nef divisor on X. Then*

$$\varepsilon(X, L; x) \geq \frac{1}{n} \tag{5.5}$$

for any very general point $x \in X$.

As always, the assertion that (5.5) holds at a very general point means that it can fail only on a countable union of proper subvarieties of X.

Remark 5.2.6. The main result of [149] is actually slightly stronger. Specifically, suppose in the situation of 5.2.5 that there exists a real number $\alpha > 0$, together with a countable union $\mathcal{B} \subseteq X$ of proper closed subvarieties, such that

$$\left(L^{\dim V} \cdot V\right) \geq (\dim V \cdot \alpha)^{\dim V}$$

for every irreducible subvariety $V \subseteq X$ with $V \not\subseteq \mathcal{B}$. Then

$$\varepsilon(X, L; x) \geq \alpha$$

for very general x. Theorem 5.2.5 follows by taking $\alpha = \frac{1}{n}$ and invoking Corollary 2.2.11 to ensure that the restriction of L to V is big (and of course nef) on V. □

Remark 5.2.7. (Lower bounds on Gromov width). Assume that X is smooth and L is ample, and denote by (X, ω_L) the corresponding symplectic manifold. In view of Theorem 5.1.22 of McDuff and Polterovich, 5.2.5 leads to the a priori lower bound

$$w_G\big(X, \omega_L\big) \ \geq \ \sqrt{\frac{1}{\pi \dim X}}$$

on the Gromov width of (X, ω_L). □

To motivate what follows, we give a quick preview of the proof of Theorem 5.2.5. Suppose that the statement fails. Then given any point $x \in X$, there exists a curve $C_x \subseteq X$ through x with

$$\operatorname{mult}_x(C_x) \ > \ n \cdot \big(L \cdot C_x\big).$$

On the other hand, consider for $k > 0$ any divisor $F \in |kL|$ that passes through x. If $C_x \nsubseteq F$, then

$$k \cdot \big(C_x \cdot L\big) \ = \ \big(C_x \cdot F\big) \ \geq \ \operatorname{mult}_x(C_x) \cdot \operatorname{mult}_x(F).$$

So to get a contradiction, it suffices to produce $F \not\supseteq C_x$ with $\operatorname{mult}_x(F) \geq \frac{k}{n}$. To this end we choose for $k \gg 0$ a divisor $E \in |kL|$ with large multiplicity at a general point $y \in X$. By what we have said, one expects that $C_y \subseteq E$. However by a "gap" argument, we are able to locate a point $x = x(y) \in E$ such that the difference between the multiplicity of E at x and at a general point of C_x is not too small. The essential idea then is to apply a suitable differential operator to the defining equation of E. This leads to a new divisor $F \in |kL|$ that does not contain C_x but still has relatively high multiplicity at x, as required.

We carry out this argument in the next two subsections. Section 5.2.B focuses on multiplicities of divisors, and in particular on the effect of differentiation in parameter directions (Proposition 5.2.13). The proof of Theorem 5.2.5 occupies 5.2.C. We conclude the present discussion by sketching the verification of Proposition 5.2.3 and indicating some extensions.

Example 5.2.8. (Outline of proof of Proposition 5.2.3). We use the characterization 5.1.5 of Seshadri constants. The main point, which was inspired by [614], is to view the question variationally. In fact, the set

$$\Big\{(C, x) \,\big|\, C \subseteq X \text{ is a reduced, irreducible curve with } \operatorname{mult}_x(C) > \big(L \cdot C\big)\Big\}$$

consists of at most countably many algebraic families. The proposition will follow if we show that each of these families is discrete, i.e. that pairs (C, x) forcing $\varepsilon\big(L; x\big) < 1$ are rigid.

Suppose to the contrary that (C_t, x_t) is a non-trivial one-parameter family, parameterized by a smooth curve T, of reduced and irreducible curves $C_t \subseteq X$ and points $x_t \in C_t$ with $\operatorname{mult}_{x_t}(C_t) > L \cdot C_t$. Let $C = C_{t^*}$ and $x = x_{t^*}$ for general t^*, and set $m = \operatorname{mult}_{x_{t^*}}(C_{t^*})$. The given deformation of C determines a section $\rho(\frac{d}{dt}) \in H^0\big(C, \mathcal{O}_C(C)\big)$, and one shows that since the deformation preserves the m-fold point of C, $\rho(\frac{d}{dt})$ vanishes to order $\geq m-1$ at x. This implies that

$$(C^2) \geq m(m-1).$$

But $(L^2) \cdot (C^2) \leq (L \cdot C)^2$ by the Hodge index theorem, and since $(L \cdot C) \leq m-1$ by assumption, we find that

$$m(m-1) \leq (L^2) \cdot (C^2) \leq (L \cdot C)^2 \leq (m-1)^2.$$

This is a contradiction when $m > 1$, and the statement follows. □

Remark 5.2.9. (Additional results on surfaces). Seshadri constants on surfaces are studied in a number of papers, including [36], [559], [39], [485]. For instance Steffens [559] shows that if (X, L) is a polarized surface having Picard number $\rho(X) = 1$, and if $x \in X$ is a very general point, then

$$\varepsilon(L; x) \geq \left[\sqrt{(L^2)}\right],$$

where $[x]$ denotes as usual the integer part of x. It follows in particular that if $(L^2) = m^2$ is a perfect square, then $\varepsilon(L; x) = m$ at a very general point. Bauer [39] obtains lower bounds for Seshadri constants at arbitrary points in terms of geometric invariants of X. Oguiso [485] studies the behavior of Seshadri constants in families, and shows for instance that if (X_t, L_t, x_t) is any family of pointed polarized surfaces, and if $\alpha < \sqrt{(L^2)}$ is any fixed number, then the set of parameter values t at which $\varepsilon(X_t, L_t; x_t) \leq \alpha$ is Zariski-closed in the parameter variety T. □

5.2.B Multiplicities of Divisors in Families

We study in this subsection some questions involving multiplicities of divisors. Our presentation closely follows the discussion in [149, §2].

Let M be a smooth variety, and F an effective divisor on M. We denote by

$$\text{mult}_x(F) = \text{mult}_x(M, F)$$

the multiplicity of F at a point $x \in M$. The function $x \mapsto \text{mult}_x(F)$ is Zariski upper-semicontinuous on M. Therefore one can define the multiplicity of F along a subvariety:

Definition 5.2.10. (Multiplicity along a subvariety). If $Z \subseteq M$ is an irreducible subvariety, then the *multiplicity of F along Z*, written $\text{mult}_Z(F)$ or $\text{mult}_Z(M, F)$, is the multiplicity $\text{mult}_x(F)$ at a general point $x \in Z$. □

We start with a result showing that multiplicities of divisors are preserved under restriction to general fibres of a mapping. As a matter of notation, given a morphism $X \longrightarrow S$ of varieties or schemes and a point $t \in S$, write X_t for the fibre of X over t. Similarly, if $F \subset X$ is an effective divisor that does not contain the fibre X_t, then F_t denotes the restriction $F \mid X_t$ of F to that fibre.

Lemma 5.2.11. (Multiplicities along fibres). *Let $p : M \longrightarrow T$ be a morphism of smooth varieties, and suppose that $V \subseteq M$ is an irreducible subvariety dominating T:*

$$\begin{array}{ccc} V & \subseteq & M \\ & \searrow \quad \swarrow_{p} & \\ & T. & \end{array}$$

Let $F \subseteq M$ be an effective divisor. Then for a general point $t \in T$, and any irreducible component $V_t' \subseteq V_t$,

$$\operatorname{mult}_{V_t'}(M_t\,,\,F_t) \;=\; \operatorname{mult}_V(M\,,\,F).$$

Proof. Since the assertion involves only a general point of T, we may suppose (by the theorem on generic smoothness) that p is a smooth morphism. Say $\operatorname{mult}_V(F) = m$. It is enough to show that for a sufficiently general point $t \in T$, and for any component $V_t' \subseteq V_t$, there is at least one point $y \in V_t'$ such that $\operatorname{mult}_y(M_t, F_t) = m$. Since V dominates T, we can choose local analytic coordinates

$$(z\,,\,x) \;=\; (z_1, \dots, z_d\,,\,x_1, \dots, x_e)$$

on M, and $z = (z_1, \dots, z_d)$ on T, in such a way that $p(z,x) = z$, V contains an open neighborhood of the "z-plane" $\{x_1 = \ldots = x_e = 0\}$, and F has multiplicity m along this plane. Let $f(z,x)$ be a local equation for F, and given a fixed point $t = (t_1, \dots, t_d) \in T$, write $\bar{f}_t(x) = f(t,x)$. It is enough to show that there exists $t \in T$ such that

$$\operatorname{ord}_0 \bar{f}_t(x) \;=\; \operatorname{ord}_{(t,0)} f(z,x).$$

But this is clear if one writes f in the form

$$f(z,x) \;=\; \sum b_I(z)\, x^I,$$

where $I = (i_1, \dots, i_e)$ is a multi-index, and $b_I(z)$ is a power series in z. Since F has multiplicity m along $\{x = 0\}$, the terms of weight $|I| < m$ must vanish, and then it is enough to choose t so that $b_I(t) \neq 0$ for some I with $|I| = m$. (A less computational argument appears in [149, §2].) □

Corollary 5.2.12. *In the situation of the lemma, let $t \in T$ be a general point. Then*

$$\operatorname{mult}_y(M, F) \;=\; \operatorname{mult}_y(M_t, F_t)$$

for every $y \in M_t$.

Proof. If not, one could construct a subvariety $V \subseteq M$, dominating T, such that

$$\operatorname{mult}_y(M_t, F_t) \;>\; \operatorname{mult}_y(F, T)$$

for every $y \in V$. But this contradicts the lemma. □

We now come to a construction that will play a central role in the proof of Theorem 5.2.5. Given a family of divisors in a fixed linear series, the idea is to "differentiate in parameter directions" to lower the multiplicities of the generic divisor in the family.

Proposition 5.2.13 (Smoothing divisors in families). *Let X and T be smooth irreducible varieties, with T affine, and suppose that*

$$Z \subseteq V \subseteq X \times T$$

are irreducible subvarieties such that V dominates X. Fix a divisor L on X, and suppose given on $X \times T$ a divisor $F \in |\mathrm{pr}_1^*(L)|$. Write

$$\ell = \mathrm{mult}_Z(F) \quad , \quad k = \mathrm{mult}_V(F).$$

Then there exists a divisor $F' \in |\mathrm{pr}_1^*(L)|$ on $X \times T$ having the property that

$$\mathrm{mult}_Z(F') \geq \ell - k \quad , \quad \text{and } V \not\subseteq \mathrm{Supp}(F').$$

Let $\sigma \in \Gamma\big(X \times T, \mathrm{pr}_1^*\mathcal{O}_X(L)\big)$ be the section defining F. The plan is to obtain F' as the divisor of the section $\sigma' = D\sigma$, where D is a general differential operator of order k on T. So we begin with some remarks about differentiating sections of the line bundle $\mathrm{pr}_1^*\mathcal{O}_X(L)$ in parameter directions.

Let D be a differential operator of order $\leq k$ on T, i.e. a section of the locally free $\mathcal{O}_T$-module $\mathcal{D}_T^k$ of all such, and let $\sigma \in \Gamma\big(X \times T, \mathrm{pr}_1^*\mathcal{O}_X(L)\big)$ be a section of $\mathrm{pr}_1^*\mathcal{O}_X(L)$. We claim that then D acts naturally on σ to determine a new section $D\sigma \in \Gamma\big(X \times T, \mathrm{pr}_1^*\mathcal{O}_X(L)\big)$ of the same bundle. In fact, choose local coordinates x and t on X and T, and let $g_{\alpha,\beta}(x)$ be the transition functions of $\mathcal{O}_X(L)$ with respect to a suitable open covering of X. Then $\sigma = \{s_\alpha(x,t)\}$ is given by a collection of functions $s_\alpha(x,t)$ such that

$$s_\alpha(x,t) = g_{\alpha,\beta}(x)\, s_\beta(x,t).$$

Viewing D locally as a differential operator in the t-variables, one has

$$Ds_\alpha(x,t) = g_{\alpha,\beta}(x)\, Ds_\beta(x,t).$$

Therefore the $\{Ds_\alpha(x,t)\}$ patch together to define a section

$$D\sigma \in \Gamma\big(X \times T, \mathrm{pr}_1^*\mathcal{O}_X(L)\big),$$

as asserted. We refer to [149, §2] for a more formal discussion.

Having said this, we turn to the:

Proof of Proposition 5.2.13. Since T is affine, the vector bundle $\mathcal{D}_T^k$ is globally generated. Choose finitely many differential operators

$$D_\alpha \in \Gamma\big(T, \mathcal{D}_T^k\big)$$

that span $\mathcal{D}_T^k$ at every point of T, and let $\sigma \in \Gamma\big(X \times T, \mathrm{pr}_1^* \mathcal{O}_X(L)\big)$ be the section defining the given divisor F. We claim that if $D \in \Gamma(T, \mathcal{D}_T^k)$ is a sufficiently general **C**-linear combination of the D_α, then

$$\sigma' \ =_{\mathrm{def}} \ D\sigma \in \Gamma\big(X \times T, \mathrm{pr}_1^* \mathcal{O}_X(L)\big)$$

does not vanish on V. Granting this for the moment, put $F' = \mathrm{div}(\sigma')$. Then F' does not contain V. On the other hand, differential operators of order $\leq k$ decrease multiplicity by at most k. Therefore $\mathrm{mult}_Z(F') \geq \ell - k$, as required.

It remains to prove that σ' does not vanish identically on V. To this end, consider the algebraic subset

$$X \times T \ \supseteq \ B \ = \ \big\{ (x,t) \mid D_\alpha \sigma(x,t) = 0 \quad \text{for all} \ \ \alpha \big\}$$

cut out by the common zeroes of all the sections $D_\alpha \sigma$. It is enough to show that $V \not\subseteq B$. For this we study the first projection

$$\mathrm{pr}_1 : X \times T \longrightarrow X.$$

Fix any point $x \in X$, and consider the fibre $F_x \subseteq T$ of F over x. Assume that $F_x \neq T$ (which will certainly hold for general x), so that F_x is a divisor on T. Given $t \in T$, it follows from the fact that the D_α generate $\mathcal{D}_T^k$ at t that

$$(x,t) \in B \quad \Longleftrightarrow \quad \mathrm{mult}_t(F_x) > k.$$

On the other hand, since V dominates X, Lemma 5.2.11 applies to pr_1 and we conclude that

$$\mathrm{mult}_t(F_x) \ = \ \mathrm{mult}_V(F) \ = \ k$$

for sufficiently general $(x,t) \in V$. This proves that V is not contained in B, and the proposition follows. □

Example 5.2.14. (Bertini's theorem with multiplicities). This circle of ideas leads to a proof of Bertini's theorem in its original form, which involves multiplicities of divisors:

> **Theorem.** *Let X be a smooth variety, L an integral divisor on X, and $V \subseteq H^0\big(X, \mathcal{O}_X(L)\big)$ a finite-dimensional subspace defining a linear series $|V| = \mathbf{P}_{\mathrm{sub}}(V)$ on X. If $D \in |V|$ is a general divisor, then*
>
> $$\mathrm{mult}_x D \ \leq \ \mathrm{mult}_x |V| + 1$$
>
> *for every $x \in X$.*

Recall (Definition 2.3.11) that the multiplicity of $|V|$ at $x \in X$ is by definition the minimum of the multiplicities $\mathrm{mult}_x(E)$ of all divisors $E \in |V|$. The theorem implies for instance that if a general divisor $D \in |V|$ is singular at

x, then x must be a base point of $|V|$. Kleiman's paper [345] contains an interesting discussion of this result and its history.

For the proof, let $F \subseteq X \times |V|$ be the universal divisor. Fix a point $0 \in |V|$ sufficiently general so that

$$\text{mult}_{(x,0)}\big(X \times |V|\,,\,F\big) \;=\; \text{mult}_x\big(X, F_0\big) \tag{*}$$

for every $x \in X$ (Corollary 5.2.12). Note that in affine coordinates $t_1, \ldots, t_r$ on $|V|$ centered at 0, F is defined by the vanishing of the section

$$s \;=_{\text{def}}\; s_0 + t_1 s_1 + \ldots + t_r s_r, \tag{**}$$

where $s_0, \ldots, s_r \in V$ is a basis with $F_0 = \text{div}(s_0)$; to lighten notation we are writing s_i in place of the more correct $\text{pr}_1^*(s_i)$ in (**). Consider now the derivatives $D_i(s) = \frac{\partial s}{\partial t_i}$ of s with respect to t_i. Since differentiation lowers multiplicity by at most one, we have

$$\text{ord}_{(x,0)}\big(\,D_i(s)\,\big) \;\geq\; \text{ord}_{(x,0)}(s) - 1 \;=\; \text{mult}_{(x,0)}(F) - 1 \tag{***}$$

for all i and every $x \in X$. On the other hand, s_0 and the restrictions $s_i = D_i(s)|_{t=0}$ together span V. Combining (*) and (***), one then finds:

$$\begin{aligned}
\text{mult}_x|V| \;&=\; \min\,\big\{\text{ord}_x(s_i)\,,\,\text{ord}_x(s_0)\big\} \\
&\geq\; \min\,\big\{\,\text{ord}_{(x,0)}(D_i(s))\,,\,\text{ord}_{(x,0)}(s)\,\big\} \\
&\geq\; \text{mult}_{(x,0)}(X \times |V|\,,\,F) - 1 \\
&=\; \text{mult}_x(X, F_0) - 1
\end{aligned}$$

for every $x \in X$. Taking $D = F_0$, the theorem is proved. □

5.2.C Proof of Theorem 5.2.5

We now outline the proof of Theorem 5.2.5. We will make several simplifying assumptions, and refer to [149] for some of the verifications.

Step 1: Reductions.

We start with some elementary simplifications. By induction on $n = \dim X$ we may assume that the result is known for all irreducible varieties of dimension $< n$. There is also no loss in assuming that X is smooth, since it is enough to prove the theorem for a resolution of singularities of X. Fix next any $\delta > 0$ and put $n' = n + \delta$. We will show that

$$\varepsilon\big(L; x\big) \;\geq\; \frac{1}{n'} \quad \text{for general } x \in X. \tag{5.6}$$

Then the theorem follows by letting $\delta \to 0$. (See [149, (3.2)–(3.4)] for details.)

Step 2: Seshadri-exceptional curves.

Aiming for a contradiction, assume that (5.6) fails. Then for each point $x \in X$ we can find a reduced and irreducible curve $C_x \ni x$ with the property that

$$\mathrm{mult}_x\big(C_x\big) \;>\; n' \cdot \big(L \cdot C_x\big). \tag{5.7}$$

We shall call C_x a *Seshadri-exceptional curve based at x*. By standard arguments with Hilbert schemes, the existence of exceptional curves based at every point is only possible if most of these curves fit together in an algebraic family. More precisely, there exist a smooth affine variety T, a dominating map $g : T \longrightarrow X$, and an irreducible subvariety

$$C \subseteq X \times T,$$

flat over T, such that for every $t \in T$ the fibre $C_t \subseteq X$ is a Seshadri-exceptional curve based at $g(t)$. In other words, C_t is a reduced and irreducible curve on X, passing through $g(t)$, with

$$\mathrm{mult}_{g(t)}\big(C_t\big) \;>\; n' \cdot \big(L \cdot C_t\big).$$

We denote by $\Gamma \subseteq X \times T$ the graph of g, so that $\Gamma \subseteq C$.

We will henceforth *make the simplifying assumption that*

$$T \subseteq X,$$

with g the inclusion. We then omit mention of g, so that from now on T is an open subset of X, $C_t \subseteq X$ is a exceptional curve based at $t \in X$, and we have a diagram

$$\begin{array}{ccc} C & \subseteq & X \times T \\ & \searrow \quad \swarrow & \\ & T. & \end{array}$$

Step 3: Subvarieties swept out by exceptional curves.

We next consider a construction analogous to one used by Kollár, Miyaoka, and Mori in [367]. Given an irreducible subvariety $W \subseteq X$, the idea is to form a new variety $CW \supseteq W$ by taking the union $\cup_{s \in W} C_s$ of all the Seshadri-exceptional curves based in W. However, we need to be able to let W vary with parameters.

Lemma 5.2.15. *Let $Z \subseteq X \times T$ be an irreducible subvariety dominating both X and T. Then one can construct an irreducible closed subvariety*

$$CZ \;\subseteq\; X \times T$$

having the following properties:

(i). $Z \subseteq CZ$ *and* $\dim CZ \leq \dim Z + 1$.

(ii). *For general $t \in T$, the fibre $(CZ)_t$ of CZ over t has the form*

$$(CZ)_t = \text{closure}\Big(\bigcup_{s \in Z_t \cap T} C_s \Big). \quad \square$$

See [149, 3.5] for the proof.

The induction hypothesis now implies that $Z \subsetneqq CZ$:

Lemma 5.2.16. *Let $Z \subseteq X \times T$ be a proper irreducible subvariety dominating both X and T, and consider the subvariety $CZ \subseteq X \times T$ constructed in the previous lemma. Then Z is a proper subvariety of CZ.*

Proof. Assume to the contrary that $CZ = Z$, and fix a very general point $t \in T$. Given a general point $s \in Z_t$, it follows from 5.2.15 (ii) that there exists a Seshadri-exceptional curve C_s based at s such that C_s lies in $Z_t = (CZ)_t$. This means that the restriction $L \,|\, Z_t$ of L to Z_t has small Seshadri constant at a general point, i.e.

$$\varepsilon\big(Z_t,\, L \,|\, Z_t; s\big) < \frac{1}{n'}$$

for a dense open set of points $s \in Z_t$. Since $\dim Z_t < n$, the induction hypothesis will give a contradiction as soon as we know that $L \,|\, Z_t$ is big for general t. But this follows from Corollary 2.2.11: since the projection $Z \longrightarrow X$ is dominating, the fibres Z_t are not all contained in any proper subvariety of X. $\square$

Step 4: A gap argument

Let $t \in T \subseteq X$ be any point. Then by an elementary dimension count, there exists for $k \gg 0$ a divisor $F_t \in |kL|$ such that

$$\text{mult}_t(F_t) > k \cdot \frac{n}{n'}$$

(Proposition 1.1.31). Moreover, an evident globalization of the proof of that proposition shows that for general $t \in T$ one can take the divisors F_t to be fibres of a global divisor

$$F \subset X \times T \quad , \quad F \in |\text{pr}_1^*(kL)|$$

(see [149], 3.8). Note that then $\text{mult}_\Gamma(F) > k\frac{n}{n'}$ thanks to Lemma 5.2.11, where as above $\Gamma \subseteq X \times T$ is the graph of $T \subseteq X$.

We claim now that there exists an irreducible subvariety $Z \subseteq X \times T$, dominating both X and T, having the property that

$$\Big(\text{mult}_Z(F) - \text{mult}_{CZ}(F)\Big) > \frac{k}{n'}. \tag{5.8}$$

In fact, set

$$Z_0 = \Gamma = \operatorname{graph}(T \subseteq X) \quad , \quad Z_1 = CZ_0 \subseteq X \times T,$$

and for $1 \leq i \leq n-1$ apply 5.2.15 inductively to form

$$Z_{i+1} = CZ_i \subseteq X \times T.$$

It follows from 5.2.16 that Z_i is a proper subvariety of Z_{i+1}, and consequently Z_i has relative dimension i over T. In particular, $Z_n = X \times T$. Thus we have a chain

$$\Gamma = Z_0 \subseteq Z_1 \subseteq \ldots \subseteq Z_{n-1} \subseteq Z_n = X \times T$$

of irreducible subvarieties of $X \times T$. But consider the multiplicities $\operatorname{mult}_{Z_i}(F)$ of F along Z_i. By construction,

$$\operatorname{mult}_{Z_0}(F) > \frac{kn}{n'} \quad \text{and} \quad \operatorname{mult}_{Z_n}(F) = 0.$$

It follows that there is at least one index $i \in [0, n-1]$ such that

$$\Big(\operatorname{mult}_{Z_i}(F) - \operatorname{mult}_{Z_{i+1}}(F)\Big) > \frac{k}{n'},$$

so we can take $Z = Z_i$.

Step 5: Exploiting jumps in multiplicity

We now come to the core of the argument. Starting with the variety Z produced in the previous step, the idea is to differentiate in parameter directions to produce a new divisor F' that does not contain CZ but still has relatively high multiplicity along Z. Then a calculation of intersection numbers gives a contradiction.

Turning to details, fix $Z \subseteq X \times T$ as in (5.8). Since Z and hence also CZ dominate X, Proposition 5.2.13 implies the existence of a divisor

$$F' \in |\operatorname{pr}_1^*(kL)|$$

on $X \times T$ with

$$\operatorname{mult}_Z(F') > \frac{k}{n'} \quad , \quad CZ \not\subseteq \operatorname{Supp}(F').$$

Now fix a general point $t \in T$, and let $x \in Z_t \cap T$ be a general point in the fibre of Z over t. Then

$$\operatorname{mult}_x(F'_t) > \frac{k}{n'}.$$

On the other hand, since $CZ \not\subseteq \operatorname{Supp}(F')$, there is a Seshadri-exceptional curve C_x through x with

$$C_x \not\subseteq \operatorname{Supp}(F'_t).$$

Now compute as in the proof of 5.1.17:

$$\begin{aligned} k\,(L\cdot C_x) &= (F'_t\cdot C_x) \\ &\geq \text{mult}_x(F'_t)\cdot \text{mult}_x(C_x) \\ &> \frac{k}{n'}\cdot \text{mult}_x(C_x). \end{aligned}$$

But $\text{mult}_x(C_x) \geq n'\cdot(L\cdot C_x)$ by (5.7), and we have the required contradiction.

Remark 5.2.17. (Variants and extensions). The paper [373] of Küchle and Steffens establishes similar lower bounds via a different approach. The argument just completed was reconsidered from a more geometric point of view by Hwang and Keum in [301]. Their idea is to view a covering family of Seshadri-exceptional curves as defining a foliation on a dense open subset of X. Among other things, they show in this manner that if X is a smooth projective threefold with Picard number 1, and if L is an ample line bundle on X, then

$$\varepsilon(L;x) \geq \frac{\sqrt[3]{(L^3)}}{3}.$$

In another direction, Nakamaye [470] obtains an improvement of Theorem 5.2.5 valid in all dimensions. □

5.3 Abelian Varieties

In this section we study local positivity on abelian varieties, where there are interesting relations with metric and arithmetic invariants. In Sections 5.3.A and 5.3.B we present a theorem of the author [393] showing that the Seshadri constant of a polarized abelian variety is bounded below in terms of the minimal length of its periods. Some generalizations and complements are considered in Section 5.3.C, where in particular we outline the result of Bauer–Szemberg [38, Appendix] and Bauer [39] that the Seshadri constants of abelian surfaces are governed by Pell's equation.

5.3.A Period Lengths and Seshadri Constants

The present subsection and the next are devoted to explaining a relation between Seshadri constants of abelian varieties and a Riemannian invariant studied by Buser and Sarnak [76]. Definitions, statements, and applications are given here, while the proof of the main theorem occupies Section 5.3.B.

Definitions. We start by setting notation and defining the invariants in question. Let A be an abelian variety of dimension g. Viewing A as a complex torus, write

$$A = V/\Lambda,$$

where $V \cong \mathbf{C}^g$ is a complex vector space of dimension g, and $\Lambda \cong \mathbf{Z}^{2g}$ is a lattice in V. We identify V with the universal covering space of A, and denote by $\pi : V \longrightarrow A$ the natural projection.

Fix now an ample divisor L on A. It is classical that L determines a positive definite Hermitian form $H = H_L$ on V. To construct it, one can note for instance that there is a unique translation-invariant Hermitian metric h_L on A whose associated Kähler form represents the cohomology class of L:

$$\omega_L =_{\text{def}} -\text{Im}(h_L) \sim c_1\big(\mathcal{O}_A(L)\big).$$

The required form on V is then realized as the pullback $H_L =_{\text{def}} \pi^* h_L$. Evidently H_L depends only on the algebraic equivalence class of L. Recall also that by diagonalizing H_L one can associate to L its *type*, a g-tuple $(d_1, \ldots, d_g)$ of positive integers with $d_1 \mid d_2 \mid \ldots \mid d_g$ that determines the algebraic equivalence class in question. See [383, Chapters 2 and 4] for details.

The plan is to use H_L to define a metric invariant of the polarized abelian variety (A, L). Specifically, the real part

$$B = B_L =_{\text{def}} \text{Re}(H_L)$$

is a Euclidean inner product on V: write $\|v\| = \|v\|_B$ for the corresponding length of a vector $v \in V$. Thus the period lattice Λ sits naturally inside the Euclidean space (V, B_L), and in particular one can discuss the lengths of lattice vectors:

Definition 5.3.1. (Buser–Sarnak invariant). The *Buser–Sarnak invariant* of (A, L) is (the square of) the minimal length of a non-zero lattice vector in Λ:

$$\begin{aligned} m(A, L) &= \min_{0 \neq \ell \in \Lambda} \|\ell\|_B^2 \\ &= \min_{0 \neq \ell \in \Lambda} H_L(\ell, \ell). \quad \square \end{aligned}$$

Equivalently, B_L determines a translation-invariant Riemannian metric on A whose volume is fixed by the type of L, and then $m(A, L)$ is the squared length of the shortest non-constant geodesic on A. For arbitrary Euclidean lattices, these invariants have been actively studied at least since Minkowski (see [484], [96]). The nice idea of Buser and Sarnak [76] is to investigate how $m(A, L)$ reflects the holomorphic geometry of (A, L).

Example 5.3.2. (Elliptic curves). We work out the story in dimension $g = 1$. Write

$$\Lambda = \Lambda_\tau = \mathbf{Z} + \mathbf{Z}\tau$$

where $\tau = a + ib$ a point in the upper half-plane. Taking the polarization L on $\mathbf{C}/\Lambda$ to have degree 1, the corresponding Hermitian form is given by

$$H(v, w) = \frac{{}^t v \cdot \overline{w}}{b}.$$

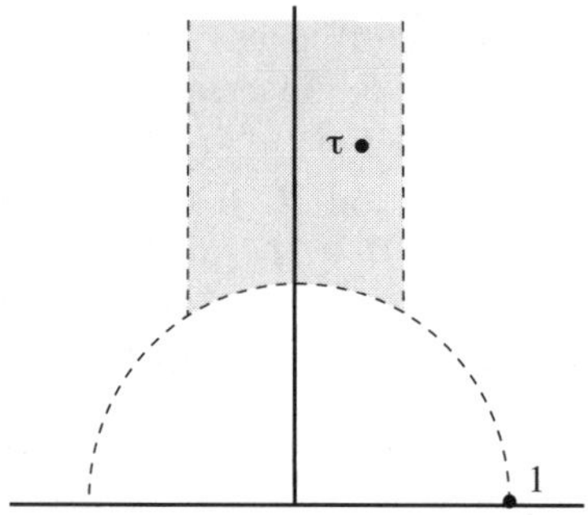

Figure 5.2. Buser–Sarnak invariant for curves

Now assume that τ lies in the classical fundamental domain for $\mathrm{SL}_2(\mathbf{Z})$, so that $|\tau| \geq 1$ and $|\mathrm{Im}(\tau)| \leq \frac{1}{2}$ (Figure 5.2). Then $1 \in \Lambda$ is the minimal vector, and

$$m(\mathbf{C}/\Lambda_\tau) \;=\; H(1,1) \;=\; \frac{1}{b}.$$

Note that this function approaches zero as τ approaches the cusp at infinity corresponding to the "boundary" of the moduli space $\mathcal{A}_1 = \mathbf{A}^1$. □

Work of Buser and Sarnak. We now summarize without proof the main results of [76]. Suppose for this that L has type $(1,\ldots,1)$, i.e. that L is a *principal polarization.* In this case $h^0\big(A, \mathcal{O}_A(L)\big) = 1$, and following classical notation we denote by $\Theta \subseteq A$ the corresponding divisor. Isomorphism classes of principally polarized abelian varieties (PPAVs) of dimension g are parameterized by a moduli space $\mathcal{A}_g$ of dimension $\frac{g(g+1)}{2}$.

Considerations of volume [76, (1.9), (1.10)] show that

$$m(A,\Theta) \;\leq\; \frac{4}{\pi}\,\sqrt[g]{g!} \;\approx\; (\text{constant}) \cdot \frac{g}{e} \tag{5.9}$$

for any PPAV (A,Θ). Here and subsequently we use the rough approximation $\sqrt[g]{g!} \approx \frac{g}{e}$ coming somewhat abusively from Stirling's formula $g! \sim (\sqrt{2\pi g})\big(\frac{g}{e}\big)^g$. The first main result of Buser–Sarnak estimates the maximum value of $m(A,\Theta)$ as (A,Θ) ranges over the moduli space $\mathcal{A}_g$:

Theorem 5.3.3. (Buser–Sarnak theorem I). *There exist principally polarized abelian varieties* (A,Θ) *for which*

$$m(A,\Theta) \;\geq\; \frac{2^{1/g}}{\pi} \cdot \sqrt[g]{g!}\;. \tag{5.10}$$

This is established by an averaging argument adapted from the geometry of numbers. In fact, Buser and Sarnak show that the set of all (A,Θ) for which (5.10) fails is parameterized by a region in $\mathcal{A}_g$ whose volume (with respect to a natural metric) decreases exponentially with g.

Remark 5.3.4. (Arbitrary polarizations). Bauer [38, Theorem 1] has generalized Theorem 5.3.3 to abelian varieties carrying a polarization L of arbitrary type $(d_1, \ldots, d_g)$. Specifically, he shows that there exist (A, L) with

$$m(A, L) \ \geq \ \frac{2^{1/g}}{\pi} \cdot \sqrt[g]{d \cdot g!}, \tag{5.11}$$

where $d = d_1 \cdot \ldots \cdot d_g$. □

The most surprising result of [76] concerns the exceptional behavior of Jacobians. Specifically, let C be a compact Riemann surface of genus g, and denote by JC the Jacobian of C with its canonical principal polarization Θ_C. Fixing a metric on C, the Buser–Sarnak invariant has the alternative interpretation [76, (3.1)]

$$m(JC, \Theta_C) \ = \ \min_w \int_C w \wedge *w,$$

the minimum being taken over all closed and non-exact real one-forms w on C with integral periods.

Theorem 5.3.5. (Buser–Sarnak theorem II). *The Buser–Sarnak invariant of JC satisfies the upper bound*

$$m(JC, \Theta_C) \ \leq \ \frac{3}{\pi} \log(4g + 3). \tag{5.12}$$

In other words, when g is large, Jacobians have periods of exceptionally short length among all PPAVs of dimension g. The proof of 5.3.5 in [76] involves some delicate constructions with the hyperbolic model of C. Buser and Sarnak also produce examples of curves C for which $m(JC, \Theta_C) \geq (\text{constant}) \cdot \log(g)$, showing that the inequality (5.12) is essentially best possible. Gromov puts these results into a broader framework in [254].

Seshadri constants. Turning to Seshadri constants, let (A, L) be a polarized abelian variety. Then $\varepsilon(A,\ L; x)$ is independent of $x \in A$ thanks to 5.1.3 and the homogeneity of A, and so we write simply $\varepsilon(A, L)$. The main result of the present section shows that these holomorphic invariants are bounded below in terms of the minimal period length of A.

Theorem 5.3.6. (Seshadri constants and period lengths). *One has the inequality*

$$\varepsilon(A, L) \ \geq \ \frac{\pi}{4} \cdot m(A, L). \tag{5.13}$$

This is actually very elementary: the proof appears in the next subsection. For the remainder of this subsection we indicate some corollaries and applications.

To begin with, we restate the theorem in more classical language. Let

$$V_k = \Gamma\big(A, \mathcal{O}_A(kL)\big)$$

be the space of sections of $\mathcal{O}_A(kL)$: one can identify V_k with a suitable space of kth order theta functions. Denote by $s_k = s(V_k; 0)$ the number of jets that V_k separates at the origin in the sense of Definition 5.1.16. Then Theorems 5.1.17 and 5.3.6 combine to yield

Corollary 5.3.7. (Jets of higher order theta functions). *One has*

$$\lim_{k\to\infty} \frac{s_k}{k} \geq \frac{\pi}{4} \cdot m(A, L). \quad \square$$

More picturesquely, we may say that if none of the periods of A are small, then higher order theta functions separate many jets.

Next, putting 5.3.6 together with the first theorem of Buser–Sarnak (and its extension (5.11) by Bauer), one obtains:

Corollary 5.3.8. *If (A, L) is a very general abelian variety with a polarization of type $(d_1, \ldots, d_g)$, then*

$$\varepsilon(A, L) \geq \frac{2^{1/g}}{4} \cdot \sqrt[g]{d \cdot g!}, \tag{5.14}$$

where as before $d = d_1 \cdot \ldots \cdot d_g$. $\square$

The hypothesis on (A, L) means that the stated inequality holds for all abelian varieties parameterized by the complement of countably many proper subvarieties of the moduli space of polarized abelian varieties of given type. Note also that the lower bound appearing in (5.14) differs from the upper bound

$$\varepsilon(L, A) \leq \sqrt[g]{d \cdot g!}$$

coming from (5.2) by a factor of less than 4.

Example 5.3.9. (Very ample polarizations). Assume that

$$d = d_1 \cdot \ldots \cdot d_g \geq \frac{(8g)^g}{2g!} \approx \frac{(8e)^g}{2}.$$

If (A, L) is a general polarized abelian variety of type $(d_1, \ldots, d_g)$, then L is very ample. (Combine (5.14) and 5.1.19.) This is due to Bauer [38, Corollary 2]: see [116] for other results along these lines. $\square$

Example 5.3.10. (Lower bounds for arbitrary abelian varieties). One has

$$\varepsilon(A, L) \geq 1$$

for any polarized abelian variety (A, L), and if A contains an elliptic curve $E \subseteq A$ with $\deg_E(L) = 1$ then equality is attained. (Argue much as in 5.1.18 using translates of divisors in $|L|$.) Nakamaye [467] shows conversely that if (A, Θ) is a PPAV with $\varepsilon(A, \Theta) = 1$, then in fact A contains such a curve. $\square$

We now come back to Jacobians. Whereas upper bounds on the Buser–Sarnak invariant $m(JC, \Theta_C)$ require serious effort, it is very easy to obtain bounds on local positivity.

Proposition 5.3.11. (Seshadri constants of Jacobians). *Let C be a smooth projective curve of genus g.*

(i). *The Seshadri constant $\varepsilon(JC, \Theta_C)$ satisfies the bound*

$$\varepsilon(JC, \Theta_C) \ \le \ \sqrt{g}.$$

(ii). *Assume that C admits a d-fold branched covering $\phi : C \longrightarrow \mathbf{P}^1$. Then*

$$\varepsilon(JC, \Theta_C) \ \le \ \frac{gd}{g+d-1}.$$

Thanks to Theorem 5.3.6, statement (i) leads to a quick and elementary proof that Jacobians have periods of small length, although the specific inequality that comes out is not as strong as (5.12). It also follows that if C is a d-sheeted covering of $\mathbf{P}^1$, then

$$m(JC, \Theta_C) \ \le \ \frac{4d}{\pi}.$$

For hyperelliptic curves (i.e. when $d = 2$), statement (ii) was established by Steffens [559].

Sketch of Proof of Proposition 5.3.11. We assume to begin with that C is non-hyperelliptic. For (i), consider the surface $S = C - C \subseteq JC$, i.e. the image of the subtraction map

$$s : C \times C \longrightarrow JC \quad , \quad (x, y) \mapsto x - y.$$

It is elementary and well known that if C is non-hyperelliptic, then s is an isomorphism outside the diagonal $\Delta \subseteq C \times C$, which blows down to the origin $0 \in S$, at which S has multiplicity $2g - 2$ (see [15, pp. 223, 263]). Moreover, the cohomology class of S is given by[4] $[S] = \frac{2}{(g-2)!} \cdot \theta^{g-2}$, where θ denotes the cohomology class of Θ_C. Therefore

$$\deg_{\Theta}(S) \ = \ \frac{2}{(g-2)!} \cdot \theta^g \ = \ 2g(g-1).$$

Applying Proposition 5.1.9 we find

[4] As explained in [15, Exercise V.D-3], when $d < \frac{g}{2}$ the cohomology class of the difference variety $V_d = C_d - C_d$ is expressed as

$$v_d \ = \ \frac{\binom{2d}{d}}{(g-2d)!} \cdot \theta^{g-2d}.$$

(Note that the actual formula stated in [15] is misprinted.)

$$\begin{aligned}\varepsilon(JC,\Theta_C) &\leq \sqrt{\frac{\deg_\Theta(S)}{\operatorname{mult}_0 S}} \\ &= \sqrt{\frac{2g(g-1)}{2g-2}} \\ &= \sqrt{g},\end{aligned}$$

as required.

Turning to (ii), denote by $f_1, f_2, \delta \in N^1(C\times C)$ respectively the classes of the fibres of the two projections $C \times C \longrightarrow C$ and the diagonal Δ_C. Then in the first place

$$s^*\big(\Theta_C\big) \ \equiv_{\text{num}} \ (g-1)(f_1+f_2)+\delta.$$

Moreover, there is an effective curve $\Gamma \subseteq C \times C$ with

$$\Gamma \ \equiv_{\text{num}} \ d(f_1+f_2)-\delta:$$

geometrically, Γ is the closure of the set $\Gamma_0 = \{\,(x,y) \mid x \neq y, \phi(x)=\phi(y)\,\}$.[5] Now if $\varepsilon = \varepsilon(JC,\Theta_C)$, then $s^*(\Theta) - \varepsilon \cdot \Delta$ is nef, and hence

$$\begin{aligned}\Gamma \cdot \big(s^*(\Theta_C) - \varepsilon\Delta\big) \ &= \ \big(d(f_1+f_2)-\delta\big)\cdot\big((g-1)(f_1+f_2)+(1-\varepsilon)\,\delta\big) \\ &\geq \ 0.\end{aligned}$$

This leads to the stated inequality. Finally, when C is hyperelliptic the only thing that needs proof is the case $d = 2$ of (ii), and this follows by looking at the image in $J(C)$ of the curve Γ just constructed. □

Remark 5.3.12. (Curves of low genus). Steffens [559] shows that if (A,Θ) is an irreducible principally polarized abelian surface, then $\varepsilon(A,\Theta) = \frac{4}{3}$, as predicted by 5.3.11 (ii). PPAVs of dimension three have been studied by Bauer and Szemberg [41]. They establish that $\varepsilon(A,\Theta) = \frac{3}{2}$ if A is a hyperelliptic Jacobian, and $\varepsilon(A,\Theta) = \frac{12}{7}$ otherwise. Debarre [114] proves that $\varepsilon(JC,\Theta_C) = 2$ for a non-hyperelliptic curve of genus 4. □

Remark 5.3.13. (Hyperelliptic Jacobians). Debarre [114], and independently Kong [369], prove that if C is a hyperelliptic curve of genus g, then $\varepsilon(JC,\Theta_C) = \frac{2g}{g+1}$, i.e. the bound in 5.3.11 (ii) is attained. It is natural to conjecture that the converse holds, i.e. that if (A,Θ) is an irreducible PPAV of dimension g with $\varepsilon(A,\Theta) \leq \frac{2g}{g+1}$, then in fact (A,Θ) must be a hyperelliptic Jacobian. This would follow from a conjecture of van Geemen and Van der Geer [581] using work of Welters [605]. Debarre points out that there exist non-Jacobian PPAVs (A,Θ) containing an elliptic curve $E \subset A$ with $\big(E\cdot\Theta\big) = 2$, for which $\varepsilon(A,\Theta) \leq 2$. So it seems that in general one can't hope to recognize non-hyperelliptic Jacobians through their Seshadri constants. □

[5] Compare the proof of Kouvidakis's Theorem 1.5.8.

Remark 5.3.14. (Prym varieties). Bauer [38] has generalized 5.3.11 to the setting of Prym varieties. He shows for example that if (P, Ξ) is the Prym variety associated to an étale double cover $\widetilde{C} \longrightarrow C$ of a curve C of genus $g \geq 3$, then

$$\varepsilon(P, \Xi) \ \leq \ \sqrt{2(g-1)} \ = \ \sqrt{2} \dim P. \quad \square$$

5.3.B Proof of Theorem 5.3.6

We now turn to the proof of Theorem 5.3.6, which is closely based on ideas from [422, §5]. In effect, we will simply write down explicitly a suitable Kähler form on the blow-up $\mathrm{Bl}_0(A)$. The presentation is adapted from [393]. A different approach is indicated in Remark 5.3.18.

We begin with a local construction on $\mathbf{C}^g$. Let

$$V \ \subseteq \ \mathbf{C}^g \times \mathbf{P}^{g-1}$$

be the blowing up of the origin $0 \in \mathbf{C}^g$, embedded in the usual way as an incidence correspondence. Write f and q respectively for the projections

$$\begin{array}{ccc} & V & \\ {\scriptstyle f} \swarrow & & \searrow {\scriptstyle q} \\ \mathbf{C}^g & & \mathbf{P}^{g-1}, \end{array}$$

so that f is the blowing up and q realizes V as the total space of the line bundle $\mathcal{O}_{\mathbf{P}^{g-1}}(-1)$. As in (5.3) let $B(\lambda) \subseteq \mathbf{C}^g$ be the ball of radius λ, and denote by $V(\lambda)$ its inverse image in V, i.e.

$$V(\lambda) \ = \ f^{-1} B(\lambda) \ \subseteq \ V.$$

Thus $V(\lambda)$ is an open neighborhood of the exceptional divisor $E = \mathbf{P}^{g-1} \subseteq V$. Finally, let $\sigma = \omega_{\mathrm{FS}}$ be the Fubini–Study Kähler form on $\mathbf{P}^{g-1}$, normalized so that $\int_{\mathbf{P}^1} \sigma = \pi$, the integral being taken over a line in $\mathbf{P}^{g-1}$. As in Example 1.2.43, this normalization is chosen so that if $S = S^{2g-1} \subseteq \mathbf{C}^g$ is the unit sphere, and if $p : S \longrightarrow \mathbf{P}^{g-1}$ is the Hopf map, then $p^*\sigma = \omega_{\mathrm{std}} \mid S$, where as in (5.4) ω_{std} is the standard Kähler form on $\mathbf{C}^g$.

The first essential point is to construct a suitable Kähler form on this blow-up:

Proposition 5.3.15. (Kähler form on blow-up of the origin, [422]). *Fix $\lambda > 0$. Given any small $\eta > 0$, there exists $0 < \delta \ll 1$, together with a Kähler form $\tau = \tau(\lambda, \eta)$ on V, such that*

$$\tau \ = \ \begin{cases} f^*(\omega_{\mathrm{std}}) & \text{on } V - \overline{V(\lambda + \eta)}; \\ f^*(\omega_{\mathrm{std}}) + \lambda^2 \cdot q^*(\sigma) & \text{on } \ V(\delta). \end{cases} \tag{5.15}$$

In other words, τ coincides with the standard form on $\mathbf{C}^g$ off a ball of radius (a tiny bit larger than) λ, whereas we are "twisting" by a form representing $\pi\lambda^2 \cdot q^* c_1\big(\mathcal{O}_{\mathbf{P}^{g-1}}(1)\big)$ in a neighborhood of E.

The plan is to construct τ in the first instance on $\mathbf{C}^g - \{0\} = V - E$, and then extend over E. So we start with some remarks concerning forms on $\mathbf{C}^g - \{0\}$.

Lemma 5.3.16. *Given a monotone increasing smooth function $\phi : (0, \infty) \longrightarrow (0, \infty)$, consider the C^∞ mapping*

$$F = F_\phi \,:\, \mathbf{C}^g - \{0\} \longrightarrow \mathbf{C}^g \quad , \quad z \mapsto \frac{\phi(|z|)}{|z|} \cdot z.$$

(i). *The two-form*

$$\eta_\phi \;=_{\text{def}}\; F_\phi^*\big(\omega_{\text{std}}\big)$$

is a Kähler form on $\mathbf{C}^g - \{0\}$.

(ii). *Fix a real number $\lambda > 0$, and consider the increasing function $\psi(r) = \sqrt{\lambda^2 + r^2}$. Then*

$$\eta_\psi \;=\; \omega_{\text{std}} \,+\, h^*\big(\lambda^2\sigma\big),$$

where $h : \mathbf{C}^g - \{0\} \longrightarrow \mathbf{P}^{g-1}$ is the canonical projection.

Proof. We follow [422, p. 425]. For (i), let $S = S^{2g-1} \subseteq \mathbf{C}^g$ be the unit sphere, and use the natural diffeomorphism

$$S \times (0, \infty) \;\cong\; \mathbf{C}^g - \{0\} \;\;,\;\; (u, r) \mapsto r \cdot u$$

to introduce "polar coordinates" (u, r) on $\mathbf{C}^g - \{0\}$. Thus $F(u, r) = \phi(r) \cdot u$. Given $x = (u, r) \in \mathbf{C}^g - \{0\}$, choose an orthonormal complex basis $v_1, \ldots, v_g \in T_x\mathbf{C}^g$ (with respect to the standard Hermitian inner product) so that v_1 points in the radial direction, and so that the complex subspace spanned by $v_2, \ldots, v_g$ is contained in T_uS. Then

$$\eta_\phi(v_\alpha, i \cdot v_\alpha) \;>\; 0$$

for $1 \leq \alpha \leq g$, whereas this form vanishes on all other pairs. It follows that η_ϕ is positive of type $(1, 1)$ (Definition 1.2.39), and of course it is closed since it is the pullback of a closed form. This proves (i). For (ii), consider the 2-form

$$\kappa \;=\; F_\psi^*(\omega_{\text{std}}) \;-\; \omega_{\text{std}}$$

on $\mathbf{C}^g - \{0\}$. Working in polar coordinates as above, one sees that κ is $\mathbf{C}^*$-invariant and degenerate in fibre directions of h, i.e. $\kappa = h^*\kappa_0$ for some closed 2-form κ_0 on $\mathbf{P}^{g-1}$. On the other hand, $F_\psi \mid S$ is multiplication by $\sqrt{\lambda^2 + 1}$, and therefore $\kappa|S = \lambda^2\omega_{\text{std}}|S$. It then follows from Example 1.2.43 that $\kappa_0 = \lambda^2\sigma$, as required. □

Proof of Proposition 5.3.15. Choose a monotone increasing smooth function $\phi(r)$ such that

$$\begin{aligned} \phi(r) &= \sqrt{\lambda^2 + r^2} \quad \text{for } 0 < r < \delta \ll 1, \\ \phi(r) &= r \quad \text{for } r > \lambda + \eta \end{aligned}$$

for some $0 < \delta \ll 1$, and let $F = F_\phi : \mathbf{C}^g - \{0\} \longrightarrow \mathbf{C}^g$ be the mapping appearing in the the previous lemma. Noting that f induces an isomorphism $V - E = \mathbf{C}^g - \{0\}$, it follows from the lemma that $\tau_0 =_{\text{def}} f^* F_\phi^* \omega_{\text{std}}$ is a Kähler form on $V - E$ which satisfies the stated properties away from E. The second equality in (5.15) then shows that τ_0 extends over E to give the required form τ on all of V. □

The proposition leads to a lower bound on Seshadri constants of bundles whose first Chern class is locally represented by ω_{std}. Specifically, let X be a smooth projective variety of dimension n, L an ample divisor on X, and ω_L a Kähler form on X representing $c_1\big(\mathcal{O}_X(L)\big)$. We view (X, ω_L) as a symplectic manifold. Given $x \in X$, define a real number $\lambda(\omega_L; x) \geq 0$ by looking for the largest radius $\lambda > 0$ for which there exists a holomorphic and symplectic embedding

$$j = j_\lambda : \big(B(\lambda)\, ,\, \omega_{\text{std}}\big) \hookrightarrow \big(X\, ,\, \omega_L\big) \ \text{ with } \ 0 \mapsto x. \tag{*}$$

More precisely, if there is no $\lambda > 0$ for which such an embedding exists, set $\lambda(\omega_L; x) = 0$. Otherwise put

$$\lambda(\omega_L; x) \;=\; \sup \big\{ \lambda > 0 \mid \exists \text{ holomorphic and symplectic } j_\lambda \text{ as in } (*) \, \big\}.$$

Proposition 5.3.17. *One has the inequality*

$$\varepsilon\big(L; x\big) \;\geq\; \pi \cdot \lambda(\omega_L\, ;\, x)^2.$$

Proof. Let $f : Y = \mathrm{Bl}_x(X) \longrightarrow X$ be the blowing-up of x, with exceptional divisor $E \subseteq Y$, and fix any $\lambda < \lambda(\omega_L; x)$. It is enough to show that the $\mathbf{R}$-divisor class $f^*L - \pi\lambda^2 E$ is ample on Y. To this end, fix $\lambda < \lambda_1 < \lambda(\omega_L; x)$. Then by definition there exists a holomorphic and symplectic embedding $B(\lambda_1) \hookrightarrow X$, and so for any $\nu < \lambda_1$ one can view the local model $V(\nu)$ of the blow-up as being embedded in Y as a neighborhood of the exceptional divisor. Proposition 5.3.15 thus guarantees the existence of a Kähler form $\overline{\omega}_L$ on Y that agrees with ω_L off $V(\lambda_2)$ for suitable $\lambda < \lambda_2 < \lambda_1$, and is given by 5.3.15 (ii) in a neighborhood $V(\delta)$ of E. The positivity of $f^*L - \pi\lambda^2 E$ will follow as soon as we show that the class of $\overline{\omega}_L$ satisfies

$$[\,\overline{\omega}_L\,] \;=\; [\,f^*\omega_L\,] \;-\; \pi\lambda^2[\,E\,]. \tag{**}$$

But $\overline{\omega}_L - f^*\omega_L$ is supported in a small tubular neighborhood of E, so (**) is a consequence of 5.3.15 (ii) and the normalization of σ. □

Theorem 5.3.6 now follows at once:

Proof of Theorem 5.3.6. Let $\pi : V \longrightarrow A$ be the universal covering, and let $H = H_L$ be the Hermitian form on V determined by L. Thus $\omega = \mathrm{Im}(H) = \pi^*\omega_L$ is the pullback of a Kähler form ω_L on A representing $c_1\big(\mathcal{O}_A(L)\big)$. Fix a basis of V with respect to which H is the standard Hermitian form $H(v,w) = {}^t v \cdot \overline{w}$ on $\mathbf{C}^g$, and let $z_\alpha = x_\alpha + iy_\alpha$ be the corresponding coordinates. Then

$$\omega \;=\; \pi^*\omega_L \;=\; \omega_{\mathrm{std}}, \tag{5.16}$$

and $H(v,v) = |v|^2$ is just the usual Euclidean length. In particular,

$$m(A,L) \;=\; \min_{0 \neq \ell \in \Lambda} |\ell|^2.$$

Now put $\lambda = \frac{\sqrt{m(A,L)}}{2}$. Then no two points of $B(\lambda)$ are congruent (modulo Λ), and consequently the composition

$$j_\lambda : B(\lambda) \hookrightarrow V \overset{\pi}{\longrightarrow} A$$

is an embedding. But j_λ is of course holomorphic, and thanks to (5.16) it is symplectic as well. Therefore $\lambda(\omega_L; 0) \geq \frac{\sqrt{m(A,L)}}{2}$, and the theorem follows from 5.3.17. □

Remark 5.3.18. (Alternative approach to Theorem 5.3.6). Hwang and To [304], and independently Demailly, observed that one can deduce Theorem 5.3.6 from an inequality concerning the volumes of analytic subvarieties of the ball. Specifically, fix $R > 0$ and denote by $B(R)$ the open ball of radius R centered at the origin $0 \in \mathbf{C}^g$. Let $W \subseteq B(R)$ be a closed analytic subvariety of $B(R)$ of pure dimension k passing through the origin. The result in question states that

$$\mathrm{vol}(W) \;\geq\; \mathrm{mult}_0(W) \cdot \frac{\pi^k R^{2k}}{k!}, \tag{*}$$

where $\mathrm{vol}(W)$ denotes the usual $2k$-dimensional Euclidean volume of V in $\mathbf{C}^g$: see [436, Theorem 9.5] or [126, Theorems 2.8 and 2.9]. Given a polarized abelian variety (A, L) and a curve $C \subseteq A$ through 0, one applies (*) to the intersection of the inverse image of C under the universal covering $\pi : V \longrightarrow A$ with the ball $B\left(\frac{\sqrt{m(A,L)}}{2}\right)$. Since the intersection number $(C \cdot L)$ is bounded below by the volume of the resulting analytic curve W, this leads to the required inequality

$$\frac{(C \cdot L)}{\mathrm{mult}_0 C} \;\geq\; \pi \cdot \left(\frac{\sqrt{m(A,L)}}{2}\right)^2 .$$

In [304], (*) is generalized to Hermitian symmetric spaces other than $\mathbf{C}^g$. □

5.3.C Complements

Here we discuss some related results, largely without proof. We begin with the work of Bauer–Szemberg [38, Appendix] and Bauer [39] computing Seshadri constants of abelian surfaces. Then we turn to analogues of Theorem 5.3.6 for quotients of bounded symmetric domains due to Hwang and To [302], [303].

Abelian surfaces. Let A be an abelian surface and L an ample divisor on A of type $(1, d)$. Thus

$$\big(L^2\big) \;=\; 2d \quad \text{and} \quad h^0\big(A, L\big) \;=\; d.$$

We will assume that A has Picard number $\rho(A) = 1$ (Definition 1.1.17): this holds when (A, L) is sufficiently general.

The inequality 5.2 implies that $\varepsilon(A, L) \le \sqrt{2d}$. When $2d$ is a perfect square, results of Steffens [559] quoted in 5.2.9 imply that in fact $\varepsilon(A, L) = \sqrt{2d}$. Matters become more interesting when $\sqrt{2d}$ is irrational. Here one considers *Pell's equation*

$$\ell^2 \;-\; 2dk^2 \;=\; 1. \tag{5.17}$$

Denote by (ℓ_0, k_0) the smallest solution to (5.17) in positive integers. Then one has:

Theorem 5.3.19. (Seshadri constants of abelian surfaces). *Assuming as above that $2d$ is not a perfect square and that $\rho(A) = 1$, the Seshadri constant of (A, L) is given by*

$$\varepsilon(A, L) \;=\; 2d \cdot \frac{k_0}{\ell_0}.$$

In particular, $\varepsilon(A, L)$ is always rational. Some values of $\varepsilon(A, L)$ as a function of d are tabulated in [39, p. 572].

Partial proof of Theorem 5.3.19. We will sketch the inequality

$$\varepsilon(A, L) \;\le\; 2d \cdot \frac{k_0}{\ell_0}$$

established by Bauer and Szemberg in the Appendix to [38]. We can assume for this that L is a symmetric divisor, and it is enough to prove the existence of a curve $D \subseteq A$ passing through the origin such that $\big(D \cdot L\big) = 4dk_0$ and $\mathrm{mult}_0(D) \ge 2\ell_0$. The idea of Bauer and Szemberg is to consider the space $\Gamma\big(A, \mathcal{O}_A(2kL)\big)^+$ of *even* sections of $\mathcal{O}_A(2kL)$: one may view these as the sections pulling back from the Kummer surface $A/\{\pm 1\}$, and the point is to exploit the fact that the constant term in Riemann–Roch is different on a $K3$ than on an abelian surface. Specifically, the space of even sections of a symmetric line bundle of type $(2k, 2kd)$ has dimension

$$h^0\big(A, \mathcal{O}_A(2kL)\big)^+ \;=\; 2 + 2dk^2. \tag{*}$$

On the other hand, an even section vanishes to even order at the origin, so that it is at most

$$1 + 3 + \ldots + (m-1) \;=\; \left(\frac{m}{2}\right)^2$$

conditions on even sections to vanish at the origin to even order m. Therefore one can find a section of $\mathcal{O}_A(2kL)$ vanishing to order $\geq m$ at the origin as soon as $\left(\frac{m}{2}\right)^2 \leq \left(2dk^2+1\right)$. Noting that $\ell_0^2 = 2dk_0^2 + 1$, we arrive at the required curve D. For the reverse inequality, Bauer [39, Theorem 6.1] shows by working on the Kummer surface $A/\{\pm 1\}$ that the curve D just constructed (or a closely related curve) is irreducible. Here one uses the hypothesis that $\rho(A) = 1$. □

Quotients of balls and bounded symmetric domains. Hwang and To [302], [303] have obtained interesting bounds for the Seshadri constants of the canonical bundle on quotients of balls and other bounded symmetric domains. We summarize without proof the simplest of their results.

Consider the unit ball $\mathbf{B} \subseteq \mathbf{C}^n$ equipped with its Poincaré metric. Fix a discrete torsion-free cocompact subgroup $\Gamma \subseteq \mathrm{PU}(1,n)$, and let $X = \mathbf{B}/\Gamma$ be the corresponding smooth compact quotient. Then X carries a canonical Kähler metric coming from the Poincaré metric on $\mathbf{B}$, which in turn allows one to define two basic metric invariants. First, the *injectivity radius* ρ_x of X at a point $x \in X$ is defined to be

$$\rho_x \;=\; \frac{1}{2} \cdot \min_{\substack{\gamma \in \Gamma \\ \gamma \neq 1}} \, d(x_0, \gamma x_0),$$

where $x_0 \in \mathbf{B}$ is a point lying over $x \in X$, and $d(\cdot,\cdot)$ is the Poincaré distance function on $\mathbf{B}$. This plays the role of the minimal period length $m(X, A)$ in the case of abelian varieties. Second, define the *diameter* of X at $x \in X$ to be $D_x = \max_{y \in X} d(x,y)$, where here d denotes the distance function associated with the Poincaré metric on X. The main result of [302] is the following:

Theorem 5.3.20. *Given X as above, together with fixed point $x \in X$, one has*

$$(n+1) \cdot \sinh^2(\rho_x) \;\leq\; \varepsilon\big(K_X; x\big) \;\leq\; (n+1) \cdot \sinh^2(D_x).$$

The lower bound is established by an analogue of symplectic blowing up, whereas the upper bound is proven by convexity arguments.

Define the global injectivity radius of the ball-quotient X to be $\rho_X = \min_{x \in X} \rho_x$. Then the theorem combines with Proposition 5.1.19 to yield

Corollary 5.3.21. *With X as above, suppose that*

$$\rho_X \ > \sinh^{-1}\sqrt{\frac{2n}{n+1}} \ .$$

Then $2K_X$ is very ample. In particular, any ball quotient X admits a finite étale cover X' on which $2K_{X'}$ is very ample.

Both the theorem and corollary are generalized in [303] to smooth compact quotients of general bounded symmetric domains. The work of Hwang and To has been extended by Yeung in [616] and [617].

5.4 Local Positivity Along an Ideal Sheaf

This section describes some results of Cutkosky, Ein, and the author [99] concerning local positivity along an arbitrary ideal sheaf. We define an analogue of the Seshadri constant in which the fixed point $x \in X$ is replaced by an ideal sheaf $\mathcal{I} \subseteq \mathcal{O}_X$. The theme loosely speaking is that the resulting invariant bounds the complexity of $\mathcal{I}$ or its powers. This picture was inspired by work of Paoletti [491], [492], who considered ideal sheaves of smooth subvarieties.

The first subsection treats the formal properties of this s-invariant, and presents some examples. The main results appear in Section 5.4.B. We follow very closely the presentation in [99].

5.4.A Definition and Formal Properties of the s-Invariant

Let X be an irreducible projective variety, let $\mathcal{I} \subseteq \mathcal{O}_X$ be an ideal sheaf, and fix an ample divisor L on X.

Definition 5.4.1. (The s-invariant of an ideal). Let

$$\mu : X' \ = \ \mathrm{Bl}_{\mathcal{I}}(X) \longrightarrow X \tag{5.18}$$

be the blowing-up of X along $\mathcal{I}$, with exceptional divisor E. The *s-invariant of $\mathcal{I}$ with respect to L* is the positive real number

$$s_L(\mathcal{I}) \ = \ \min\{ s \in \mathbf{R} \mid \mu^*(sL) - E \text{ is a nef } \mathbf{R}\text{-divisor on } \mathrm{Bl}_{\mathcal{I}}(X) \}. \quad \square$$

One can think of $s_L(\mathcal{I})$ as measuring how many times one has to twist $\mathcal{I}$ by L in order to render it positive. The main case to keep in mind is that in which L is a hyperplane divisor on $X = \mathbf{P}^n$. A related invariant for vector bundles is introduced in Example 6.2.14.

Example 5.4.2. (Relation to Seshadri constants). Let $x \in X$ be a fixed point, with maximal ideal $\mathfrak{m} \subseteq \mathcal{O}_X$. Then $\varepsilon\big(L; x\big) = 1/s_L(\mathfrak{m})$. $\square$

Remark 5.4.3. (Seshadri constant of an ideal sheaf). By analogy with Definition 5.1.1, it would be natural to define the Seshadri constant $\varepsilon(L;\mathcal{I})$ of L along $\mathcal{I}$ to be the largest real number $\varepsilon > 0$ such that $\mu^*L - \varepsilon \cdot E$ is nef. Then $\varepsilon(L;\mathcal{I}) = \frac{1}{s_L(\mathcal{I})}$. However, for present purposes $s_L(\mathcal{I})$ is more convenient to work with than its reciprocal. □

Our immediate goal is to study the formal properties of this invariant, and to give some examples. It is useful to start with an extension of Definition 1.8.37.

Definition 5.4.4. (Generating degree of an ideal with respect to an ample divisor). With notation as above, the *generating degree* of $\mathcal{I}$ with respect to L is the natural number

$$d_L(\mathcal{I}) \;=\; \min\big\{\, d \in \mathbf{N} \mid \mathcal{I} \otimes \mathcal{O}_X(dL) \text{ is globally generated} \,\big\}. \quad \square$$

The first observation is that the generating degree of an ideal bounds its s-invariant.

Proposition 5.4.5. (Generating degree and s-invariant). *Let $\mathcal{I} \subseteq \mathcal{O}_X$ be an ideal sheaf on the projective variety X, and let L be any ample divisor on X. Then $s_L(\mathcal{I}) \le d_L(\mathcal{I})$. More generally,*

$$s_L(\mathcal{I}) \;\le\; \frac{d_L(\mathcal{I}^p)}{p}$$

for every integer $p \ge 1$.

We will see later (Theorem 5.4.22) that if L is globally generated, then in fact $s_L(\mathcal{I}) = \lim \frac{d_L(\mathcal{I}^p)}{p}$.

Proof of Proposition 5.4.5. Keeping the notation of (5.18), assume that $\mathcal{I}^p \otimes \mathcal{O}_X(dL)$ is generated by its global sections. This sheaf pulls back on X' to the line bundle $\mathcal{O}_{X'}\big(\mu^*(dL) - pE\big)$, and hence $\mu^*(dL) - pE$ is free. In particular, $s_L(\mathcal{I}) \le \frac{d}{p}$, as claimed. □

The next examples present some additional properties.

Example 5.4.6. (Schemes cut out by quadrics). Take $X = \mathbf{P}^n$ and L a hyperplane, and suppose that $\mathcal{I} \otimes \mathcal{O}_{\mathbf{P}^n}(2)$ is globally generated. Assume moreover that the zero-locus $Z = \text{Zeroes}(\mathcal{I})$ is not a linear space. Then $s_L(\mathcal{I}) = 2$. (By the previous example $s_L(\mathcal{I}) \le 2$, and examining the proper transform of a suitable secant line to Z gives the reverse inequality.) □

Example 5.4.7. (Irrational s-invariants). It is not hard to find examples in which the s-invariant of an ideal is irrational. In fact, let $X = B \times B$ be the product of an elliptic curve with itself, so that $\text{Nef}(X)$ is a circular cone (Example 1.5.4). Choose very ample divisors $L, C \subseteq X$, and consider the ideal sheaf $\mathcal{I} = \mathcal{O}_X(-C)$ of C. Then the s-invariant of $\mathcal{I}$ with respect to L is the largest root of the quadratic polynomial $q(t) = \big((tL - C)^2 \big)$, and for general choices of L and C this will be irrational. Observe that given L, one can take C so that the corresponding s-invariant is arbitrarily large. We will use this construction in Example 5.4.16 to produce a curve $C \subseteq \mathbf{P}^n$ whose ideal sheaf has irrational s-invariant. □

Example 5.4.8. For any ideal sheaf $\mathcal{I} \subseteq \mathcal{O}_X$ and ample line bundle L, the s-invariant $s_L(\mathcal{I})$ is an algebraic integer of degree $\leq \dim X$. (Set $s = s_L(\mathcal{I})$. Then on the blow-up $X' = \mathrm{Bl}_{\mathcal{I}}(X)$, the class $sL - E$ is nef but not ample. In this case, the Campana–Peternell theorem (Theorem 2.3.18) implies that there exists a subvariety $V' \subseteq X'$ such that $\big((sL - E)^{\dim V'} \cdot V'\big) = 0$. But this exhibits $s_L(\mathcal{I})$ as the root of a monic polynomial equation of degree $= \dim V'$.) □

Example 5.4.9. (Numerical nature and homogeneity). The s-invariant $s_L(\mathcal{I})$ depends only on the numerical equivalence class of L. For any natural number $m > 0$, one has $s_{mL}(\mathcal{I}) = \frac{1}{m} \cdot s_L(\mathcal{I})$. □

Example 5.4.10. (Alternative computation of s-invariants). Let $f : Y \longrightarrow X$ be a surjective morphism of projective varieties with the property that $\mathcal{I} \cdot \mathcal{O}_Y = \mathcal{O}_Y(-F)$ for some effective Cartier divisor F on Y. Then

$$s_L(\mathcal{I}) \ = \ \min \big\{ s \in \mathbf{R} \mid f^*(sL) - F \text{ is nef } \big\}.$$

(The hypothesis implies that f factors through the blowing-up of $\mathcal{I}$.) □

Example 5.4.11. (Sums and products of ideals). With notation as above, consider ideal sheaves $\mathcal{I}_1, \mathcal{I}_2 \subseteq \mathcal{O}_X$. Then

$$\begin{aligned} s_L\big(\mathcal{I}_1 \cdot \mathcal{I}_2\big) \ &\leq \ s_L(\mathcal{I}_1) + s_L(\mathcal{I}_2), \\ s_L\big(\mathcal{I}_1 + \mathcal{I}_2\big) \ &\leq \ \max \big\{ s_L(\mathcal{I}_1) \, , \, s_L(\mathcal{I}_2) \big\}. \end{aligned}$$

(Write $s_1 = s_L(\mathcal{I}_1)$ and $s_2 = s_L(\mathcal{I}_2)$, and let $f : Y \longrightarrow X$ be a surjective mapping from an irreducible projective variety Y that dominates the blowings-up of X along $\mathcal{I}_1$, $\mathcal{I}_2$, and $\mathcal{I}_1 + \mathcal{I}_2$. Then Y carries effective Cartier divisors F_1, F_2, and F_{12} such that

$$\mathcal{I}_1 \cdot \mathcal{O}_Y = \mathcal{O}_Y(-F_1) \ \ , \ \ \mathcal{I}_2 \cdot \mathcal{O}_Y = \mathcal{O}_Y(-F_2) \ \ , \ \ \big(\mathcal{I}_1 + \mathcal{I}_2\big) \cdot \mathcal{O}_Y = \mathcal{O}_Y(-F_{12}),$$

and $f^*(s_1 L) - F_1$ and $f^*(s_2 L) - F_2$ are nef. The first inequality follows from the observation that $\big(\mathcal{I}_1 \mathcal{I}_2\big) \cdot \mathcal{O}_Y = \mathcal{O}_Y\big(-(F_1 + F_2)\big)$. For the second, observe that one has a surjective map

$$\mathcal{O}_Y(-F_1) \oplus \mathcal{O}_Y(-F_2) \longrightarrow \mathcal{O}_Y(-F_{12})$$

of vector bundles on Y. If $s \geq \max\{s_1, s_2\}$, then the bundle on the left becomes nef when twisted (in the sense of Section 6.2) by the $\mathbf{Q}$-divisor $f^*(sL)$. Since quotients of nef $\mathbf{Q}$-twisted bundles are nef (Theorem 6.2.12), this implies that $f^*(sL) - E_{12}$ is nef.) □

Remark 5.4.12. (s-invariant of the integral closure of an ideal). Assuming that X is normal, denote by $\overline{\mathcal{I}}$ the integral closure of an ideal $\mathcal{I} \subseteq \mathcal{O}_X$ (Definition 9.6.2). Then it is established in [99, Proposition 1.12] that

$$s_L(\mathcal{I}) \ = \ s_L(\overline{\mathcal{I}}). \quad \square$$

Suppose now that X is non-singular, and let $Y \subseteq X$ be a smooth subvariety with ideal sheaf $\mathcal{I} = \mathcal{I}_{X/Y}$. We discuss next a geometric interpretation, due to Paoletti [491, p.487], for the s-invariant of $\mathcal{I}$. Paoletti's result involves data associated to non-constant mappings $f : C \longrightarrow X$ from a smooth curve C to X. Suppose first that $f(C) \not\subseteq Y$. Then $f^{-1}\mathcal{I} \subset \mathcal{O}_C$ is an ideal of finite colength in $\mathcal{O}_C$, and we define

$$s'_L(\mathcal{I}) \ = \ \sup_{\substack{f:C \longrightarrow X \\ f(C)\not\subseteq Y}} \Big\{ \frac{\text{colength}(f^{-1}\mathcal{I})}{(C \cdot_f L)} \Big\},$$

where $(C \cdot_f L)$ denotes the degree of the divisor f^*L on C. Next, write $N = N_{Y/X}$ for the normal bundle to Y in X and put

$$s''_L(\mathcal{I}) \ = \ \inf_{\substack{f:C \longrightarrow X \\ f(C)\subseteq Y}} \Big\{ t \in \mathbf{Q}^{>0} \ \Big| \ \begin{array}{c} \deg M + t \cdot (C \cdot_f L) \geq 0 \\ \text{for all rank 1 quotients } M \text{ of } f^*N^* \end{array} \Big\}.$$

Proposition 5.4.13. (Paoletti, [491]). *With the notation and hypotheses just introduced, one has*

$$s_L(\mathcal{I}) \ = \ \max \big\{ s'_L(\mathcal{I}) \, , \, s''_L(\mathcal{I}) \big\}.$$

Remark 5.4.14. When $\mathcal{I} = \mathfrak{m}$ is the maximal ideal of a point $x \in X$, the mappings appearing in the definition of $s''_L(\mathfrak{m})$ do not occur. The resulting equality $s_L(\mathfrak{m}) = s'_L(\mathfrak{m})$ is then a restatement of Proposition 5.1.5. $\square$

Sketch of Proof of Proposition 5.4.13. Consider a mapping $f : C \longrightarrow X$ with $f(C) \not\subseteq Y$, and let $f' : C \longrightarrow X' = \text{Bl}_Y(X)$ be the proper transform of f. Then $\text{colength}(f^{-1}\mathcal{I}) = \big(C \cdot_{f'} E\big)$, where as above E is the exceptional divisor of the blowing-up $\mu : \text{Bl}_Y(X) \longrightarrow X$. Therefore $s'_L(\mathcal{I})$ is the least real number $t > 0$ such that $\mu^*(tL) - E$ has non-negative degree on every curve not lying in the exceptional divisor $E \subset X'$. Similarly, $s''_L(\mathcal{I})$ is the infimum of all rational numbers $t > 0$ such that the $\mathbf{Q}$-twisted vector bundle $N^*<tL>$ is nef in the sense of Section 6.2 (Proposition 6.1.18, Remark 6.2.9). But this is equivalent to asking that $\big(\mu^*(tL) - E\big)$ be a nef divisor class on $E = \mathbf{P}(N^*)$, and the proposition follows. $\square$

For the purpose of constructing examples, it is useful to understand something about how the s-invariant behaves in chains.

Proposition 5.4.15. *Consider smooth projective varieties*

$$Z \subseteq Y \subseteq X,$$

with ideal sheaves

$$\mathcal{I}_{Z/X} \,,\, \mathcal{I}_{Y/X} \subseteq \mathcal{O}_X \,,\, \mathcal{I}_{Z/Y} \subseteq \mathcal{O}_Y.$$

Fix an ample divisor L *on* X*, and write (somewhat abusively)* $s_L(\mathcal{I}_{Z/Y})$ *for the s-invariant calculated by blowing up* Y *along* Z *and computing with the restriction of* L *to* Y*. If* $s_L(I_{Y/X}) < s_L(\mathcal{I}_{Z/X})$ *then*

$$s_L(\mathcal{I}_{Z/X}) = s_L(\mathcal{I}_{Z/Y}).$$

Sketch of Proof. Keep the notation introduced before Proposition 5.4.13. It is evident from the definitions that

$$s'_L(\mathcal{I}_{Z/X}) \leq \max\{ s'_L(\mathcal{I}_{Z/Y}) \,,\, s'_L(\mathcal{I}_{Y/X}) \},$$

and using the conormal bundle sequence

$$0 \to N^*_{Y/X}|Z \to N^*_{Z/X} \to N^*_{Z/Y} \to 0$$

it follows from Theorem 6.2.12, (ii), (v), that

$$s''_L(\mathcal{I}_{Z/X}) \leq \max\{ s''_L(\mathcal{I}_{Z/Y}) \,,\, s''_L(\mathcal{I}_{Y/X}) \}.$$

The assertion is then a consequence of 5.4.13. □

We conclude this subsection by giving a painless construction of a curve $C \subseteq \mathbf{P}^n$ with irrational s-invariant. More explicit examples appear in the paper [98] of Cutkosky.

Example 5.4.16. (Projective curve with irrational s-invariant). Start with the product $A = B \times B$ of an elliptic curve with itself, and fix a very ample divisor on A defining an embedding $A \subseteq \mathbf{P}^n$ in which A is cut out by quadrics.[6] Then Example 5.4.6 shows that $s_L(\mathcal{I}_{A/\mathbf{P}^n}) = 2$, L being the hyperplane divisor on $\mathbf{P}^n$. Choose next a curve $C \subseteq A$ having the property that $s_L\big(\mathcal{O}_A(-C)\big)$ is an irrational number $s > 2$ (Example 5.4.7). Applying the previous proposition to the chain $C \subseteq A \subseteq \mathbf{P}^n$, we find that $s_L(\mathcal{I}_{C/\mathbf{P}^n}) = s$ is irrational. □

Example 5.4.17. (An irrational polyhedral nef cone). Let $C \subseteq \mathbf{P}^n$ be the curve constructed in the previous example, and let $X = \mathrm{Bl}_C(\mathbf{P}^n)$ be the blowing-up of $\mathbf{P}^n$ along C. Then X has Picard number two, and the nef cone $\mathrm{Nef}(X) \subseteq N^1(X)_{\mathbf{R}} = \mathbf{R}^2$ is an irrational polyhedron. (Write E and H respectively for the exceptional divisor and the pullback of the hyperplane class L on $\mathbf{P}^n$. Then $\mathrm{Nef}(X)$ is spanned by the classes of H and $sH - E$, where $s = s_L(\mathcal{I}_{C/\mathbf{P}^n})$.) □

[6] For example, one can take the sum of the pullbacks under each projection of a divisor of degree 4 on B.

5.4.B Complexity Bounds

We now present two results illustrating the philosophy that the s-invariant of an ideal sheaf bounds its complexity. The first is an inequality of Bézout-type that will be useful in connection with the effective Nullstellensatz (Chapter 10.5). The second theorem asserts that the s-invariant of an ideal computes its asymptotic Castelnuovo–Mumford regularity. These (and related statements from [99]) provide a geometric perspective on some of the results and questions from Bayer and Mumford's survey [42]. As in that paper, we speak of complexity in a purely informal sense.

One can view the number and degrees of the components of an ideal's zero-locus as a first measure of its complexity. The s-invariant gives some control over these:

Theorem 5.4.18. (Degree inequality). *Let X be an irreducible projective variety, $\mathcal{I} \subseteq \mathcal{O}_X$ an ideal sheaf, and L an ample divisor on X. Denote by*

$$Y_1 \, , \, \ldots \, , Y_p \ \subseteq \ X$$

the irreducible components (with their reduced scheme structures) of the zero-locus $Y = \mathrm{Zeroes}(\mathcal{I})$ of $\mathcal{I}$, and write $s = s_L(\mathcal{I})$. Then

$$\sum_{j=1}^{p} s^{\dim Y_j} \cdot \deg_L(Y_j) \ \leq \ s^{\dim X} \cdot \deg_L(X). \tag{5.19}$$

Given an irreducible subvariety $V \subseteq X$, by $\deg_L(V)$ we understand the intersection number $\big(L^{\dim V} \cdot V\big)$. All of these degrees are positive (L being ample), and so the theorem limits the number and degrees of the components in question:

Corollary 5.4.19. *Setting $s_+ = \max\{1, s\}$, one has*

$$\sum_{j=1}^{p} \deg_L(Y_j) \ \leq \ s_+^{\dim X} \cdot \deg_L(X). \quad \square$$

Proof of Theorem 5.4.18. Let X'' be the normalization of the blow-up $X' = \mathrm{Bl}_{\mathcal{I}}(X)$, and write

$$\nu : X'' \longrightarrow X$$

for the natural map. Denote by F the exceptional divisor of ν, so that F is an effective Cartier divisor on X'' with $\mathcal{I} \cdot \mathcal{O}_{X''} = \mathcal{O}_{X''}(-F)$. Now F determines an effective Weil divisor $[F]$ on X'', say

$$[F] \ = \ \sum_{i=1}^{t} r_i \cdot [F_i],$$

where the F_i are the irreducible components of the support of F, and $r_i \geq 1$ is a positive integer. Put

$$Z_i \;=\; \nu(F_i) \;\subseteq\; X,$$

so that Z_i is a reduced and irreducible subvariety of X. Following [208], the Z_i are called the *distinguished subvarieties* of $\mathcal{I}$.[7] Then evidently $Y = \cup Z_i$, so in particular each irreducible component Y_j of Y occurs as one of the distinguished subvarieties Z_i. Therefore it is enough to prove that the Z_i satisfy the degree bound

$$\sum_{i=1}^{t} r_i \cdot s^{\dim Z_i} \cdot \deg_L(Z_i) \;\leq\; s^{\dim X}\, \deg_L(X). \tag{5.20}$$

For this, set $n = \dim X$ and denote by $l, m \in N^1(X'')_{\mathbf{R}}$ the classes of $\nu^* L$ and $\nu^*(sL) - F$ respectively. Then by definition m is nef, and so

$$\int_{X''} m^n \;\geq\; 0 \quad \text{ and } \quad \int_{F_i} l^j \cdot m^{n-1-j} \;\geq\; 0$$

for all i and j. Moreover, $F \equiv_{\text{num}} s \cdot l - m$. One then finds that

$$\begin{aligned}
s^n \cdot \deg_L(X) \;&=\; \int_{X''} (s \cdot l)^n \\
&\geq\; \int_{X''} \big((s \cdot l)^n - m^n\big) \\
&=\; \int_{X''} \Big((s \cdot l) - m\Big)\Big(\sum_{j=0}^{n-1} (s \cdot l)^j \cdot m^{n-1-j}\Big) \\
&=\; \int_{[F]} \Big(\sum_{j=0}^{n-1} (s \cdot l)^j \cdot m^{n-1-j}\Big) \\
&\geq\; \sum_{i=1}^{t} r_i \cdot \int_{F_i} (s \cdot l)^{\dim Z_i} \cdot m^{n-1-\dim Z_i}.
\end{aligned}$$

On the other hand, $\mathcal{O}_{X''}(-F)$ — being a finite pullback of the ideal of the exceptional divisor of a blow-up — is ample for ν. Therefore m has positive degree on every fibre of ν, and in particular

$$\int_{F_i} l^{\dim Z_i} \cdot m^{n-1-\dim Z_i} \;\geq\; \deg_L(Z_i).$$

All told, we find that

---

[7] Note that several of the F_i may have the same image in X, in which case there will be repetitions among the Z_i. However this doesn't cause any problems.

$$\begin{aligned} s^n \cdot \deg_L(X) &\geq \sum_{i=1}^{t} r_i \cdot \int_{F_i} (s \cdot l)^{\dim Z_i} \cdot m^{n-1-\dim Z_i} \\ &\geq \sum_{i=1}^{t} r_i \cdot s^{\dim Z_i} \cdot \deg_L(Z_i), \end{aligned}$$

as required. □

Remark 5.4.20. Some variants of Theorem 5.4.18 are discussed in [99, §2]. In particular, it is observed there that if $\mathcal{I}$ is integrally closed then one can include in the left-hand side of (5.19) the subvarieties defined by all associated primes of $\mathcal{I}$, minimal or embedded. On the other hand, [99, Example 2.8] gives a construction to show that the number of embedded components cannot be bounded in terms of the s-invariant for ideals that are not integrally closed. □

Assume now that L is free (as well as ample). Then one can discuss the Castelnuovo–Mumford regularity with respect to L of any coherent sheaf on X (Definition 1.8.4), and by analogy with Definition 1.8.17 we make

Definition 5.4.21. The *L-regularity* $\mathrm{reg}_L(\mathcal{I})$ of an ideal $\mathcal{I} \subseteq \mathcal{O}_X$ is the least integer m such that $\mathcal{I}$ is m-regular with respect to $\mathcal{O}_X(L)$. □

We refer to Section 1.8 for a survey of the theory of Castelnuovo–Mumford regularity, and in particular its role in measuring the complexity of an ideal.

The next result shows that the s-invariant of an ideal computes asymptotically the regularity and generating degree of its powers:

Theorem 5.4.22. (Asymptotic regularity and s-invariant). *Let X be an irreducible projective variety, and fix an ample and free divisor L on X. If $\mathcal{I} \subseteq \mathcal{O}_X$ is any ideal sheaf, then*

$$\liminf_{p \to \infty} \frac{\mathrm{reg}_L(\mathcal{I}^p)}{p} \;=\; \lim_{p \to \infty} \frac{d_L(\mathcal{I}^p)}{p} \;=\; s_L(\mathcal{I}). \tag{5.21}$$

One can view this as a geometric analogue of the result of Cutkosky–Herzog–Trung and Kodiyalam (Theorem 1.8.49) concerning the asymptotic regularity of powers of a homogeneous ideal.

Remark 5.4.23. With a little more effort, one can show that the lim inf appearing in (5.21) is actually a limit: see [99, §3]. □

The crucial input to the theorem is a lemma asserting that one can compute cohomology of large powers of $\mathcal{I}$ in terms of the exceptional divisor on the blow-up.

Lemma 5.4.24. *As before, let $\mu : X' = \mathrm{Bl}_{\mathcal{I}}(X) \longrightarrow X$ be the blowing up of X, with exceptional divisor E. Then there exists an integer $p_0 = p_0(\mathcal{I})$ with the property that if $p \geq p_0$ then*

$$\mu_*\big(\mathcal{O}_{X'}(-pE)\big) \;=\; \mathcal{I}^p, \tag{5.22}$$

and moreover for any divisor D on X,

$$H^i\big(X, \mathcal{I}^p(D)\big) \;=\; H^i\big(X', \mathcal{O}_{X'}(\mu^*D - pE)\big) \tag{5.23}$$

for all $i \geq 0$.

Remark 5.4.25. Note that if X is non-singular, and $\mathcal{I}$ defines a smooth subvariety of X, then the conclusion of the lemma holds for all $p \geq 0$ (Lemma 4.3.16). However, in general one has to be content with large values of p. □

Proof of Lemma 5.4.24. Since $\mathcal{O}_{X'}(-E)$ is ample for μ, Grothendieck–Serre vanishing implies that

$$R^j\mu_*\big(\mathcal{O}_{X'}(-pE)\big) \;=\; 0 \quad \text{for } j > 0 \text{ and } p \gg 0.$$

The isomorphism on global cohomology groups is then a consequence of (5.22) thanks to the Leray spectral sequence. As for (5.22), we can suppose that X is affine and that $\mathcal{I}$ is generated by functions $g_1, \ldots, g_r \in \mathcal{O}_X$. The choice of generators gives a surjection $\mathcal{O}_X^r \longrightarrow \mathcal{I}$, which in turn determines an embedding $X' \subseteq \mathbf{P}(\mathcal{O}_X^r) \;=\; \mathbf{P}_X^{r-1}$ in such a way that $\mathcal{O}_{\mathbf{P}_X^{r-1}}(1)\,|\,X' = \mathcal{O}_{X'}(-E)$. Write $\pi : \mathbf{P}_X^{r-1} = X \times \mathbf{P}^{r-1} \longrightarrow X$ for the projection. Serre vanishing for π, applied to the ideal sheaf of X' in $\mathbf{P}_X^{r-1}$, shows that if $p \gg 0$ then the natural homomorphism

$$\pi_*\big(\mathcal{O}_{\mathbf{P}_X^{r-1}}(p)\big) \longrightarrow \pi_*\big(\mathcal{O}_{X'}(-pE)\big) \tag{*}$$

is surjective. But recalling that $\pi_*\big(\mathcal{O}_{\mathbf{P}_X^{r-1}}(k)\big) = \mathrm{Sym}^k(\mathcal{O}_X^r)$ for every k, one sees that the image of the mapping in (*) is exactly $\mathcal{I}^p$. The stated equality (5.22) follows. □

Proof of Theorem 5.4.22. Set $d_p \;=\; d_L(\mathcal{I}^p)$ and $r_p \;=\; \mathrm{reg}_L(\mathcal{I}^p)$. Observe to begin with that $d_{\ell+m} \;\leq\; d_\ell + d_m$, from which it follows that $\lim \frac{d_p}{p}$ exists. Note next that $s_L(\mathcal{I}) \leq \frac{d_p}{p}$ for all p by Proposition 5.4.5, while $d_p \leq r_p$ thanks to Mumford's Theorem 1.8.5. Therefore it is enough to show that given any rational number $s' > s$, one has $r_p \leq ps'$ for all sufficiently large and divisible p.

But if $s' > s$ then $s'L - E$ is ample. So if we take p to be sufficiently large and divisible — so that in particular $ps' \in \mathbf{N}$ — then by Serre vanishing we can suppose that

$$H^i\big(X', \mathcal{O}_{X'}\big((ps' - k)L - pE\big)\big) \;=\; 0 \quad \text{for all} \quad i > 0 \text{ and } 1 \leq k \leq \dim X.$$

Assuming as we may that p is also sufficiently large so that the conclusion of Lemma 5.4.24 holds, it follows from (5.23) that $\mathcal{I}$ is ps'-regular with respect to L. Therefore $r_p \leq ps'$, as required. □

Notes

Seshadri constants were introduced in [123, §6]. Demailly seems to have been motivated here by Fujita's conjecture (see Section 10.4.A), but the examples of Miranda soon made it clear that one could not expect any simple application in this direction. However, in the intervening years it has become increasingly clear that Seshadri constants are fundamental and interesting invariants in their own right.

Most of the foundational material in Section 5.1 appears in [123]. Section 5.2 generally follows [149], while Section 5.3 is adapted from [393]. Section 5.4 is taken from [99].

Appendices

A

Projective Bundles

We recall here some basic facts about projective bundles.

Definition and construction. Let X be an algebraic variety or scheme, and let E be a vector bundle of rank e on X. We denote by

$$\pi : \mathbf{P}(E) \longrightarrow X$$

the projective bundle of one-dimensional quotients of X. Thus a point in $\mathbf{P}(E)$ is determined by specifying a point $x \in X$ together with a one-dimensional quotient of the fibre $E(x)$ of E at x. More algebraically, $\mathbf{P}(E)$ is realized as the scheme

$$\mathbf{P}(E) \ = \ \mathrm{Proj}_{\mathcal{O}_X}\big(\mathrm{Sym}(E)\big),$$

where $\mathrm{Sym}(E) = \oplus\, S^m E$ denotes the symmetric algebra of E.[1]

Serre line bundle. The projective bundle $\mathbf{P}(E)$ carries a line bundle $\mathcal{O}_{\mathbf{P}(E)}(1)$, arising as a "tautological" quotient of π^*E:

$$\pi^*E \longrightarrow \mathcal{O}_{\mathbf{P}(E)}(1) \longrightarrow 0. \tag{A.1}$$

If L is a line bundle, then $\mathbf{P}(E) \cong \mathbf{P}(E \otimes L)$ via an isomorphism under which $\mathcal{O}_{\mathbf{P}(E\otimes L)}(1)$ corresponds to $\mathcal{O}_{\mathbf{P}(E)}(1) \otimes \pi^*L$.

For $m \geq 0$ one has

$$\pi_*\mathcal{O}_{\mathbf{P}(E)}(m) \ = \ S^m E \tag{A.2a}$$

$$R^{e-1}\pi_*\mathcal{O}_{\mathbf{P}(E)}(-e-m) \ = \ \big(S^m E\big)^* \otimes \det E^*, \tag{A.2b}$$

and all other direct images vanish.

Maps to P(E). Let $p : Y \longrightarrow X$ be a variety or scheme mapping to X. Then giving a line bundle quotient $p^*E \twoheadrightarrow L$ of the pullback of E is equivalent to specifying a map $f : Y \longrightarrow \mathbf{P}(E)$ over X:

[1] Here and elsewhere we do not distinguish between E and the corresponding locally free sheaf of sections.

$$\begin{array}{ccc} Y & \xrightarrow{\ f\ } & \mathbf{P}(E) \\ & {}_{p}\searrow \quad \swarrow_{\pi} & \\ & X. & \end{array}$$

Under this correspondence $L = f^*\mathcal{O}_{\mathbf{P}(E)}(1)$, and the given quotient is the pullback of (A.1). In particular, for each $\ell > 0$ there is a natural Veronese embedding of $\mathbf{P}(E)$ into $\mathbf{P}(S^\ell E)$:

$$\begin{array}{ccc} \mathbf{P}(E) & \overset{\nu_\ell}{\hookrightarrow} & \mathbf{P}(S^\ell E) \\ & {}_{p}\searrow \quad \swarrow_{\pi} & \\ & X. & \end{array}$$

It is determined by the quotient $S^\ell \pi^* E \twoheadrightarrow \mathcal{O}_{\mathbf{P}(E)}(\ell)$ arising by taking ℓ^{th} symmetric products in (A.1).

Néron–Severi group. Assume that X is projective. Then the Néron–Severi group of $\mathbf{P}(E)$ is determined by

$$N^1(\mathbf{P}(E)) \;=\; \pi^* N^1(X) \oplus \mathbf{Z} \cdot \xi_E,$$

where ξ_E is the class of a divisor representing $\mathcal{O}_{\mathbf{P}(E)}(1)$. Writing ξ_E also for the corresponding class in $H^2\big(X;\mathbf{Z}\big)$, the integral cohomology $H^*\big(\mathbf{P}(E);\mathbf{Z}\big)$ of $\mathbf{P}(E)$ is generated as an $H^*\big(X;\mathbf{Z}\big)$-algebra by ξ_E, subject to the Grothendieck relation

$$\xi_E^e - \pi^* c_1(E) \cdot \xi_E^{e-1} + \pi^* c_2(E) \cdot \xi_E^{e-2} + \ldots + (-1)^e \pi^* c_e(E) \;=\; 0.$$

One–dimensional subspaces. On a few occasions it will be preferable to consider the bundle $\mathbf{P}_{\text{sub}}(E)$ of one–dimensional *subspaces* of E. This is related to the bundles of quotients by the isomorphism $\mathbf{P}_{\text{sub}}(E) = \mathbf{P}(E^*)$.

B

Cohomology and Complexes

We collect in this appendix several constructions and facts of a homological nature. The first subsection deals with techniques for analyzing the cohomology of a sheaf. The second focuses on some complexes associated to vector bundle maps.

B.1 Cohomology

This section lays out a collection of tools that can be useful in computing the cohomology of a coherent sheaf on a variety or scheme.

Cohomology and direct images. Let

$$f : Y \longrightarrow X$$

be a morphism of varieties or schemes, and consider a sheaf $\mathcal{F}$ on Y. It is frequently important to compute the cohomology of $\mathcal{F}$ on Y in terms of data — specifically, the direct images of $\mathcal{F}$ — on X. The basic mechanism to this end is the Leray spectral sequence

$$E_2^{p,q} \;=\; H^p\big(X, R^q f_* \mathcal{F}\big) \;\Longrightarrow\; H^{p+q}\big(Y, \mathcal{F}\big).$$

However, one often needs only a special case, in which the sequence degenerates.

Proposition B.1.1. *In the situation just described, assume that all the higher direct images of $\mathcal{F}$ vanish, i.e. suppose that $R^j f_* \mathcal{F} = 0$ for every $j > 0$. Then*

$$H^i\big(Y, \mathcal{F}\big) \;=\; H^i\big(X, f_*\mathcal{F}\big) \quad \textit{for all} \quad i. \quad \square$$

A direct proof (avoiding spectral sequences) is possible: see [280, Exercise III.8.1]. We frequently use this equality without explicit mention.

Resolutions. Let X be a projective (or complete) algebraic variety or scheme. One sometimes attempts to understand the cohomology of a coherent sheaf $\mathcal{F}$ on X in terms of a resolution, or something close to a resolution. Thus consider a complex of coherent sheaves

$$F_\bullet : \qquad 0 \longrightarrow F_\ell \longrightarrow \ldots \longrightarrow F_1 \longrightarrow F_0 \stackrel{\varepsilon}{\longrightarrow} \mathcal{F} \longrightarrow 0,$$

with ε surjective. We record some simple criteria, in increasing order of generality, that allow one to conclude the vanishing of $H^k(X, \mathcal{F})$ in terms of data involving $F_\bullet$.

Proposition B.1.2. (Chasing through complexes). *Given an integer $k \geq 0$, assume that*

$$H^k(X, F_0) \;=\; H^{k+1}(X, F_1) \;=\; \ldots \;=\; H^{k+\ell}(X, F_\ell) \;=\; 0. \tag{B.1}$$

(i). *If $F_\bullet$ is exact, then $H^k(X, \mathcal{F}) = 0$.*

(ii). *The same conclusion holds assuming that $F_\bullet$ is exact off a subset of X having dimension $\leq k$.*

(iii). *More generally still, for $i \geq 0$ denote by $\mathcal{H}_i$ the homology sheaf*

$$\mathcal{H}_i = \frac{\ker\left(F_i \longrightarrow F_{i-1}\right)}{\operatorname{im}\left(F_{i+1} \longrightarrow F_i\right)}$$

(with $F_{-1} = \mathcal{F}$). Still assuming (B.1)*, the vanishing $H^k(X, \mathcal{F}) = 0$ holds provided that*

$$H^{k+1}(X, \mathcal{H}_0) = H^{k+2}(X, \mathcal{H}_1) = \cdots = H^{k+\ell+1}(X, \mathcal{H}_\ell) = 0.$$

The hypothesis in (ii) means that all the homology sheaves of $F_\bullet$ should be supported on an algebraic set of dimension $\leq k$.

Proof of Proposition B.1.2. This is most easily checked by chopping $F_\bullet$ into short exact sequences in the usual way, and chasing through the resulting diagram. □

Example B.1.3. (Surjectivity on global sections). One can also ask for criteria to guarantee that the map

$$H^0(X, F_0) \longrightarrow H^0(X, \mathcal{F}) \tag{*}$$

determined by ε is surjective. Assume that

$$H^1(X, F_1) \;=\; H^2(X, F_2) \;=\; \ldots \;=\; H^\ell(X, F_\ell) \;=\; 0.$$

Then the homomorphism in (*) is surjective provided either that $F_\bullet$ is exact off a set of dimension 0, or that one has the case $k = 0$ of the vanishings in B.1.2 (iii). □

Remark B.1.4. (Infinite complexes). Note that the vanishing

$$H^{k+i}\big(X, F_i\big) \;=\; 0$$

appearing in (B.1) is automatic if $k + i > \dim X$, and similarly for the hypotheses in (iii). Therefore one can always replace a complex $F_\bullet$ of length $\ell \gg 0$ by a suitable truncation of length $\leq \dim X + 1$. In particular, the proposition applies also to possibly infinite resolutions of $\mathcal{F}$. Example B.1.3 extends in a similar fashion. □

Spectral sequences. It is sometimes most efficient to organize cohomological data into a spectral sequence. Suppose for example that the complex $F_\bullet$ above is exact, and denote by $F_\bullet^+$ the truncated complex

$$F_\bullet^+ : \qquad 0 \longrightarrow F_\ell \longrightarrow \ldots \longrightarrow F_1 \longrightarrow F_0 \longrightarrow 0.$$

Thus $F_\bullet^+$ is quasi-isomorphic to the single sheaf $\mathcal{F}$, and hence the hypercohomology of $F_\bullet^+$ coincides with the cohomology of $\mathcal{F}$ itself:

$$\mathbf{H}^k\big(X, F_\bullet^+\big) \;=\; H^k\big(X, \mathcal{F}\big).$$

In this case the cohomology of $\mathcal{F}$ is computed via the hypercohomology spectral sequence associated to $F_\bullet^+$:

Lemma B.1.5. *Assuming that $F_\bullet$ is exact, there is a (second quadrant) spectral sequence*

$$E_1^{p,q} \;=\; H^q\big(X, F_{-p}\big) \Longrightarrow H^{p+q}\big(X, \mathcal{F}\big)$$

abutting to the cohomology of $\mathcal{F}$. □

Example B.1.6. In the situation of the lemma, suppose that there are positive integers r and s such that all the groups $H^i\big(X, F_j\big)$ vanish except when $i = 0$ and $j \leq s$ or when $i = r$ and $j \geq s+r+1$. Then the cohomology groups of $\mathcal{F}$ are computed as the cohomology of a complex

$$\ldots \longrightarrow H^r\big(X, F_{s+r+2}\big) \longrightarrow H^r\big(X, F_{s+r+1}\big) \longrightarrow H^0\big(X, F_s\big) \\ \longrightarrow H^0\big(X, F_{s-1}\big) \longrightarrow \ldots . \quad \square \qquad \text{(B.2)}$$

Filtrations. Instead of a "resolution" such as $F_\bullet$, one sometimes wishes to infer something about the cohomology of a sheaf from a filtration. Suppose then that $\mathcal{F}$ is a coherent sheaf on X, and that one is given a filtration $K^\bullet$ of $\mathcal{F}$ by coherent subsheaves:

$$0 \;=\; K^{p+1} \;\subseteq\; K^p \;\subseteq\; \ldots \;\subseteq\; K^1 \;\subseteq\; K^0 \;=\; \mathcal{F}.$$

As usual, we set $\mathrm{Gr}^j = K^j/K^{j+1}$. In general, the cohomology of $\mathcal{F}$ is most efficiently analyzed via the spectral sequence of a filtered sheaf (cf. [248, Chapter 3, §5]), but in the simplest situations it is just as easy to work directly with the evident short exact sequences linking these data.

Lemma B.1.7. *Suppose that there is an integer $k \geq 0$ such that all the Gr^j have vanishing cohomology in all degrees $i \neq k$:*

$$H^i\big(X, \mathrm{Gr}^j\big) \;=\; 0 \quad \textit{for all} \;\; i \neq k \;\; \textit{and all} \;\; 0 \leq j \leq p.$$

Then $H^i\big(X, \mathcal{F}\big) = 0$ for $i \neq k$, and $H^k\big(X, \mathcal{F}\big)$ has a filtration with graded pieces isomorphic to $H^k\big(X, \mathrm{Gr}^j\big)$. □

Direct images. Everything here generalizes to the setting of a proper morphism $f : X \longrightarrow Y$ (without X or Y necessarily being complete), with cohomology groups being replaced by direct image sheaves. We leave the details to the reader.

B.2 Complexes

We next discuss some useful complexes. Throughout this section, X is an irreducible variety or scheme of dimension n.

Koszul complex. Let E be a vector bundle on X of rank e, and let

$$s \;\in\; \Gamma\big(X, E\big)$$

be a section of E. This section gives rise to a zero scheme

$$Z \;=\; \mathrm{Zeroes}(s) \;\subseteq\; X :$$

in terms of a local trivialization of E, s is expressed by a vector $s = (s_1, \ldots, s_e)$ of regular functions, and the ideal sheaf of Z is locally generated by the component functions s_i. These data determine a *Koszul complex* $K_\bullet(E, s)$ of sheaves on X:

$$0 \longrightarrow \Lambda^e E^* \longrightarrow \Lambda^{e-1} E^* \longrightarrow \ldots \longrightarrow \Lambda^2 E^* \longrightarrow E^* \longrightarrow \mathcal{O}_X \longrightarrow \mathcal{O}_Z \longrightarrow 0.$$

The map $\Lambda^i E^* \longrightarrow \Lambda^{i-1} E^*$ is defined via contraction with s. The complex $K_\bullet(E, s)$ is always exact off Z in the sense that its homology sheaves are supported on Z. It is globally exact if X is locally Cohen–Macaulay (e.g. non–singular) and if $\mathrm{codim}_X Z = \mathrm{rank} E$.

One may deduce several useful variants by transposition and twisting by line bundles. For instance, when A is a line bundle, a morphism $u : E \longrightarrow A$ gives rise to a complex

$$0 \to \Lambda^e E \otimes A^{\otimes -e} \to \cdots \to \Lambda^2 E \otimes A^{\otimes -2} \to E \otimes A^* \to \mathcal{O}_X \to \mathcal{O}_Z \to 0$$

with analogous exactness properties. Similarly, starting with a section $\mathcal{O}_X \longrightarrow E$, one can also form the complex

$$0 \longrightarrow \mathcal{O}_X \longrightarrow E \longrightarrow \Lambda^2 E \longrightarrow \ldots \longrightarrow \Lambda^e E \longrightarrow \Lambda^e E \otimes \mathcal{O}_Z \longrightarrow 0.$$

Example B.2.1. Given any vector bundle E on X and any positive integer k, there is a long exact sequence

$$\ldots \longrightarrow S^{k-2}E \otimes \Lambda^2 E \longrightarrow S^{k-1} \otimes E \longrightarrow S^k E \longrightarrow 0$$

of bundles on X. (Form the Koszul complex associated to the tautological quotient $\pi^* E \longrightarrow \mathcal{O}_{\mathbf{P}(E)}(1)$ on $\mathbf{P}(E)$, twist by $\mathcal{O}_{\mathbf{P}(E)}(k-1)$ and take direct images.) □

Complexes of Eagon–Northcott type. The Koszul complex is a special case of a collection of complexes determined by a map

$$u : E \longrightarrow F$$

of bundles of ranks e and f respectively on X. Assume (as we may by transposing if necessary) that $e \geq f$, and let $Z = D_{f-1}(u)$ denote the top degeneracy locus of u, i.e. the subscheme

$$D_{f-1}(u) \;=\; \{x \in X \mid \text{rank } u(x) \leq f-1\}$$

locally defined by the $f \times f$ minors of a matrix for u. Recall that Z has expected codimension $e - f + 1$ in X (while its actual codimension might be smaller).

The first of the complexes in question is due to Eagon and Northcott, and provides (in good situations) a resolution of $\mathcal{O}_Z$. The second was constructed by Buchsbaum and Rim, and resolves (generically) the cokernel of u. The family as a whole was described independently by Buchsbaum and Eisenbud and by Kirby. We refer to Eisenbud's book [164, A.2.6] for a discussion from an algebraic point of view. Following an approach pioneered by Kempf, we sketch a geometric derivation.

Kempf's idea is to pass to the projectivization $\pi : \mathbf{P}(F) \longrightarrow X$ of F, and to form the indicated composition v:

$$\begin{array}{ccc} \pi^* E & \xrightarrow{\pi^* u} & \pi^* F \\ & \searrow^{v} & \downarrow \\ & & \mathcal{O}_{\mathbf{P}(F)}(1). \end{array}$$

View v as a section of $\pi^* E^* \otimes \mathcal{O}_{\mathbf{P}(F)}(1)$, and let

$$Y \;=\; \text{Zeroes}(v) \;\subseteq\; \mathbf{P}(F)$$

denote the zero scheme of v. Note that Y projects onto Z, and given $x \in Z$ the fibre of Y over x is $\mathbf{P}(\text{coker } u(x))$. It follows by a dimension count that if each of the degeneracy loci

$$D_{f-\ell}(u) \;=\; \{x \in X \mid \text{rank}\, u(x) \leq f-\ell\,\}$$

has codimension $\geq e-f+\ell$ in X — so that in particular $Z = D_{f-1}(u)$ has the expected codimension $e-f+1$ — then Y achieves the expected codimension e in $\mathbf{P}(F)$, i.e. $\dim Y = n+(f-1)-e$.

The next step is to form the Koszul complex $K_\bullet = K_\bullet(\pi^*E^* \otimes \mathcal{O}_{\mathbf{P}(F)}(1), v)$ determined by v. Twisting by $\mathcal{O}_{\mathbf{P}(F)}(k)$ for $k \geq 0$ yields a complex $K_\bullet(k)$:

$$0 \to \pi^*\Lambda^e E(k-e) \to \cdots \to \pi^*\Lambda^2 E(k-2) \to \pi^*E(k-1) \to \mathcal{O}_{\mathbf{P}(F)}(k) \to 0.$$

Here we are writing $\pi^*\Lambda^k E(\ell)$ for $\pi^*(\Lambda^k E) \otimes \mathcal{O}_{\mathbf{P}(F)}(\ell)$, and we have omitted the cokernel $\mathcal{O}_Y(k)$ of the last map on the right.

The hyper-direct images of $K_\bullet(k)$ are computed by a (second quadrant) spectral sequence

$$E_1^{p,q} \;=\; R^q\pi_*\big((\pi^*\Lambda^{-p}E^*)(k+p)\big) \;\Longrightarrow\; \mathbf{R}^{p+q}\,\pi_*\big(K_\bullet(k)\big). \tag{B.3}$$

On the other hand, observe that

$$R^i\pi_*\left(\pi^*\Lambda^j E^* \otimes \mathcal{O}_{\mathbf{P}(F)}(\ell)\right) \;=\; \Lambda^j E^* \otimes R^i\pi_*\mathcal{O}_{\mathbf{P}(F)}(\ell).$$

Bearing in mind that the direct images on the right vanish unless either $i=0$ and $\ell \geq 0$ or $i = f-1$ and $\ell \leq -f$, one finds as in Example B.1.6 that the spectral sequence (B.3) gives rise to a single complex (EN_k) of sheaves on X. Note that (EN_k) is exact off Z, since $K_\bullet(k)$ is exact away from $\pi^{-1}(Z)$.

Now assume that X — and hence also $\mathbf{P}(F)$ — is locally Cohen–Macaulay (e.g. smooth), and that Y has the expected codimension e in $\mathbf{P}(F)$. Then $K_\bullet(k)$ is acyclic, i.e. the only homology is the cokernel of the right-most map, and so it is a resolution of $\mathcal{O}_Y(k)$. Therefore the hyper-direct images of $K_\bullet(k)$ simply compute the direct images of $\mathcal{O}_Y(k)$:

$$\mathbf{R}^i\pi_*\big(K_\bullet(k)\big) \;=\; R^i\pi_*\mathcal{O}_Y(k).$$

But it follows from the spectral sequence (B.3) for reasons of degree that these can be non–zero only when $i=0$. Thus under the stated hypotheses, (EN_k) is itself acyclic.

Mirroring Eisenbud's presentation in [164, Appendix A2.6], we next write down the complexes explicitly.[1] Note that if $0 \leq k \leq e-f+1$ then (EN_k) has length $e-f+1$. The length of each of the subsequent complexes grows by one until it reaches length e when $k \geq e$:

$$0 \to \begin{matrix} \Lambda^e E \\ \otimes \\ (S^{e-f}F)^* \otimes \Lambda^f F^* \end{matrix} \to \cdots \to \begin{matrix} \Lambda^{f+1} E \\ \otimes \\ F^* \otimes \Lambda^f F^* \end{matrix} \to \begin{matrix} \Lambda^f E \\ \otimes \\ \Lambda^f F^* \end{matrix} \to \mathcal{O}_X \to 0 \tag{EN_0}$$

[1] The table that follows differs from Figure A2.6 in [164] only in that we keep track of the twists by line bundles that arise in the global setting.

$$0 \to \begin{array}{c} \Lambda^e E \\ \otimes \\ (S^{e-f-1}F)^* \otimes \Lambda^f F^* \end{array} \to \cdots \to \begin{array}{c} \Lambda^{f+2} E \\ \otimes \\ F^* \otimes \Lambda^f F^* \end{array} \to \begin{array}{c} \Lambda^{f+1} E \\ \otimes \\ \Lambda^f F^* \end{array} \to E \to F \to 0 \quad (\mathrm{EN}_1)$$

$$0 \to \begin{array}{c} \Lambda^e E \\ \otimes \\ (S^{e-f-2}F)^* \otimes \Lambda^f F^* \end{array} \to \cdots \to \begin{array}{c} \Lambda^{f+2} E \\ \otimes \\ \Lambda^f F^* \end{array} \to \Lambda^2 E \to E \otimes F \to S^2 F \to 0 \quad (\mathrm{EN}_2)$$

• • •

$$0 \to \begin{array}{c} \Lambda^e E \\ \otimes \\ \Lambda^f F^* \end{array} \to \Lambda^{e-f} E \to \cdots \to \begin{array}{c} \Lambda^2 E \\ \otimes \\ S^{e-f-2} F \end{array} \to E \otimes S^{e-f-1} F \longrightarrow S^{e-f} F \longrightarrow 0 \quad (\mathrm{EN}_{e-f})$$

$$0 \to \Lambda^{e-f+1} E \to \begin{array}{c} \Lambda^{e-f} E \\ \otimes \\ F \end{array} \to \cdots \to \begin{array}{c} \Lambda^2 E \\ \otimes \\ S^{e-f-1} F \end{array} \to E \otimes S^{e-f} F \to S^{e-f+1} F \to 0 \quad (\mathrm{EN}_{e-f+1})$$

• • •

$$0 \to \Lambda^e E \to \begin{array}{c} \Lambda^{e-1} E \\ \otimes \\ F \end{array} \to \cdots \to \begin{array}{c} \Lambda^2 E \\ \otimes \\ S^{e-2} F \end{array} \to E \otimes S^{e-1} F \to S^e F \to 0 \quad (\mathrm{EN}_e)$$

$$0 \to \begin{array}{c} \Lambda^e E \\ \otimes \\ F \end{array} \to \begin{array}{c} \Lambda^{e-1} E \\ \otimes \\ S^2 F \end{array} \to \cdots \to \begin{array}{c} \Lambda^2 E \\ \otimes \\ S^{e-1} F \end{array} \to E \otimes S^e F \to S^{e+1} F \to 0 \quad (\mathrm{EN}_{e+1})$$

• • •

The next statement summarizes this discussion.

Theorem B.2.2. (Eagon–Northcott complexes). *Let*

$$u : E \longrightarrow F$$

be a homomorphism of vector bundles of ranks $e \geq f$ *on an irreducible variety or scheme* X *of dimension* n*, and let* $Z = D_{f-1}(u) \subseteq X$ *be the locus of points where* u *drops rank.*

(i). *For each* $k \geq 0$*, the complex* (EN_k) *is exact off* Z.

(ii). *Assume that* X *is smooth (or locally Cohen–Macaulay), and that*

$$\operatorname{codim}_X D_{f-\ell}(u) \ \geq \ e - f + \ell$$

for every $1 \leq \ell \leq f$*. Then* (EN_k) *is acyclic for every* $k \geq 0$.

(iii). *Assume that* X *is smooth (or locally Cohen–Macaulay). If* $k \leq e - f + 1$ *then* (EN_k) *is acyclic provided only that* Z *has the expected codimension* $e - f + 1$*. If* $k \geq e - f + 2$ *then* (EN_k) *is acyclic as soon as*

$$\operatorname{codim}_X D_{f-\ell}(u) \ \geq \ e - f + \ell$$

whenever $1 \leq \ell \leq \min(k, e) - (e - f)$.

Of course (ii) is a special case of (iii): we state it separately because — unlike (iii) — it follows directly from what we have said.

Proof of Theorem B.2.2. The codimension hypotheses in (ii) imply that Y has the expected codimension e in $\mathbf{P}(F)$, and as we have noted this implies the acyclicity of (EN_k). The stronger assertion in (iii) follows from [164, Theorem A2.10]. □

References

1. Shreeram Abhyankar, *Tame coverings and fundamental groups of algebraic varieties. I. Branch loci with normal crossings; Applications: Theorems of Zariski and Picard*, Amer. J. Math. **81** (1959), 46–94.
2. Dan Abramovich and A. Johan de Jong, *Smoothness, semistability, and toroidal geometry*, J. Algebraic Geom. **6** (1997), no. 4, 789–801.
3. Allen Altman and Steven Kleiman, *Introduction to Grothendieck Duality Theory*, Lect. Notes in Math., vol. 146, Springer-Verlag, Berlin, 1970.
4. A. Alzati and G. Ottaviani, *A linear bound on the t-normality of codimension two subvarieties of* $\mathbf{P}^n$, J. Reine Angew. Math. **409** (1990), 35–40.
5. Marco Andreatta and Andrew J. Sommese, *On the projective normality of the adjunction bundles*, Comment. Math. Helv. **66** (1991), no. 3, 362–367.
6. Marco Andreatta and Jarosław A. Wiśniewski, *On manifolds whose tangent bundle contains an ample subbundle*, Invent. Math. **146** (2001), no. 1, 209–217.
7. Aldo Andreotti and Theodore Frankel, *The Lefschetz theorem on hyperplane sections*, Ann. of Math. (2) **69** (1959), 713–717.
8. Aldo Andreotti and Hans Grauert, *Théorème de finitude pour la cohomologie des espaces complexes*, Bull. Soc. Math. France **90** (1962), 193–259.
9. Aldo Andreotti and Alan Mayer, *On period relations for abelian integrals on algebraic curves*, Ann. Scuola Norm. Sup. Pisa (3) **21** (1967), 189–238.
10. Urban Angehrn and Yum-Tong Siu, *Effective freeness and point separation for adjoint bundles*, Invent. Math. **122** (1995), no. 2, 291–308.
11. Flavio Angelini, *An algebraic version of Demailly's asymptotic Morse inequalities*, Proc. Amer. Math. Soc. **124** (1996), no. 11, 3265–3269.
12. Marian Aprodu and Claire Voisin, *Green–Lazarsfeld's conjecture for generic curves of large gonality*, C. R. Math. Acad. Sci. Paris **336** (2003), no. 4, 335–339.
13. Donu Arapura, *Higgs line bundles, Green–Lazarsfeld sets, and maps of Kähler manifolds to curves*, Bull. Amer. Math. Soc. (N.S.) **26** (1992), no. 2, 310–314.

14. ———, *Frobenius amplitude and strong vanishing theorems for vector bundles*, Duke Math. J. **121** (2004), no. 2, 231–267, With an appendix by Dennis S. Keeler.

15. Enrico Arbarello, Maurizio Cornalba, Phillip Griffiths, and Joe Harris, *Geometry of Algebraic Curves. Vol. I*, Grundlehren Math. Wiss., vol. 267, Springer-Verlag, New York, 1985.

16. Enrico Arbarello and Corrado De Concini, *Another proof of a conjecture of S. P. Novikov on periods of abelian integrals on Riemann surfaces*, Duke Math. J. **54** (1987), no. 1, 163–178.

17. V. I. Arnol′d, *Mathematical Methods of Classical Mechanics*, Graduate Texts in Mathematics, vol. 60, Springer-Verlag, New York, 1989.

18. V. I. Arnol′d, S. M. Guseĭn-Zade, and A. N. Varchenko, *Singularities of Differentiable Maps. Vol. II: Monodromy and Asymptotics of Integrals*, Birkhäuser Boston Inc., Boston, MA, 1988.

19. Michael Atiyah and Jurgen Berndt, *Projective planes, Severi varieties and spheres*, preprint.

20. Lucian Bădescu, *Projective Geometry and Formal Geometry*, forthcoming book.

21. ———, *Special chapters of projective geometry*, Rend. Sem. Mat. Fis. Milano **69** (1999/00), 239–326 (2001).

22. ———, *Algebraic Surfaces*, Universitext, Springer-Verlag, New York, 2001.

23. Edoardo Ballico, *Spanned and ample vector bundles with low Chern numbers*, Pacific J. Math. **140** (1989), no. 2, 209–216.

24. ———, *On ample and spanned rank-3 bundles with low Chern numbers*, Manuscripta Math. **68** (1990), no. 1, 9–16.

25. Daniel Barlet, *À propos d'une conjecture de R. Hartshorne*, J. Reine Angew. Math. **374** (1987), 214–220.

26. Daniel Barlet, L. Doustaing, and J. Magnússon, *La conjecture de R. Hartshorne pour les hypersurfaces lisses de* $\mathbf{P}^n$, J. Reine Angew. Math. **457** (1994), 189–202.

27. Daniel Barlet, Thomas Peternell, and Michael Schneider, *On two conjectures of Hartshorne's*, Math. Ann. **286** (1990), no. 1-3, 13–25.

28. Wolf Barth, *Verallgemeinerung des Bertinischen Theorems in abelschen Mannigfaltigkeiten*, Ann. Scuola Norm. Sup. Pisa (3) **23** (1969), 317–330.

29. ———, *Der Abstand von einer algebraischen Mannigfaltigkeit im komplex-projektiven Raum*, Math. Ann. **187** (1970), 150–162.

30. ———, *Transplanting cohomology classes in complex-projective space*, Amer. J. Math. **92** (1970), 951–967.

31. ———, *Larsen's theorem on the homotopy groups of projective manifolds of small embedding codimension*, Algebraic Geometry – Arcata 1974, Proc. Symp. Pure Math., Amer. Math. Soc., Providence, R.I., 1975, pp. 307–313.

32. Wolf Barth and Mogens Larsen, *On the homotopy groups of complex projective algebraic manifolds*, Math. Scand. **30** (1972), 88–94.

33. Wolf Barth and A. Van de Ven, *A decomposability criterion for algebraic 2-bundles on projective spaces*, Invent. Math. **25** (1974), 91–106.

34. Charles M. Barton, *Tensor products of ample vector bundles in characteristic p*, Amer. J. Math. **93** (1971), 429–438.

35. Alexander Barvinok, *A Course in Convexity*, Graduate Studies in Mathematics, vol. 54, American Mathematical Society, Providence, RI, 2002.

36. Thomas Bauer, *Seshadri constants of quartic surfaces*, Math. Ann. **309** (1997), no. 3, 475–481.

37. ______, *On the cone of curves of an abelian variety*, Amer. J. Math. **120** (1998), no. 5, 997–1006.

38. ______, *Seshadri constants and periods of polarized abelian varieties*, Math. Ann. **312** (1998), no. 4, 607–623, With an appendix by the author and Tomasz Szemberg.

39. ______, *Seshadri constants on algebraic surfaces*, Math. Ann. **313** (1999), no. 3, 547–583.

40. Thomas Bauer, Alex Küronya, and Tomasz Szemberg, *Zariski chambers, volume and stable base loci*, preprint, 2003.

41. Thomas Bauer and Tomasz Szemberg, *Local positivity of principally polarized abelian threefolds*, J. Reine Angew. Math. **531** (2001), 191–200.

42. David Bayer and David Mumford, *What can be computed in algebraic geometry?*, Computational Algebraic Geometry and Commutative Algebra (Cortona, 1991), Cambridge Univ. Press, Cambridge, 1993, pp. 1–48.

43. David Bayer and Michael Stillman, *A criterion for detecting m-regularity*, Invent. Math. **87** (1987), no. 1, 1–11.

44. ______, *On the complexity of computing syzygies*, J. Symbolic Comput. **6** (1988), no. 2–3, 135–147.

45. Arnaud Beauville, *Annulation du H^1 et systèmes paracanoniques sur les surfaces*, J. Reine Angew. Math. **388** (1988), 149–157.

46. ______, *Annulation du H^1 pour les fibrés en droites plats*, Complex Algebraic Varieties (Bayreuth, 1990), Lect. Notes in Math., vol. 1507, Springer, Berlin, 1992, pp. 1–15.

47. ______, *Vector bundles on curves and generalized theta functions: recent results and open problems*, Current Topics in Complex Algebraic Geometry, Cambridge Univ. Press, Cambridge, 1995, pp. 17–33.

48. Alexander Beilinson, Joseph Bernstein, and Pierre Deligne, *Faisceaux Pervers*, Astérisque, vol. 100, Soc. Math. France, Paris, 1982.

49. Mauro Beltrametti, Michael Schneider, and Andrew Sommese, *Chern inequalities and spannedness of adjoint bundles*, Proceedings of the Hirzebruch 65 Conference on Algebraic Geometry (Ramat Gan, 1993), 1996, pp. 97–107.

50. Mauro Beltrametti and Andrew Sommese, *The Adjunction Theory of Complex Projective Varieties*, Walter de Gruyter & Co., Berlin, 1995.

51. Carlos Berenstein, Roger Gay, Alekos Vidras, and Alain Yger, *Residue Currents and Bezout Identities*, Progr. Math., vol. 114, Birkhäuser Verlag, Basel, 1993.

52. P. Berthelot, Alexander Grothendieck, and Luc Illusie, *Théorie des Intersections et Théorème de Riemann–Roch* (SGA6), Lect. Notes in Math., vol. 225, Springer-Verlag, Berlin, 1971.
53. Aaron Bertram, *An existence theorem for Prym special divisors*, Invent. Math. **90** (1987), no. 3, 669–671.
54. Aaron Bertram, Lawrence Ein, and Robert Lazarsfeld, *Surjectivity of Gaussian maps for line bundles of large degree on curves*, Algebraic Geometry (Chicago, 1989), Lect. Notes in Math., vol. 1479, Springer, Berlin, 1991, pp. 15–25.
55. ______, *Vanishing theorems, a theorem of Severi, and the equations defining projective varieties*, J. Amer. Math. Soc. **4** (1991), no. 3, 587–602.
56. Edward Bierstone and Pierre D. Milman, *Canonical desingularization in characteristic zero by blowing up the maximum strata of a local invariant*, Invent. Math. **128** (1997), no. 2, 207–302.
57. Paul Biran, *Constructing new ample divisors out of old ones*, Duke Math. J. **98** (1999), no. 1, 113–135.
58. ______, *A stability property of symplectic packing*, Invent. Math. **136** (1999), no. 1, 123–155.
59. ______, *From symplectic packing to algebraic geometry and back*, European Congress of Mathematics, Vol. II (Barcelona, 2000), Progr. in Math., vol. 202, Birkhäuser, Basel, 2001, pp. 507–524.
60. Spencer Bloch and David Gieseker, *The positivity of the Chern classes of an ample vector bundle*, Invent. Math. **12** (1971), 112–117.
61. Fedor Bogomolov, *Holomorphic tensors and vector bundles on projective manifolds*, Izv. Akad. Nauk SSSR Ser. Mat. **42** (1978), no. 6, 1227–1287, 1439.
62. ______, *Unstable vector bundles and curves on surfaces*, Proceedings of the International Congress of Mathematicians (Helsinki), Acad. Sci. Fennica, 1980, pp. 517–524.
63. Fedor Bogomolov and Bruno De Oliveira, *Convexity of coverings of algebraic varieties and vanishing theorems*, Preprint, 2003.
64. Fedor Bogomolov and Tony Pantev, *Weak Hironaka theorem*, Math. Res. Lett. **3** (1996), no. 3, 299–307.
65. Ciprian Borcea, *Homogeneous vector bundles and families of Calabi–Yau threefolds. II*, Several Complex Variables and Complex Geometry, Part 2, Proc. Symp. Pure Math., vol. 52, Amer. Math. Soc., Providence, RI, 1991, pp. 83–91.
66. Sébastien Boucksom, *On the volume of a line bundle*, Internat. J. Math. **13** (2002), no. 10, 1043–1063.
67. ______, *Higher dimensional Zariski decompositions*, preprint, 2003.
68. Sébastien Boucksom, Jean-Pierre Demailly, Mihai Paun, and Thomas Peternall, *The pseudo-effective cone of a compact Kähler manifold and varieties of negative Kodaira dimension*, preprint, 2004.
69. Nicolas Bourbaki, *Commutative Algebra. Chapters 1–7*, Elements of Mathematics, Springer-Verlag, Berlin, 1998.

70. Ana Bravo and Orlando Villamayor, *A strong desingularization theorem*, preprint.

71. Glen E. Bredon, *Topology and Geometry*, Graduate Texts in Mathematics, vol. 139, Springer-Verlag, New York, 1993.

72. Markus Brodmann and Rodney Sharp, *Local Cohomology: An Algebraic Introduction with Geometric Applications*, Cambridge studies in advanced mathematics, vol. 60, Cambridge University Press, Cambridge, 1998.

73. W. Dale Brownawell, *Bounds for the degrees in the Nullstellensatz*, Ann. of Math. (2) **126** (1987), no. 3, 577–591.

74. Winfried Bruns and Jürgen Herzog, *Cohen–Macaulay Rings*, Cambridge Studies in Advanced Mathematics, vol. 39, Cambridge University Press, Cambridge, 1993.

75. Nero Budur and Morihiko Saito, *Multiplier ideals, V-filtration, and spectrum*, preprint, 2003.

76. P. Buser and Peter Sarnak, *On the period matrix of a Riemann surface of large genus*, Invent. Math. **117** (1994), no. 1, 27–56.

77. Frédéric Campana, *Ensembles de Green–Lazarsfeld et quotients résolubles des groupes de Kähler*, J. Algebraic Geom. **10** (2001), no. 4, 599–622.

78. Frédéric Campana and Thomas Peternell, *Algebraicity of the ample cone of projective varieties*, J. Reine Angew. Math. **407** (1990), 160–166.

79. ______, *Projective manifolds whose tangent bundles are numerically effective*, Math. Ann. **289** (1991), no. 1, 169–187.

80. ______, *On the second exterior power of tangent bundles of threefolds*, Compositio Math. **83** (1992), no. 3, 329–346.

81. ______, *Rational curves and ampleness properties of the tangent bundle of algebraic varieties*, Manuscripta Math. **97** (1998), no. 1, 59–74.

82. James Carlson, Mark Green, Phillip Griffiths, and Joe Harris, *Infinitesimal variations of Hodge structure. I*, Compositio Math. **50** (1983), no. 2-3, 109–205.

83. James Carlson, Stefan Müller-Stach, and Chris Peters, *Period Mappings and Period Domains*, Cambridge Studies in Advanced Mathematics, vol. 85, Cambridge University Press, Cambridge, 2003.

84. Karen Chandler, *Regularity of the powers of an ideal*, Comm. Algebra **25** (1997), no. 12, 3773–3776.

85. Pierre-Emmanuel Chaput, *Severi varieties*, Math. Z. **240** (2002), no. 2, 451–459.

86. Marc Chardin and Clare D'Cruz, *Castelnuovo-Mumford regularity: examples of curves and surfaces*, J. Algebra **270** (2003), no. 1, 347–360.

87. Marc Chardin and Bernd Ulrich, *Liaison and Castelnuovo–Mumford regularity*, Amer. J. Math. **124** (2002), no. 6, 1103–1124.

88. Jungkai Chen and Christopher Hacon, *Characterization of abelian varieties*, Invent. Math. **143** (2001), no. 2, 435–447.

89. ______, *Pluricanonical maps of varieties of maximal Albanese dimension*, Math. Ann. **320** (2001), no. 2, 367–380.

90. Jaydeep V. Chipalkatti, *A generalization of Castelnuovo regularity to Grassmann varieties*, Manuscripta Math. **102** (2000), no. 4, 447–464.

91. Koji Cho and Ei-ichi Sato, *Smooth projective varieties with the ample vector bundle $\lambda^2 T_X$ in any characteristic*, J. Math. Kyoto Univ. **35** (1995), no. 1, 1–33.

92. Ciro Ciliberto and Alexis Kouvidakis, *On the symmetric product of a curve with general moduli*, Geom. Dedicata **78** (1999), no. 3, 327–343.

93. Herbert Clemens, *Curves on generic hypersurfaces*, Ann. Sci. École Norm. Sup. (4) **19** (1986), no. 4, 629–636.

94. ———, *A local proof of Petri's conjecture at the general curve*, J. Differential Geom. **54** (2000), no. 1, 139–176.

95. Herbert Clemens and Christopher Hacon, *Deformations of the trivial line bundle and vanishing theorems*, Amer. J. Math. **124** (2002), no. 4, 769–815.

96. John Conway and Neil Sloane, *Sphere Packings, Lattices and Groups*, third ed., Springer-Verlag, New York, 1999.

97. S. Dale Cutkosky, *Zariski decomposition of divisors on algebraic varieties*, Duke Math. J. **53** (1986), no. 1, 149–156.

98. ———, *Irrational asymptotic behaviour of Castelnuovo–Mumford regularity*, J. Reine Angew. Math. **522** (2000), 93–103.

99. S. Dale Cutkosky, Lawrence Ein, and Robert Lazarsfeld, *Positivity and complexity of ideal sheaves*, Math. Ann. **321** (2001), no. 2, 213–234.

100. S. Dale Cutkosky, Jürgen Herzog, and Ngô Viêt Trung, *Asymptotic behaviour of the Castelnuovo–Mumford regularity*, Compositio Math. **118** (1999), no. 3, 243–261.

101. S. Dale Cutkosky and V. Srinivas, *On a problem of Zariski on dimensions of linear systems*, Ann. of Math. (2) **137** (1993), no. 3, 531–559.

102. ———, *Periodicity of the fixed locus of multiples of a divisor on a surface*, Duke Math. J. **72** (1993), no. 3, 641–647.

103. Mark de Cataldo, *Singular Hermitian metrics on vector bundles*, J. Reine Angew. Math. **502** (1998), 93–122.

104. Mark de Cataldo and Luca Migliorini, *The hard Lefschetz theorem and the topology of semismall maps*, Ann. Sci. École Norm. Sup. (4) **35** (2002), no. 5, 759–772.

105. ———, *The Hodge theory of algebraic maps*, preprint, 2003.

106. Tommaso de Fernex, Lawrence Ein, and Mircea Mustaţă, *Multiplicities and log canonical threshold*, J. Alg. Geometry, to appear.

107. ———, *Bounds for log canonical thresholds with applications to birational rigidity*, Math. Res. Lett. **10** (2003), no. 2–3, 219–236.

108. A. Johan de Jong, *Smoothness, semi-stability and alterations*, Inst. Hautes Études Sci. Publ. Math. (1996), no. 83, 51–93.

109. Olivier Debarre, *Fulton–Hansen and Barth–Lefschetz theorems for subvarieties of abelian varieties*, J. Reine Angew. Math. **467** (1995), 187–197.

110. ______, *Théorèmes de connexité et variétés abéliennes*, Amer. J. Math. **117** (1995), no. 3, 787–805.

111. ______, *Théorèmes de connexité pour les produits d'espaces projectifs et les grassmanniennes*, Amer. J. Math. **118** (1996), no. 6, 1347–1367.

112. ______, *Tores et Variétés Abéliennes Complexes*, Société Mathématique de France, Paris, 1999.

113. ______, *Théorèmes de Lefschetz pour les lieux de dégénérescence*, Bull. Soc. Math. France **128** (2000), no. 2, 283–308.

114. ______, *Higher-Dimensional Algebraic Geometry*, Springer-Verlag, New York, 2001.

115. ______, *Varieties with ample cotangent bundle*, preprint, 2003.

116. Olivier Debarre, Klaus Hulek, and J. Spandaw, *Very ample linear systems on abelian varieties*, Math. Ann. **300** (1994), no. 2, 181–202.

117. Pierre Deligne, *Cohomologie Étale* (SGA4.5), Lect. Notes in Math., vol. 569, Springer-Verlag, Berlin, 1977.

118. ______, *Le groupe fondamental du complément d'une courbe plane n'ayant que des points doubles ordinaires est abélien (d'après W. Fulton)*, Séminaire Bourbaki, Année 1979/80, Lect. Notes in Math., vol. 842, Springer, Berlin, 1981, pp. 1–10.

119. Jean-Pierre Demailly, *Complex Analytic and Algebraic Geometry*, book in preparation.

120. ______, *Champs magnétiques et inégalités de Morse pour la d''-cohomologie*, C. R. Acad. Sci. Paris Sér. I Math. **301** (1985), no. 4, 119–122.

121. ______, *Vanishing theorems for tensor powers of an ample vector bundle*, Invent. Math. **91** (1988), no. 1, 203–220.

122. ______, *Regularization of closed positive currents and intersection theory*, J. Algebraic Geom. **1** (1992), no. 3, 361–409.

123. ______, *Singular Hermitian metrics on positive line bundles*, Complex Algebraic Varieties (Bayreuth, 1990), Lect. Notes in Math., vol. 1507, 1992, pp. 87–104.

124. ______, *A numerical criterion for very ample line bundles*, J. Differential Geom. **37** (1993), no. 2, 323–374.

125. ______, *Effective bounds for very ample line bundles*, Invent. Math. **124** (1996), no. 1-3, 243–261.

126. ______, *L^2 vanishing theorems for positive line bundles and adjunction theory*, Transcendental Methods in Algebraic Geometry (Cetraro, 1994), Lect. Notes in Math., vol. 1464, Springer, Berlin, 1996, pp. 1–97.

127. ______, *Algebraic criteria for Kobayashi hyperbolic projective varieties and jet differentials*, Algebraic Geometry, Santa Cruz 1995, Proc. Symp. Pure Math., vol. 62, Amer. Math. Soc., Providence, RI, 1997, pp. 285–360.

128. ______, *Méthodes L^2 et résultats effectifs en géométrie algébrique*, Séminaire Bourbaki, Année 1998/99, Exp. No. 852, no. 266, 2000, pp. 59–90.

129. ______, *Multiplier ideal sheaves and analytic methods in algebraic geometry*, School on Vanishing Theorems and Effective Results in Algebraic Geometry

(Trieste, 2000), ICTP Lect. Notes, vol. 6, Abdus Salam Int. Cent. Theoret. Phys., Trieste, 2001, pp. 1–148.

130. Jean-Pierre Demailly, Lawrence Ein, and Robert Lazarsfeld, *A subadditivity property of multiplier ideals*, Michigan Math. J. **48** (2000), 137–156.

131. Jean-Pierre Demailly and János Kollár, *Semi-continuity of complex singularity exponents and Kähler–Einstein metrics on Fano orbifolds*, Ann. Sci. École Norm. Sup. (4) **34** (2001), no. 4, 525–556.

132. Jean-Pierre Demailly and Mihai Paun, *Numerical characterization of the Kähler cone of a compact Kähler manifold*, to appear.

133. Jean-Pierre Demailly, Thomas Peternell, and Michael Schneider, *Compact complex manifolds with numerically effective tangent bundles*, J. Algebraic Geom. **3** (1994), no. 2, 295–345.

134. Jean-Pierre Demailly and Henri Skoda, *Relations entre les notions de positivités de P. A. Griffiths et de S. Nakano pour les fibrés vectoriels*, Séminaire Pierre Lelong–Henri Skoda. Année 1978/79, Lect. Notes in Math, vol. 822, 1980, pp. 304–309.

135. Michel Demazure, *Caractérisations de l'espace projectif (conjectures de Hartshorne et de Frankel) (d'après Shigefumi Mori)*, Séminaire Bourbaki, Année 1979/80, Lect. Notes in Math., vol. 842, Springer, Berlin, 1981, pp. 11–19.

136. Jan Denef and François Loeser, *Germs of arcs on singular algebraic varieties and motivic integration*, Invent. Math. **135** (1999), no. 1, 201–232.

137. Harm Derksen and Jessica Sidman, *A sharp bound for the Castelnuovo–Mumford regularity of subspace arrangements*, Adv. Math. **172** (2002), no. 2, 151–157.

138. Mireille Deschamps, *Courbes de genre géométrique borné sur une surface de type général (d'après F. A. Bogomolov)*, Séminaire Bourbaki, Année 1977/78, Lect. Notes in Math., vol. 710, Springer, 1979, pp. 233–247.

139. S. K. Donaldson, *Anti self-dual Yang–Mills connections over complex algebraic surfaces and stable vector bundles*, Proc. London Math. Soc. (3) **50** (1985), no. 1, 1–26.

140. Lawrence Ein, *An analogue of Max Noether's theorem*, Duke Math. J. **52** (1985), no. 3, 689–706.

141. ———, *Varieties with small dual varieties. II*, Duke Math. J. **52** (1985), no. 4, 895–907.

142. ———, *Varieties with small dual varieties. I*, Invent. Math. **86** (1986), no. 1, 63–74.

143. ———, *Subvarieties of generic complete intersections*, Invent. Math. **94** (1988), no. 1, 163–169.

144. ———, *Unpublished notes on multiplier ideals*, 1995.

145. ———, *Multiplier ideals, vanishing theorems and applications*, Algebraic Geometry—Santa Cruz 1995, Proc. Symp. Pure Math., vol. 62, Amer. Math. Soc., Providence, RI, 1997, pp. 203–219.

146. ———, *Notes from lectures at Catania*, unpublished, 1999.

147. ______, *Linear systems with removable base loci*, Comm. Algebra **28** (2000), no. 12, 5931–5934.

148. Lawrence Ein, Bo Ilic, and Robert Lazarsfeld, *A remark on projective embeddings of varieties with non-negative cotangent bundles*, Complex Analysis and Algebraic Geometry, de Gruyter, Berlin, 2000, pp. 165–171.

149. Lawrence Ein, Oliver Küchle, and Robert Lazarsfeld, *Local positivity of ample line bundles*, J. Differential Geom. **42** (1995), no. 2, 193–219.

150. Lawrence Ein and Robert Lazarsfeld, *Stability and restrictions of Picard bundles, with an application to the normal bundles of elliptic curves*, Complex Projective Geometry (Trieste, 1989/Bergen, 1989), London Math. Soc. Lecture Note Series, vol. 179, Cambridge Univ. Press, Cambridge, 1992, pp. 149–156.

151. ______, *Global generation of pluricanonical and adjoint linear series on smooth projective threefolds*, J. Amer. Math. Soc. **6** (1993), no. 4, 875–903.

152. ______, *Seshadri constants on smooth surfaces*, Journées de Géométrie Algébrique d'Orsay (Orsay, 1992), Astérisque, no. 218, 1993, pp. 177–186.

153. ______, *Syzygies and Koszul cohomology of smooth projective varieties of arbitrary dimension*, Invent. Math. **111** (1993), no. 1, 51–67.

154. ______, *Singularities of theta divisors and the birational geometry of irregular varieties*, J. Amer. Math. Soc. **10** (1997), no. 1, 243–258.

155. ______, *A geometric effective Nullstellensatz*, Invent. Math. **137** (1999), no. 2, 427–448.

156. Lawrence Ein, Robert Lazarsfeld, and Mircea Mustaţă, *Contact loci in arc spaces*, to appear.

157. Lawrence Ein, Robert Lazarsfeld, Mircea Mustaţă, Michael Nakamaye, and Mihnea Popa, *Asymptotic invariants of base loci*, preprint, 2003.

158. Lawrence Ein, Robert Lazarsfeld, and Michael Nakamaye, *Zero-estimates, intersection theory, and a theorem of Demailly*, Higher-Dimensional Complex Varieties (Trento, 1994), de Gruyter, Berlin, 1996, pp. 183–207.

159. Lawrence Ein, Robert Lazarsfeld, and Karen E. Smith, *Uniform bounds and symbolic powers on smooth varieties*, Invent. Math. **144** (2001), no. 2, 241–252.

160. ______, *Uniform approximation of Abhyankar valuation ideals in smooth function fields*, Amer. J. Math. **125** (2003), no. 2, 409–440.

161. Lawrence Ein, Robert Lazarsfeld, Karen E. Smith, and Dror Varolin, *Jumping coefficients of multiplier ideals*, to appear.

162. Lawrence Ein, Mircea Mustaţă, and Takehiko Yasuda, *Jet schemes, log discrepancies and inversion of adjunction*, Invent. Math. **153** (2003), no. 3, 519–535.

163. David Eisenbud, *The Geometry of Syzygies*, forthcoming.

164. ______, *Commutative Algebra, with a View Toward Algebraic Geometry*, Graduate Texts in Math, no. 150, Springer-Verlag, New York, 1995.

165. David Eisenbud and Shiro Goto, *Linear free resolutions and minimal multiplicity*, J. Algebra **88** (1984), no. 1, 89–133.

166. Santiago Encinas and Orlando Villamayor, *Good points and constructive resolution of singularities*, Acta Math. **181** (1998), no. 1, 109–158.

167. ______, *A course on constructive desingularization and equivariance*, Resolution of Singularities (Obergurgl, 1997), Progr. Math., vol. 181, Birkhäuser, Basel, 2000, pp. 147–227.

168. Hélène Esnault and Eckart Viehweg, *Sur une minoration du degré d'hypersurfaces s'annulant en certains points*, Math. Ann. **263** (1983), no. 1, 75–86.

169. ______, *Dyson's lemma for polynomials in several variables (and the theorem of Roth)*, Invent. Math. **78** (1984), no. 3, 445–490.

170. ______, *Logarithmic de Rham complexes and vanishing theorems*, Invent. Math. **86** (1986), no. 1, 161–194.

171. ______, *Rêvetements cycliques. II (autour du théorème d'annulation de J. Kollár)*, Géométrie algébrique et applications, II (La Rábida, 1984), Travaux en cours, no. 23, Hermann, Paris, 1987, pp. 81–96.

172. ______, *Effective bounds for semipositive sheaves and for the height of points on curves over complex function fields*, Compositio Math. **76** (1990), no. 1-2, 69–85, Algebraic geometry (Berlin, 1988).

173. ______, *Ample sheaves on moduli schemes*, Algebraic Geometry and Analytic Geometry (Tokyo, 1990), ICM-90 Satell. Conf. Proc., Springer, Tokyo, 1991, pp. 53–80.

174. ______, *Lectures on Vanishing Theorems*, DMV Seminar, vol. 20, Birkhäuser Verlag, Basel, 1992.

175. E. Graham Evans and Phillip Griffith, *The syzygy problem*, Ann. of Math. (2) **114** (1981), no. 2, 323–333.

176. Gerd Faltings, *Formale Geometrie und homogene Räume*, Invent. Math. **64** (1981), no. 1, 123–165.

177. ______, *Verschwindungssätze und Untermannigfaltigkeiten kleiner Kodimension des komplex-projektiven Raumes*, J. Reine Angew. Math. **326** (1981), 136–151.

178. ______, *Diophantine approximation on abelian varieties*, Ann. of Math. (2) **133** (1991), no. 3, 549–576.

179. ______, *The general case of S. Lang's conjecture*, Barsotti Symposium in Algebraic Geometry (Abano Terme, 1991), Academic Press, San Diego, CA, 1994, pp. 175–182.

180. ______, *Mumford-Stabilität in der algebraischen Geometrie*, Proceedings of the International Congress of Mathematicians, (Zürich, 1994), Birkhäuser, 1995, pp. 648–655.

181. Gerd Faltings and Gisbert Wüstholz, *Diophantine approximations on projective spaces*, Invent. Math. **116** (1994), no. 1-3, 109–138.

182. Fuquan Fang, Sérgio Mendonça, and Xiaochung Rong, *A connectedness principle in the geometry of positive curvature*, preprint.

183. Gavril Farkas and Angela Gibney, *The Mori cones of moduli spaces of pointed curves of small genus*, Trans. Amer. Math. Soc. **355** (2003), no. 3, 1183–1199.

184. Charles Favre and Mattias Jonsson, *Valuative criterion for integrability*, preprint.

185. Guillermo Fernández del Busto, *A Matsusaka-type theorem on surfaces*, J. Algebraic Geom. **5** (1996), no. 3, 513–520.

186. Hubert Flenner, Liam O'Carroll, and Wolfgang Vogel, *Joins and Intersections*, Monographs in Mathematics, Springer-Verlag, Berlin, 1999.

187. Hubert Flenner and Wolfgang Vogel, *Joins, tangencies and intersections*, Math. Ann. **302** (1995), no. 3, 489–505.

188. Jens Franke, Yuri Manin, and Yuri Tschinkel, *Rational points of bounded height on Fano varieties*, Invent. Math. **95** (1989), no. 2, 421–435.

189. Klaus Fritzsche, *Pseudoconvexity properties of complements of analytic subvarieties*, Math. Ann. **230** (1977), no. 2, 107–122.

190. Osamu Fujino, *Notes on toric varieties from Mori theoretic viewpoint*, Tohoku Math. J. (2) **55** (2003), no. 4, 551–564.

191. Takao Fujita, *Defining equations for certain types of polarized varieties*, Complex Analysis and Algebraic Geometry, Cambridge University Press, 1977, pp. 165–173.

192. ———, *On Kähler fiber spaces over curves*, J. Math. Soc. Japan **30** (1978), no. 4, 779–794.

193. ———, *On Zariski problem*, Proc. Japan Acad. Ser. A Math. Sci. **55** (1979), no. 3, 106–110.

194. ———, *Semipositive line bundles*, J. Fac. Sci. Univ. Tokyo Sect. IA Math. **30** (1983), no. 2, 353–378.

195. ———, *Vanishing theorems for semipositive line bundles*, Algebraic Geometry (Tokyo/Kyoto, 1982), Lect. Notes in Math., no. 1016, Springer, Berlin, 1983, pp. 519–528.

196. ———, *Triple covers by smooth manifolds*, J. Fac. Sci. Univ. Tokyo Sect. IA Math. **35** (1988), no. 1, 169–175.

197. ———, *Classification Theories of Polarized Varieties*, London Math. Soc. Lect. Note Series, vol. 155, Cambridge University Press, Cambridge, 1990.

198. ———, *The Kodaira energy and the classification of polarized manifolds*, Sūgaku **45** (1993), no. 3, 244–255.

199. ———, *Approximating Zariski decomposition of big line bundles*, Kodai Math. J. **17** (1994), no. 1, 1–3.

200. ———, *On Kodaira energy of polarized log varieties*, J. Math. Soc. Japan **48** (1996), no. 1, 1–12.

201. William Fulton, *Ample vector bundles, Chern classes, and numerical criteria*, Invent. Math. **32** (1976), no. 2, 171–178.

202. ———, *On the fundamental group of the complement of a node curve*, Ann. of Math. (2) **111** (1980), no. 2, 407–409.

203. ———, *On the topology of algebraic varieties*, Algebraic Geometry, Bowdoin, 1985, Proc. Symp. Pure Math., vol. 46, Amer. Math. Soc., Providence, RI, 1987, pp. 15–46.

204. ———, *Flags, Schubert polynomials, degeneracy loci, and determinantal formulas*, Duke Math. J. **65** (1992), no. 3, 381–420.

205. ______, *Introduction to Toric Varieties*, Annals of Math. Studies, vol. 131, Princeton University Press, Princeton, NJ, 1993.

206. ______, *Positive polynomials for filtered ample vector bundles*, Amer. J. Math. **117** (1995), no. 3, 627–633.

207. ______, *Young Tableaux*, London Mathematical Society Student Texts, vol. 35, Cambridge University Press, Cambridge, 1997.

208. ______, *Intersection Theory*, second ed., Ergebnisse der Math. und ihrer Grenzgebiete (3), vol. 2, Springer-Verlag, Berlin, 1998.

209. William Fulton and Johan Hansen, *A connectedness theorem for projective varieties, with applications to intersections and singularities of mappings*, Ann. of Math. (2) **110** (1979), no. 1, 159–166.

210. William Fulton and Joe Harris, *Representation Theory, a First Course*, Graduate Texts in Math., vol. 129, Springer-Verlag, New York, 1991.

211. William Fulton and Robert Lazarsfeld, *Connectivity and its applications in algebraic geometry*, Algebraic Geometry (Chicago, 1980), Lect. Notes in Math., vol. 862, Springer, Berlin, 1981, pp. 26–92.

212. ______, *On the connectedness of degeneracy loci and special divisors*, Acta Math. **146** (1981), no. 3-4, 271–283.

213. ______, *Positivity and excess intersection*, Enumerative Geometry and Classical Algebraic Geometry (Nice, 1981), Prog. Math., vol. 24, Birkhäuser Boston, Mass., 1982, pp. 97–105.

214. ______, *Positive polynomials for ample vector bundles*, Ann. of Math. (2) **118** (1983), no. 1, 35–60.

215. William Fulton and Piotr Pragacz, *Schubert Varieties and Degeneracy Loci*, Lect. Notes in Math., vol. 1689, Springer-Verlag, Berlin, 1998.

216. Terence Gaffney and Robert Lazarsfeld, *On the ramification of branched coverings of* $\mathbf{P}^n$, Invent. Math. **59** (1980), no. 1, 53–58.

217. Francisco Gallego and B. P. Purnaprajna, *Projective normality and syzygies of algebraic surfaces*, J. Reine Angew. Math. **506** (1999), 145–180.

218. ______, *Syzygies of projective surfaces: an overview*, J. Ramanujan Math. Soc. **14** (1999), no. 1, 65–93.

219. Jørgen Anders Geertsen, *Push-forward of degeneracy classes and ampleness*, Proc. Amer. Math. Soc. **129** (2001), no. 7, 1885–1890.

220. Anthony V. Geramita, Alessandro Gimigliano, and Yves Pitteloud, *Graded Betti numbers of some embedded rational n-folds*, Math. Ann. **301** (1995), no. 2, 363–380.

221. Franco Ghione, *Un problème du type Brill–Noether pour les fibrés vectoriels*, Algebraic Geometry—Open Problems (Ravello, 1982), Lecture Notes in Math., vol. 997, Springer, Berlin, 1983, pp. 197–209.

222. Angela Gibney, Seán Keel, and Ian Morrison, *Towards the ample cone of* $\overline{M}_{g,n}$, J. Amer. Math. Soc. **15** (2002), no. 2, 273–294.

223. David Gieseker, *p-ample bundles and their Chern classes*, Nagoya Math. J. **43** (1971), 91–116.

224. ______, *On a theorem of Bogomolov on Chern classes of stable bundles*, Amer. J. Math. **101** (1979), no. 1, 77–85.

225. ______, *Stable curves and special divisors: Petri's conjecture*, Invent. Math. **66** (1982), no. 2, 251–275.

226. Jacob Eli Goodman, *Affine open subsets of algebraic varieties and ample divisors*, Ann. of Math. (2) **89** (1969), 160–183.

227. Mark Goresky and Robert MacPherson, *Stratified Morse Theory*, Ergebnisse der Math. und ihrer Grenzgebiete (3), vol. 14, Springer-Verlag, Berlin, 1988.

228. Gerd Gotzmann, *Eine Bedingung für die Flachheit und das Hilbertpolynom eines graduierten Ringes*, Math. Z. **158** (1978), no. 1, 61–70.

229. Russell Goward, *A simple algorithm for principalization of monomial ideals*, preprint, 2003.

230. William Graham, *Non-emptiness of skew symmetric degeneracy loci*, preprint, 2003.

231. ______, *Non-emptiness of symmetric degeneracy loci*, preprint, 2003.

232. Hans Grauert, *Über Modifikationen und exzeptionelle analytische Mengen*, Math. Ann. **146** (1962), 331–368.

233. ______, *Mordells Vermutung über rationale Punkte auf algebraischen Kurven und Funktionenkörper*, Inst. Hautes Études Sci. Publ. Math. (1965), no. 25, 131–149.

234. Hans Grauert and Oswald Riemenschneider, *Verschwindungssätze für analytische Kohomologiegruppen auf komplexen Räumen*, Invent. Math. **11** (1970), 263–292.

235. Mark Green, *Koszul cohomology and the geometry of projective varieties*, J. Differential Geom. **19** (1984), no. 1, 125–171.

236. ______, *A new proof of the explicit Noether–Lefschetz theorem*, J. Differential Geom. **27** (1988), no. 1, 155–159.

237. ______, *Koszul cohomology and geometry*, Lectures on Riemann surfaces (Trieste, 1987), World Sci. Publishing, Teaneck, NJ, 1989, pp. 177–200.

238. ______, *Restrictions of linear series to hyperplanes, and some results of Macaulay and Gotzmann*, Algebraic Curves and Projective Geometry (Trento, 1988), Lect. Notes in Math., vol. 1389, Springer, Berlin, 1989, pp. 76–86.

239. ______, *Infinitesimal methods in Hodge theory*, Algebraic Cycles and Hodge Theory (Torino, 1993), Lect. Notes in Math., vol. 1594, Springer, Berlin, 1994, pp. 1–92.

240. ______, *Generic initial ideals*, Six Lectures on Commutative Algebra (Bellaterra, 1996), Progr. Math., vol. 166, Birkhäuser, Basel, 1998, pp. 119–186.

241. Mark Green and Robert Lazarsfeld, *On the projective normality of complete linear series on an algebraic curve*, Invent. Math. **83** (1985), no. 1, 73–90.

242. ______, *Deformation theory, generic vanishing theorems, and some conjectures of Enriques, Catanese and Beauville*, Invent. Math. **90** (1987), no. 2, 389–407.

243. ______, *Some results on the syzygies of finite sets and algebraic curves*, Compositio Math. **67** (1988), no. 3, 301–314.

244. ______, *Higher obstructions to deforming cohomology groups of line bundles*, J. Amer. Math. Soc. **4** (1991), no. 1, 87–103.

245. Phillip Griffiths, *Hermitian differential geometry and the theory of positive and ample holomorphic vector bundles*, J. Math. Mech. **14** (1965), 117–140.

246. ______, *The extension problem in complex analysis. II. Embeddings with positive normal bundle*, Amer. J. Math. **88** (1966), 366–446.

247. ______, *Hermitian differential geometry, Chern classes, and positive vector bundles*, Global Analysis (Papers in Honor of K. Kodaira), Univ. Tokyo Press, Tokyo, 1969, pp. 185–251.

248. Phillip Griffiths and Joseph Harris, *Principles of Algebraic Geometry*, Wiley-Interscience, New York, 1978.

249. ______, *Algebraic geometry and local differential geometry*, Ann. Sci. École Norm. Sup. (4) **12** (1979), no. 3, 355–452.

250. ______, *On the variety of special linear systems on a general algebraic curve*, Duke Math. J. **47** (1980), no. 1, 233–272.

251. Mikhael Gromov, *Pseudoholomorphic curves in symplectic manifolds*, Invent. Math. **82** (1985), no. 2, 307–347.

252. ______, *Partial Differential Relations*, Ergebnisse der Math. und ihrer Grenzgebiete (3), vol. 9, Springer-Verlag, Berlin, 1986.

253. ______, *Convex sets and Kähler manifolds*, Advances in Differential Geometry and Topology, World Sci. Publishing, Teaneck, NJ, 1990, pp. 1–38.

254. ______, *Systoles and intersystolic inequalities*, Actes de la Table Ronde de Géométrie Différentielle (Luminy, 1992), Soc. Math. France, Paris, 1996, pp. 291–362.

255. Alexander Grothendieck, *Éléments de Géométrie Algébrique. II. Étude Globale Élémentaire de Quelques Classes de Morphismes*, Inst. Hautes Études Sci. Publ. Math. (1961), no. 8, 222.

256. ______, *Éléments de Géométrie Algébrique. III. Étude Cohomologique des Faisceaux Cohérents. I*, Inst. Hautes Études Sci. Publ. Math. (1961), no. 11, 167.

257. ______, *Éléments de Géométrie Algébrique. IV. Étude Locale des Schémas et des Morphismes de Schémas IV*, Inst. Hautes Études Sci. Publ. Math. (1967), no. 32, 361.

258. ______, *Cohomologie Locale des Faisceaux Cohérents et Théorèmes de Lefschetz Locaux et Globaux* (SGA 2), Advanced Stud. in Pure Math, vol. 2, North-Holland, Amsterdam, 1968.

259. Laurent Gruson, Robert Lazarsfeld, and Christian Peskine, *On a theorem of Castelnuovo, and the equations defining space curves*, Invent. Math. **72** (1983), no. 3, 491–506.

260. Christopher Hacon, *Examples of spanned and ample vector bundles with small numerical invariants*, C. R. Acad. Sci. Paris Sér. I Math. **323** (1996), no. 9, 1025–1029.

261. ______, *Divisors on principally polarized abelian varieties*, Compositio Math. **119** (1999), no. 3, 321–329.

262. ______, *Fourier transforms, generic vanishing theorems and polarizations of abelian varieties*, Math. Z. **235** (2000), no. 4, 717–726.

263. ______, *A derived category approach to generic vanishing*, preprint, 2003.

264. Helmut Hamm, *Lokale topologische Eigenschaften komplexer Räume*, Math. Ann. **191** (1971), 235–252.

265. ______, *On the vanishing of local homotopy groups for isolated singularities of complex spaces*, J. Reine Angew. Math. **323** (1981), 172–176.

266. Helmut Hamm and Lê Dung Tráng, *Rectified homotopical depth and Grothendieck conjectures*, The Grothendieck Festschrift, Vol. II, Birkhäuser Boston, Boston, MA, 1990, pp. 311–351.

267. ______, *Relative homotopical depth and a conjecture of Deligne*, Math. Ann. **296** (1993), no. 1, 87–101.

268. ______, *Vanishing theorems for constructible sheaves. I*, J. Reine Angew. Math. **471** (1996), 115–138.

269. Johan Hansen, *A connectedness theorem for flagmanifolds and Grassmannians*, Amer. J. Math. **105** (1983), no. 3, 633–639.

270. Nobuo Hara, *Geometric interpretation of tight closure and test ideals*, Trans. Amer. Math. Soc. **353** (2001), no. 5, 1885–1906.

271. Nobuo Hara and Ken-Ichi Yoshida, *A generalization of tight closure and mutiplier ideals*, Trans. Amer. Math. Soc. **355** (2003), no. 8, 3143–3174.

272. Joe Harris, *On the Severi problem*, Invent. Math. **84** (1986), no. 3, 445–461.

273. Joe Harris and Loring W. Tu, *The connectedness of symmetric degeneracy loci: odd ranks*, Topics in Algebra, Part 2 (Warsaw, 1988), PWN, Warsaw, 1990, pp. 249–256.

274. Robin Hartshorne, *Ample vector bundles*, Inst. Hautes Études Sci. Publ. Math. No. **29** (1966), 63–94.

275. ______, *Cohomological dimension of algebraic varieties*, Ann. of Math. (2) **88** (1968), 403–450.

276. ______, *Ample Subvarieties of Algebraic Varieties*, Lect. Notes in Math., vol. 156, Springer-Verlag, Berlin, 1970.

277. ______, *Ample vector bundles on curves*, Nagoya Math. J. **43** (1971), 73–89.

278. ______, *Equivalence relations on algebraic cycles and subvarieties of small codimension*, Proc. Symp. Pure Math., vol. 29, pp. 129–164, Amer. Math. Soc., Providence, R.I., 1973.

279. ______, *Varieties of small codimension in projective space*, Bull. Amer. Math. Soc. **80** (1974), 1017–1032.

280. ______, *Algebraic Geometry*, Graduate Texts in Math., vol. 52, Springer-Verlag, New York, 1977.

281. Robin Hartshorne (ed.), *Algebraic Geometry, Arcata 1974*, Proc. Symp. Pure Math., vol. 29, Providence, R.I., American Mathematical Society, 1979.

282. Brendan Hassett, Hui-Wen Lin, and Chin-Lung Wang, *The weak Lefschetz principle is false for ample cones*, Asian J. Math. **6** (2002), no. 1, 95–99.

283. Brendan Hassett and Yuri Tschinkel, *Rational curves on holomorphic symplectic fourfolds*, Geom. Funct. Anal. **11** (2001), no. 6, 1201–1228.

284. ______, *On the effective cone of the moduli space of pointed rational curves*, Topology and geometry: commemorating SISTAG, Contemp. Math., vol. 314, Amer. Math. Soc., Providence, RI, 2002, pp. 83–96.

285. Stefan Helmke, *On Fujita's conjecture*, Duke Math. J. **88** (1997), no. 2, 201–216.

286. ______, *On global generation of adjoint linear systems*, Math. Ann. **313** (1999), no. 4, 635–652.

287. Michel Hickel, *Solution d'une conjecture de C. Berenstein–A. Yger et invariants de contact à l'infini*, Ann. Inst. Fourier (Grenoble) **51** (2001), no. 3, 707–744.

288. Heisuke Hironaka, *Resolution of singularities of an algebraic variety over a field of characteristic zero. I, II*, Ann. of Math. (2) 79 (1964), 109–203; ibid. (2) **79** (1964), 205–326.

289. André Hirschowitz, *Une conjecture pour la cohomologie des diviseurs sur les surfaces rationnelles génériques*, J. Reine Angew. Math. **397** (1989), 208–213.

290. Melvin Hochster and Craig Huneke, *Comparison of symbolic and ordinary powers of ideals*, Invent. Math. **147** (2002), no. 2, 349–369.

291. Audun Holme and Michael Schneider, *A computer aided approach to codimension 2 subvarieties of* $\mathbf{P}_n$, $n \geq 6$, J. Reine Angew. Math. **357** (1985), 205–220.

292. Jason Howald, *Computations with monomial ideals*, Ph.D. thesis, University of Michigan, 2001.

293. ______, *Multiplier ideals of monomial ideals*, Trans. Amer. Math. Soc. **353** (2001), no. 7, 2665–2671.

294. ______, *Multiplier ideals of sufficiently general polynomials*, preprint, 2003.

295. Reinhold Hübl and Irena Swanson, *Discrete valuations centered on local domains*, J. Pure Appl. Algebra **161** (2001), no. 1-2, 145–166.

296. Craig Huneke, *Uniform bounds in Noetherian rings*, Invent. Math. **107** (1992), no. 1, 203–223.

297. ______, *Tight Closure and Its Applications*, CBMS Regional Conference Series in Math., vol. 88, Conference Board of the Mathematical Sciences, Washington, DC, 1996.

298. Daniel Huybrechts, *Compact hyper-Kähler manifolds: basic results*, Invent. Math. **135** (1999), no. 1, 63–113.

299. ______, *Erratum: "Compact hyper-Kähler manifolds: basic results"*, Invent. Math. **152** (2003), no. 1, 209–212.

300. Daniel Huybrechts and Manfred Lehn, *The Geometry of Moduli Spaces of Sheaves*, Aspects of Mathematics, E31, Friedr. Vieweg & Sohn, Braunschweig, 1997.

301. Jun-Muk Hwang and Jong Hae Keum, *Seshadri-exceptional foliations*, Math. Ann. **325** (2003), no. 2, 287–297.

302. Jun-Muk Hwang and Wing-Keung To, *On Seshadri constants of canonical bundles of compact complex hyperbolic spaces*, Compositio Math. **118** (1999), no. 2, 203–215.

303. ______, *On Seshadri constants of canonical bundles of compact quotients of bounded symmetric domains*, J. Reine Angew. Math. **523** (2000), 173–197.

304. ______, *Volumes of complex analytic subvarieties of Hermitian symmetric spaces*, Amer. J. Math. **124** (2002), no. 6, 1221–1246.

305. Eero Hyry and Karen E. Smith, *On a Non-Vanishing Conjecture of Kawamata and the Core of an Ideal*, preprint, 2002.

306. Bo Ilic and J. M. Landsberg, *On symmetric degeneracy loci, spaces of symmetric matrices of constant rank and dual varieties*, Math. Ann. **314** (1999), no. 1, 159–174.

307. Shuzo Izumi, *A measure of integrity for local analytic algebras*, Publ. Res. Inst. Math. Sci. **21** (1985), no. 4, 719–735.

308. Jean-Pierre Jouanolou, *Théorèmes de Bertini et Applications*, Progr. Math., vol. 42, Birkhäuser Boston, Boston, MA, 1983.

309. Morris Kalka, Bernard Shiffman, and Bun Wong, *Finiteness and rigidity theorems for holomorphic mappings*, Michigan Math. J. **28** (1981), no. 3, 289–295.

310. M. Kapranov, *Veronese curves and Grothendieck–Knudsen moduli space $\overline{M}_{0,n}$*, J. Algebraic Geom. **2** (1993), no. 2, 239–262.

311. K. K. Karčjauskas, *A generalized Lefschetz theorem*, Funkcional. Anal. i Priložen. **11** (1977), no. 4, 80–81.

312. ______, *Homotopy properties of algebraic sets*, Zap. Nauchn. Sem. Leningrad. Otdel. Mat. Inst. Steklov. (LOMI) **83** (1979), 67–72, 103, Studies in topology, III.

313. Masaki Kashiwara, *B-functions and holonomic systems. Rationality of roots of B-functions*, Invent. Math. **38** (1976/77), no. 1, 33–53.

314. Ludger Kaup, *Eine topologische Eigenschaft Steinscher Räume*, Nachr. Akad. Wiss. Göttingen (1966), 213–224.

315. Yujiro Kawamata, *Characterization of abelian varieties*, Compositio Math. **43** (1981), no. 2, 253–276.

316. ______, *A generalization of Kodaira–Ramanujam's vanishing theorem*, Math. Ann. **261** (1982), no. 1, 43–46.

317. ______, *The cone of curves of algebraic varieties*, Ann. of Math. (2) **119** (1984), no. 3, 603–633.

318. ______, *Pluricanonical systems on minimal algebraic varieties*, Invent. Math. **79** (1985), no. 3, 567–588.

319. ______, *The Zariski decomposition of log-canonical divisors*, Algebraic Geometry, Bowdoin, 1985, Amer. Math. Soc., Providence, RI, 1987, pp. 425–433.

320. ______, *On Fujita's freeness conjecture for 3-folds and 4-folds*, Math. Ann. **308** (1997), no. 3, 491–505.

321. ______, *Subadjunction of log canonical divisors for a subvariety of codimension 2*, Birational Algebraic Geometry (Baltimore, 1996), Contemp. Math., vol. 207, Amer. Math. Soc., Providence, RI, 1997, pp. 79–88.

322. ______, *Subadjunction of log canonical divisors. II*, Amer. J. Math. **120** (1998), no. 5, 893–899.

323. ———, *Deformations of canonical singularities*, J. Amer. Math. Soc. **12** (1999), no. 1, 85–92.

324. ———, *On the extension problem of pluricanonical forms*, Algebraic Geometry: Hirzebruch 70 (Warsaw, 1998), Amer. Math. Soc., Providence, RI, 1999, pp. 193–207.

325. ———, *On a relative version of Fujita's freeness conjecture*, Complex Geometry (Göttingen, 2000), Springer, Berlin, 2002, pp. 135–146.

326. Yujiro Kawamata, Katsumi Matsuda, and Kenji Matsuki, *Introduction to the minimal model problem*, Algebraic Geometry, Sendai, 1985, North-Holland, Amsterdam, 1987, pp. 283–360.

327. Yujiro Kawamata and Eckart Viehweg, *On a characterization of an abelian variety in the classification theory of algebraic varieties*, Compositio Math. **41** (1980), no. 3, 355–359.

328. Seán Keel, *Intersection theory of moduli space of stable n-pointed curves of genus zero*, Trans. Amer. Math. Soc. **330** (1992), no. 2, 545–574.

329. ———, *Basepoint freeness for nef and big line bundles in positive characteristic*, Ann. of Math. (2) **149** (1999), no. 1, 253–286.

330. ———, *Polarized pushouts over finite fields*, Comm. Algebra **31** (2003), no. 8, 3955–3982, Special issue in honor of Steven L. Kleiman.

331. Dennis Keeler, *Ample filters of invertible sheaves*, J. Algebra **259** (2003), no. 1, 243–283.

332. George Kempf, *Varieties with rational singularities*, The Lefschetz Centennial Conference, Part I, Comtemp. Math., vol. 58, Amer. Math. Soc., Providence, R.I., 1986, pp. 179–182.

333. ———, *Projective coordinate rings of abelian varieties*, Algebraic Analysis, Geometry, and Number Theory, Johns Hopkins Univ. Press, Baltimore, MD, 1989, pp. 225–235.

334. A. G. Khovanskiĭ, *Analogues of Aleksandrov–Fenchel inequalities for hyperbolic forms*, Dokl. Akad. Nauk SSSR **276** (1984), no. 6, 1332–1334.

335. Meeyoung Kim, *A Barth–Lefschetz type theorem for branched coverings of Grassmannians*, J. Reine Angew. Math. **470** (1996), 109–122.

336. ———, *On branched coverings of quadrics*, Arch. Math. (Basel) **67** (1996), no. 1, 76–79.

337. Meeyoung Kim and Laurent Manivel, *On branched coverings of some homogeneous spaces*, Topology **38** (1999), no. 5, 1141–1160.

338. Meeyoung Kim and Andrew J. Sommese, *Two results on branched coverings of Grassmannians*, J. Math. Kyoto Univ. **38** (1998), no. 1, 21–27.

339. Meeyoung Kim and Jon Wolfson, *Theorems of Barth–Lefschetz type on Kähler manifolds of non-negative bisectional curvature*, Forum Math. **15** (2003), no. 2, 261–273.

340. Steven Kleiman, *A note on the Nakai–Moisezon test for ampleness of a divisor*, Amer. J. Math. **87** (1965), 221–226.

341. ———, *Toward a numerical theory of ampleness*, Ann. of Math. (2) **84** (1966), 293–344.

342. ______, *Ample vector bundles on algebraic surfaces*, Proc. Amer. Math. Soc. **21** (1969), 673–676.

343. ______, *The transversality of a general translate*, Compositio Math. **28** (1974), 287–297.

344. ______, *Misconceptions about K_X*, Enseign. Math. (2) **25** (1979), no. 3-4, 203–206.

345. ______, *Bertini and his two fundamental theorems*, Rend. Circ. Mat. Palermo (2) Suppl. (1998), no. 55, 9–37.

346. ______, *Cartier divisors versus invertible sheaves*, Comm. Algebra **28** (2000), no. 12, 5677–5678.

347. Steven Kleiman and Dan Laksov, *On the existence of special divisors*, Amer. J. Math. **94** (1972), 431–436.

348. ______, *Another proof of the existence of special divisors*, Acta Math. **132** (1974), 163–176.

349. Steven Kleiman and Anders Thorup, *A geometric theory of the Buchsbaum–Rim multiplicity*, J. Algebra **167** (1994), no. 1, 168–231.

350. ______, *Mixed Buchsbaum–Rim multiplicities*, Amer. J. Math. **118** (1996), no. 3, 529–569.

351. Shoshichi Kobayashi, *Negative vector bundles and complex Finsler structures*, Nagoya Math. J. **57** (1975), 153–166.

352. ______, *Differential Geometry of Complex Vector Bundles*, Princeton University Press, Princeton, NJ, 1987.

353. K. Kodaira, *On a differential-geometric method in the theory of analytic stacks*, Proc. Nat. Acad. Sci. U.S.A. **39** (1953), 1268–1273.

354. Vijay Kodiyalam, *Asymptotic behaviour of Castelnuovo–Mumford regularity*, Proc. Amer. Math. Soc. **128** (2000), no. 2, 407–411.

355. Shoji Koizumi, *Theta relations and projective normality of Abelian varieties*, Amer. J. Math. **98** (1976), no. 4, 865–889.

356. János Kollár, *Higher direct images of dualizing sheaves. I*, Ann. of Math. (2) **123** (1986), no. 1, 11–42.

357. ______, *Higher direct images of dualizing sheaves. II*, Ann. of Math. (2) **124** (1986), no. 1, 171–202.

358. ______, *Subadditivity of the Kodaira dimension: fibers of general type*, Algebraic Geometry, Sendai, 1985, North-Holland, Amsterdam, 1987, pp. 361–398.

359. ______, *Sharp effective Nullstellensatz*, J. Amer. Math. Soc. **1** (1988), no. 4, 963–975.

360. ______, *Flips and Abundance for Algebraic Threefolds*, Astérisque, vol. 211, 1992.

361. ______, *Shafarevich maps and plurigenera of algebraic varieties*, Invent. Math. **113** (1993), no. 1, 177–215.

362. ______, *Shafarevich Maps and Automorphic Forms*, Princeton University Press, Princeton, NJ, 1995.

363. ______, *Rational Curves on Algebraic Varieties*, Ergebnisse der Math. und ihrer Grenzgebiete (3), vol. 32, Springer-Verlag, Berlin, 1996.

364. ______, *Singularities of pairs*, Algebraic Geometry, Santa Cruz 1995, Proc. Symp. Pure Math., vol. 62, Amer. Math. Soc., Providence, RI, 1997, pp. 221–287.

365. ______, *Effective Nullstellensatz for arbitrary ideals*, J. Eur. Math. Soc. (JEMS) **1** (1999), no. 3, 313–337.

366. János Kollár and T. Matsusaka, *Riemann–Roch type inequalities*, Amer. J. Math. **105** (1983), no. 1, 229–252.

367. János Kollár, Yoichi Miyaoka, and Shigefumi Mori, *Rational curves on Fano varieties*, Classification of Irregular Varieties (Trento, 1990), Lect. Notes in Math., vol. 1515, Springer, Berlin, 1992, pp. 100–105.

368. János Kollár and Shigefumi Mori, *Birational Geometry of Algebraic Varieties*, Cambridge Tracts in Math., vol. 134, Cambridge University Press, Cambridge, 1998.

369. Jian Kong, *Seshadri constants on Jacobian of curves*, Trans. Amer. Math. Soc. **355** (2003), no. 8, 3175–3180.

370. Alexis Kouvidakis, *Divisors on symmetric products of curves*, Trans. Amer. Math. Soc. **337** (1993), no. 1, 117–128.

371. Sándor J. Kovács, *The cone of curves of a $K3$ surface*, Math. Ann. **300** (1994), no. 4, 681–691.

372. Henrik Kratz, *Compact complex manifolds with numerically effective cotangent bundles*, Doc. Math. **2** (1997), 183–193.

373. Oliver Küchle and Andreas Steffens, *Bounds for Seshadri constants*, New Trends in Algebraic Geometry (Warwick, 1996), London Math. Soc. Lect. Note Series, vol. 264, Cambridge Univ. Press, Cambridge, 1999, pp. 235–254.

374. Alex Küronya, *A divisorial valuation with irrational volume*, J. Algebra **262** (2003), no. 2, 413–423.

375. Sijong Kwak, *Castelnuovo regularity for smooth subvarieties of dimensions* 3 *and* 4, J. Algebraic Geom. **7** (1998), no. 1, 195–206.

376. Klaus Lamotke, *The topology of complex projective varieties after S. Lefschetz*, Topology **20** (1981), no. 1, 15–51.

377. J. M. Landsberg, *Differential-geometric characterizations of complete intersections*, J. Differential Geom. **44** (1996), no. 1, 32–73.

378. ______, *On degenerate secant and tangential varieties and local differential geometry*, Duke Math. J. **85** (1996), no. 3, 605–634.

379. ______, *Algebraic Geometry and Projective Differential Geometry*, Seoul National University Research Institute of Mathematics Global Analysis Research Center, Seoul, 1999.

380. Serge Lang, *Hyperbolic and Diophantine analysis*, Bull. Amer. Math. Soc. (N.S.) **14** (1986), no. 2, 159–205.

381. ______, *Introduction to Complex Hyperbolic Spaces*, Springer-Verlag, New York, 1987.

382. ———, *Number Theory, III. Diophantine Geometry*, Encyclopedia of Math. Sciences, vol. 60, Springer-Verlag, Berlin, 1991.

383. Herbert Lange and Christina Birkenhake, *Complex Abelian Varieties*, Grundlehren Math. Wiss., vol. 302, Springer-Verlag, Berlin, 1992.

384. Mogens Larsen, *On the topology of complex projective manifolds*, Invent. Math. (1973), 251–260.

385. F. Laytimi, *On degeneracy loci*, Internat. J. Math. **7** (1996), no. 6, 745–754.

386. F. Laytimi and W. Nahm, *A generalization of Le Potier's vanishing theorem*, preprint, 2002.

387. Robert Lazarsfeld, *A Barth-type theorem for branched coverings of projective space*, Math. Ann. **249** (1980), no. 2, 153–162.

388. ———, *Branched coverings of projective space*, Ph.D. thesis, Brown University, 1980.

389. ———, *Some applications of the theory of positive vector bundles*, Complete Intersections (Acireale, 1983), Lect. Notes in Math, vol. 1092, Springer, Berlin, 1984, pp. 29–61.

390. ———, *Brill–Noether–Petri without degenerations*, J. Differential Geom. **23** (1986), no. 3, 299–307.

391. ———, *A sharp Castelnuovo bound for smooth surfaces*, Duke Math. J. **55** (1987), no. 2, 423–429.

392. ———, *A sampling of vector bundle techniques in the study of linear series*, Lectures on Riemann Surfaces (Trieste, 1987), World Sci. Publishing, Teaneck, NJ, 1989, pp. 500–559.

393. ———, *Lengths of periods and Seshadri constants of abelian varieties*, Math. Res. Lett. **3** (1996), no. 4, 439–447.

394. ———, *Lectures on linear series*, Complex Algebraic Geometry (Park City, UT, 1993), IAS/Park City Math. Series, vol. 3, Amer. Math. Soc., Providence, RI, 1997, pp. 161–219.

395. Robert Lazarsfeld and A. Van de Ven, *Topics in the Geometry of Projective Space: Recent Work of F. L. Zak*, DMV Seminar, vol. 4, Birkhäuser Verlag, Basel, 1984.

396. Joseph Le Potier, *"Vanishing theorem" pour un fibré vectoriel holomorphe positif de rang quelconque*, Journées de Géométrie Analytique (Poitiers, 1972), Soc. Math. France, Paris, 1974, pp. 107–119. Bull. Soc. Math. France Suppl. Mém., No. 38.

397. ———, *Annulation de la cohomologie à valeurs dans un fibré vectoriel holomorphe positif de rang quelconque*, Math. Ann. **218** (1975), no. 1, 35–53.

398. Marc Levine, *Pluri-canonical divisors on Kähler manifolds*, Invent. Math. **74** (1983), no. 2, 293–303.

399. Anatoly Libgober, *Alexander invariants of plane algebraic curves*, Singularities, Part 2 (Arcata, Calif., 1981), Proc. Symp. Pure Math., vol. 40, Amer. Math. Soc., Providence, RI, 1983, pp. 135–143.

400. Ben Lichtin, *Poles of* $|f(z,w)|^{2s}$ *and roots of the b-function*, Ark. Mat. **27** (1989), no. 2, 283–304.

401. David Lieberman and David Mumford, *Matsusaka's big theorem*, Algebraic Geometry – Arcata 1974, Proc. Symp. Pure Math., vol. 29, Amer. Math. Soc., 1975, pp. 513–530.

402. Joseph Lipman, *Unique factorization in complete local rings*, Algebraic Geometry – Arcata, 1974, Proc. Symp. Pure Math., vol. 29, Amer. Math. Soc., Providence, R.I., 1975, pp. 531–546.

403. ———, *Adjoints of ideals in regular local rings*, Math. Res. Lett. **1** (1994), no. 6, 739–755.

404. Joseph Lipman and Bernard Teissier, *Pseudorational local rings and a theorem of Briançon–Skoda about integral closures of ideals*, Michigan Math. J. **28** (1981), no. 1, 97–116.

405. Joseph Lipman and Kei-ichi Watanabe, *Integrally closed ideals in two-dimensional regular local rings are multiplier ideals*, Math. Res. Lett. **10** (2003), no. 4, 423–434.

406. François Loeser and Michel Vaquié, *Le polynôme d'Alexander d'une courbe plane projective*, Topology **29** (1990), no. 2, 163–173.

407. Eduard Looijenga, *Cohomology and intersection homology of algebraic varieties*, Complex Algebraic Geometry (Park City, UT, 1993), IAS/Park City Math. Series, vol. 3, Amer. Math. Soc., Providence, RI, 1997, pp. 221–263.

408. ———, *Motivic measures*, Séminaire Bourbaki, Année 1999/2000, Exp. 276, 2002, pp. 267–297.

409. Martin Lübke, *Beweis einer Vermutung von Hartshorne für den Fall homogener Mannigfaltigkeiten*, J. Reine Angew. Math. **316** (1980), 215–220.

410. Tie Luo, *Riemann–Roch type inequalities for nef and big divisors*, Amer. J. Math. **111** (1989), no. 3, 457–487.

411. ———, *A note on the Hodge index theorem*, Manuscripta Math. **67** (1990), no. 1, 17–20.

412. Gennady Lyubeznik, *Étale cohomological dimension and the topology of algebraic varieties*, Ann. of Math. (2) **137** (1993), no. 1, 71–128.

413. B. Malgrange, *Sur les polynômes de I. N. Bernstein*, Séminaire Goulaouic–Schwartz 1973–1974: Équations aux dérivées partielles et analyse fonctionnelle, Exp. No. 20, Centre de Math., École Polytech., Paris, 1974, p. 10.

414. ———, *Le polynôme de Bernstein d'une singularité isolée*, Fourier Integral Operators and Partial Differential Equations, Lect. Notes in Math., vol. 459, Springer, Berlin, 1975, pp. 98–119.

415. Yuri Manin, *Rational points on algebraic curves over function fields*, Izv. Akad. Nauk SSSR Ser. Mat. **27** (1963), 1395–1440.

416. Laurent Manivel, *Théorèmes d'annulation sur certaines variétés projectives*, Comment. Math. Helv. **71** (1996), no. 3, 402–425.

417. ———, *Vanishing theorems for ample vector bundles*, Invent. Math. **127** (1997), no. 2, 401–416.

418. Mireille Martin-Deschamps, *Propriétés de descente des variétés à fibré cotangent ample*, Ann. Inst. Fourier (Grenoble) **34** (1984), no. 3, 39–64.

419. Kenji Matsuki, *Introduction to the Mori Program*, Universitext, Springer-Verlag, New York, 2002.

420. T. Matsusaka, *Polarized varieties with a given Hilbert polynomial*, Amer. J. Math. **94** (1972), 1027–1077.

421. Arthur Mattuck, *Symmetric products and Jacobians*, Amer. J. Math. **83** (1961), 189–206.

422. Dusa McDuff and Leonid Polterovich, *Symplectic packings and algebraic geometry*, Invent. Math. **115** (1994), no. 3, 405–434.

423. Dusa McDuff and Dietmar Salamon, *Introduction to Symplectic Topology*, Oxford University Press, New York, 1995.

424. John Milnor, *Morse Theory*, Annals of Math. Studies, vol. 51, Princeton University Press, Princeton, N.J., 1963.

425. Rick Miranda, *Triple covers in algebraic geometry*, Amer. J. Math. **107** (1985), no. 5, 1123–1158.

426. ———, *Linear systems of plane curves*, Notices Amer. Math. Soc. **46** (1999), no. 2, 192–201.

427. Yoichi Miyaoka, *On the Mumford–Ramanujam vanishing theorem on a surface*, Journées de Géometrie Algébrique d'Angers, Sijthoff & Noordhoff, Alphen aan den Rijn, 1980, pp. 239–247.

428. ———, *Algebraic surfaces with positive indices*, Classification of Algebraic and Analytic Manifolds (Katata, 1982), Birkhäuser Boston, Boston, Mass., 1983, pp. 281–301.

429. ———, *The Chern classes and Kodaira dimension of a minimal variety*, Algebraic Geometry, Sendai, 1985, North-Holland, Amsterdam, 1987, pp. 449–476.

430. ———, *Deformations of a morphism along a foliation and applications*, Algebraic Geometry, Bowdoin, 1985, Proc. Symp. Pure Math., vol. 46, Amer. Math. Soc., Providence, RI, 1987, pp. 245–268.

431. Yoichi Miyaoka and Shigefumi Mori, *A numerical criterion for uniruledness*, Ann. of Math. (2) **124** (1986), no. 1, 65–69.

432. Yoichi Miyaoka and Thomas Peternell, *Geometry of Higher-Dimensional Algebraic Varieties*, DMV Seminar, vol. 26, Birkhäuser Verlag, Basel, 1997.

433. Boris Moishezon, *Projective imbeddings of algebraic manifolds*, Dokl. Akad. Nauk SSSR **141** (1961), 555–557.

434. ———, *Remarks on projective imbeddings of algebraic manifolds*, Dokl. Akad. Nauk SSSR **145** (1962), 996–999.

435. Ngaiming Mok, *The uniformization theorem for compact Kähler manifolds of nonnegative holomorphic bisectional curvature*, J. Differential Geom. **27** (1988), no. 2, 179–214.

436. Frank Morgan, *Geometric Measure Theory*, second ed., Academic Press Inc., San Diego, CA, 1995.

437. Shigefumi Mori, *Projective manifolds with ample tangent bundles*, Ann. of Math. (2) **110** (1979), no. 3, 593–606.

438. ———, *Threefolds whose canonical bundles are not numerically effective*, Ann. of Math. (2) **116** (1982), no. 1, 133–176.

439. ______, *Classification of higher-dimensional varieties*, Algebraic Geometry, Bowdoin, 1985, Proc. Symp. Pure Math., vol. 46, Amer. Math. Soc., Providence, RI, 1987, pp. 245–268.

440. ______, *Flip theorem and the existence of minimal models for* 3*-folds*, J. Amer. Math. Soc. **1** (1988), no. 1, 117–253.

441. Atsushi Moriwaki, *Geometric height inequality on varieties with ample cotangent bundles*, J. Algebraic Geom. **4** (1995), no. 2, 385–396.

442. ______, *Remarks on rational points of varieties whose cotangent bundles are generated by global sections*, Math. Res. Lett. **2** (1995), no. 1, 113–118.

443. Christophe Mourougane and Francesco Russo, *Some remarks on nef and good divisors on an algebraic variety*, C. R. Acad. Sci. Paris Sér. I Math. **325** (1997), no. 5, 499–504.

444. Shigeru Mukai, *Duality between* $D(X)$ *and* $D(\hat{X})$ *with its application to Picard sheaves*, Nagoya Math. J. **81** (1981), 153–175.

445. David Mumford, *Lectures on Curves on an Algebraic Surface*, Annals of Math. Studies, vol. 59, Princeton University Press, Princeton, N.J., 1966.

446. ______, *Pathologies. III*, Amer. J. Math. **89** (1967), 94–104.

447. ______, *Abelian Varieties*, Tata Institute Studies in Mathematics, vol. 5, Oxford University Press, 1970.

448. ______, *Varieties defined by quadratic equations*, Questions on Algebraic Varieties (C.I.M.E., III Ciclo, Varenna, 1969), Edizioni Cremonese, Rome, 1970, pp. 29–100.

449. ______, *Theta characteristics of an algebraic curve*, Ann. Sci. École Norm. Sup. (4) **4** (1971), 181–192.

450. ______, *Curves and Their Jacobians*, The University of Michigan Press, Ann Arbor, Mich., 1975, (reprinted in The Red Book of Varieties and Schemes, 2nd Edition, Lect. Notes in Math., Volume 1358).

451. ______, *Hilbert's fourteenth problem–the finite generation of subrings such as rings of invariants*, Mathematical Developments Arising from Hilbert Problems, Proc. Symp. Pure Math., vol. 28, Amer. Math. Soc.., 1976, pp. 431–444.

452. ______, *Some footnotes to the work of C. P. Ramanujam*, C. P. Ramanujam—a tribute, Springer, Berlin, 1978, pp. 247–262.

453. ______, *Algebraic Geometry, I. Complex Projective Varieties*, Grundlehren Math. Wiss., vol. 221, Springer-Verlag, Berlin, 1981.

454. Vicente Muñoz and Francisco Presas, *Semipositive bundles and Brill–Noether theory*, Bull. London Math. Soc. **35** (2003), no. 2, 179–190.

455. Mircea Mustaţă, *Jet schemes of locally complete intersection canonical singularities*, Invent. Math. **145** (2001), no. 3, 397–424.

456. ______, *The multiplier ideals of a sum of ideals*, Trans. Amer. Math. Soc. **354** (2002), no. 1, 205–217.

457. ______, *On multiplicities of graded sequences of ideals*, J. Algebra **256** (2002), no. 1, 229–249.

458. ______, *Singularities of pairs via jet schemes*, J. Amer. Math. Soc. **15** (2002), no. 3, 599–615.

459. ———, *Vanishing theorems on toric varieties*, Tohoku Math. J. (2) **54** (2002), no. 3, 451–470.

460. Alan Nadel, *Multiplier ideal sheaves and existence of Kähler–Einstein metrics of positive scalar curvature*, Proc. Nat. Acad. Sci. U.S.A. **86** (1989), no. 19, 7299–7300.

461. ———, *Multiplier ideal sheaves and Kähler–Einstein metrics of positive scalar curvature*, Ann. of Math. (2) **132** (1990), no. 3, 549–596.

462. ———, *The boundedness of degree of Fano varieties with Picard number one*, J. Amer. Math. Soc. **4** (1991), no. 4, 681–692.

463. Masayoshi Nagata, *On the* 14-*th problem of Hilbert*, Amer. J. Math. **81** (1959), 766–772.

464. Yoshikazu Nakai, *Non-degenerate divisors on an algebraic surface*, J. Sci. Hiroshima Univ. Ser. A **24** (1960), 1–6.

465. ———, *A criterion of an ample sheaf on a projective scheme*, Amer. J. Math. **85** (1963), 14–26.

466. ———, *Some fundamental lemmas on projective schemes*, Trans. Amer. Math. Soc. **109** (1963), 296–302.

467. Michael Nakamaye, *Multiplicity estimates and the product theorem*, Bull. Soc. Math. France **123** (1995), no. 2, 155–188.

468. ———, *Stable base loci of linear series*, Math. Ann. **318** (2000), no. 4, 837–847.

469. ———, *Base loci of linear series are numerically determined*, Trans. Amer. Math. Soc. **355** (2003), no. 2, 551–566.

470. ———, *Seshadri constants at very general points*, preprint, 2003.

471. Noboru Nakayama, *Invariance of plurigenera of algebraic varieties*, preprint.

472. ———, *Zariski-decomposition and abundance*, unpublished preprint, 1997.

473. Terrence Napier and Mohan Ramachandran, *The* L^2 $\overline{\partial}$*-method, weak Lefschetz theorems, and the topology of Kähler manifolds*, J. Amer. Math. Soc. **11** (1998), no. 2, 375–396.

474. M. S. Narasimhan and C. S. Seshadri, *Stable and unitary vector bundles on a compact Riemann surface*, Ann. of Math. (2) **82** (1965), 540–567.

475. Raghavan Narasimhan, *On the homology groups of Stein spaces*, Invent. Math. **2** (1967), 377–385.

476. Lei Ne and Jon Wolfson, *The Lefschetz theorem for CR submanifolds and the nonexistence of real analytic Levi flat submanifolds*, preprint, 2003.

477. P. E. Newstead, *Introduction to Moduli Problems and Orbit Spaces*, Tata Institute of Fundamental Research, Bombay, 1978.

478. Peter A. Nielsen, *A sharper Riemann–Roch inequality for big, semi-ample divisors*, Japan. J. Math. (N.S.) **17** (1991), no. 2, 267–284.

479. Junjiro Noguchi, *A higher-dimensional analogue of Mordell's conjecture over function fields*, Math. Ann. **258** (1981/82), no. 2, 207–212.

480. Junjiro Noguchi and Toshikazu Sunada, *Finiteness of the family of rational and meromorphic mappings into algebraic varieties*, Amer. J. Math. **104** (1982), no. 4, 887–900.

481. Madhav V. Nori, *Zariski's conjecture and related problems*, Ann. Sci. École Norm. Sup. (4) **16** (1983), no. 2, 305–344.

482. ______, *Algebraic cycles and Hodge-theoretic connectivity*, Invent. Math. **111** (1993), no. 2, 349–373.

483. ______, *Constructible sheaves*, Algebra, Arithmetic and Geometry, Part I, II (Mumbai, 2000), Tata Inst. Fund. Res. Stud. Math., vol. 16, Tata Inst. Fund. Res., Bombay, 2002, pp. 471–491.

484. Joseph Oesterlé, *Empilements de sphères*, Astérisque (1990), no. 189-190, Exp. No. 727, 375–397, Séminaire Bourbaki, Vol. 1989/90.

485. Keiji Oguiso, *Seshadri constants in a family of surfaces*, Math. Ann. **323** (2002), no. 4, 625–631.

486. Arthur Ogus, *Local cohomological dimension of algebraic varieties*, Ann. of Math. (2) **98** (1973), 327–365.

487. Christian Okonek, *Barth–Lefschetz theorems for singular spaces*, J. Reine Angew. Math. **374** (1987), 24–38.

488. Christian Okonek, Michael Schneider, and Heinz Spindler, *Vector Bundles on Complex Projective Spaces*, Progr. Math., vol. 3, Birkhäuser Boston, Mass., 1980.

489. Andrei Okounkov, *Why would multiplicities be log-concave?*, preprint.

490. Gianluca Pacienza, *On the nef cone of symmetric products of a generic curve*, Amer. J. Math. **125** (2003), no. 5, 1117–1135.

491. Roberto Paoletti, *Seshadri constants, gonality of space curves, and restriction of stable bundles*, J. Differential Geom. **40** (1994), no. 3, 475–504.

492. ______, *Free pencils on divisors*, Math. Ann. **303** (1995), no. 1, 109–123.

493. ______, *The asymptotic growth of equivariant sections of positive and big line bundles*, preprint, 2002.

494. Kapil H. Paranjape, *The Bogomolov–Pantev resolution, an expository account*, New Trends in Algebraic Geometry (Warwick, 1996), London Math. Soc. Lecture Note Ser., vol. 264, Cambridge Univ. Press, Cambridge, 1999, pp. 347–358.

495. Giuseppe Pareschi, *Syzygies of abelian varieties*, J. Amer. Math. Soc. **13** (2000), no. 3, 651–664.

496. Giuseppe Pareschi and Mihnea Popa, *Regularity on abelian varieties. I*, J. Amer. Math. Soc. **16** (2003), no. 2, 285–302.

497. ______, *Regularity on abelian varieties. II. Basic results on linear series and defining equations*, J. Algebraic Geom. **13** (2004), no. 1, 167–193.

498. Sam Payne, *Fujita's very ampleness conjecture for singular toric varieties*, preprint, 2004.

499. Irena Peeva and Bernd Sturmfels, *Syzygies of codimension 2 lattice ideals*, Math. Z. **229** (1998), no. 1, 163–194.

500. Mathias Peternell, *Ein Lefschetz-Satz für Schnitte in projektiv-algebraischen Mannigfaltigkeiten*, Math. Ann. **264** (1983), no. 3, 361–388.

501. ______, *Algebraische Varietäten und q-vollständige komplexe Räume*, Math. Z. **200** (1989), no. 4, 547–581.

502. Thomas Peternell, Joseph Le Potier, and Michael Schneider, *Vanishing theorems, linear and quadratic normality*, Invent. Math. **87** (1987), no. 3, 573–586.

503. Thomas Peternell and Andrew Sommese, *Ample vector bundles and branched coverings, II*, preprint.

504. ______, *Ample vector bundles and branched coverings*, Comm. Algebra **28** (2000), no. 12, 5573–5599.

505. Henry Pinkham, *A Castelnuovo bound for smooth surfaces*, Invent. Math. **83** (1986), no. 2, 321–332.

506. Piotr Pragacz, *Symmetric polynomials and divided differences in formulas of intersection theory*, Parameter Spaces (Warsaw, 1994), Polish Acad. Sci., Warsaw, 1996, pp. 125–177.

507. Yuri Prokhorov, *On Zariski decomposition problem*, preprint, 2002.

508. S. Ramanan, *A note on C. P. Ramanujam*, C. P. Ramanujam—a Tribute, Tata Inst. Fund. Res. Studies in Math., vol. 8, Springer, Berlin, 1978, pp. 11–13.

509. C. P. Ramanujam, *Remarks on the Kodaira vanishing theorem*, J. Indian Math. Soc. (N.S.) **36** (1972), 41–51.

510. Ziv Ran, *On projective varieties of codimension* 2, Invent. Math. **73** (1983), no. 2, 333–336.

511. ______, *Local differential geometry and generic projections of threefolds*, J. Differential Geom. **32** (1990), no. 1, 131–137.

512. Ziv Ran and Herbert Clemens, *A new method in Fano geometry*, Internat. Math. Res. Notices (2000), no. 10, 527–549.

513. M. S. Ravi, *Regularity of ideals and their radicals*, Manuscripta Math. **68** (1990), no. 1, 77–87.

514. David Rees and Rodney Sharp, *On a theorem of B. Teissier on multiplicities of ideals in local rings*, J. London Math. Soc. (2) **18** (1978), no. 3, 449–463.

515. Miles Reid, *Canonical 3-folds*, Journées de Géometrie Algébrique d'Angers, Sijthoff & Noordhoff, Alphen aan den Rijn, 1980, pp. 273–310.

516. ______, *Young person's guide to canonical singularities*, Algebraic Geometry, Bowdoin, 1985, Proc. Symp. Pure Math., vol. 46, Amer. Math. Soc., Providence, RI, 1987, pp. 345–414.

517. ______, *Twenty-five years of 3-folds—an old person's view*, Explicit Birational Geometry of 3-Folds, London Math. Soc. Lecture Note Ser., vol. 281, Cambridge Univ. Press, Cambridge, 2000, pp. 313–343.

518. Igor Reider, *Vector bundles of rank 2 and linear systems on algebraic surfaces*, Ann. of Math. (2) **127** (1988), no. 2, 309–316.

519. Paul C. Roberts, *Multiplicities and Chern Classes in Local Algebra*, Cambridge Tracts in Mathematics, vol. 133, Cambridge University Press, Cambridge, 1998.

520. Francesco Russo, *A characterization of nef and good divisors by asymptotic multiplier ideals*, unpublished manuscript.

521. Bernard Saint-Donat, *Projective models of K-3 surfaces*, Amer. J. Math. **96** (1974), 602–639.

522. Michael Schneider, *Ein einfacher Beweis des Verschwindungssatzes für positive holomorphe Vektorraumbündel*, Manuscripta Math. **11** (1974), 95–101.

523. ______, *Symmetric differential forms as embedding obstructions and vanishing theorems*, J. Algebraic Geom. **1** (1992), no. 2, 175–181.

524. Michael Schneider and Alessandro Tancredi, *Almost-positive vector bundles on projective surfaces*, Math. Ann. **280** (1988), no. 4, 537–547.

525. Michael Schneider and Jörg Zintl, *The theorem of Barth–Lefschetz as a consequence of Le Potier's vanishing theorem*, Manuscripta Math. **80** (1993), no. 3, 259–263.

526. Richard Schoen and Jon Wolfson, *Theorems of Barth–Lefschetz type and Morse theory on the space of paths*, Math. Z. **229** (1998), no. 1, 77–89.

527. Frank-Olaf Schreyer, *Green's conjecture for general p-gonal curves of large genus*, Algebraic Curves and Projective Geometry (Trento, 1988), Lect. Notes in Math., vol. 1389, Springer, Berlin, 1989, pp. 254–260.

528. Stefan Schröer, *Remarks on the existence of Cartier divisors*, Arch. Math. (Basel) **75** (2000), no. 1, 35–38.

529. Tsutomu Sekiguchi, *On the normal generation by a line bundle on an Abelian variety*, Proc. Japan Acad. Ser. A Math. Sci. **54** (1978), no. 7, 185–188.

530. Jean-Pierre Serre, *Faisceaux algébriques cohérents*, Ann. of Math. (2) **61** (1955), 197–278.

531. ______, *Revêtements ramifiés du plan projectif (d'après S. Abhyankar)*, Séminaire Bourbaki, Année 1959/60, Exp. 204, Soc. Math. France, Paris, 1995, pp. 483–489.

532. Igor R. Shafarevich, *Basic Algebraic Geometry. 1. Varieties in Projective Space*, second ed., Springer-Verlag, Berlin, 1994.

533. Bernard Shiffman and Andrew Sommese, *Vanishing Theorems on Complex Manifolds*, Progr. Math., vol. 56, Birkhäuser Boston Inc., Boston, MA, 1985.

534. V. V. Shokurov, *A nonvanishing theorem*, Izv. Akad. Nauk SSSR Ser. Mat. **49** (1985), no. 3, 635–651.

535. Jessica Sidman, *On the Castelnuovo–Mumford regularity of products of ideal sheaves*, Adv. Geom. **2** (2002), no. 3, 219–229.

536. Carlos Simpson, *Subspaces of moduli spaces of rank one local systems*, Ann. Sci. École Norm. Sup. (4) **26** (1993), no. 3, 361–401.

537. Yum-Tong Siu, *An effective Matsusaka big theorem*, Ann. Inst. Fourier (Grenoble) **43** (1993), no. 5, 1387–1405.

538. ______, *Effective very ampleness*, Invent. Math. **124** (1996), no. 1–3, 563–571.

539. ______, *Invariance of plurigenera*, Invent. Math. **134** (1998), no. 3, 661–673.

540. ______, *Extension of twisted pluricanonical sections with plurisubharmonic weight and invariance of semipositively twisted plurigenera for manifolds not necessarily of general type*, Complex Geometry (Göttingen, 2000), Springer, Berlin, 2002, pp. 223–277.

541. ______, *A new bound for the effective Matsusaka big theorem*, Houston J. Math. **28** (2002), no. 2, 389–409, Special issue for S. S. Chern.

542. Yum-Tong Siu and Shing-Tung Yau, *Compact Kähler manifolds of positive bisectional curvature*, Invent. Math. **59** (1980), no. 2, 189–204.

543. Karen E. Smith, *Fujita's freeness conjecture in terms of local cohomology*, J. Algebraic Geom. **6** (1997), no. 3, 417–429.

544. ______, *The multiplier ideal is a universal test ideal*, Comm. Algebra **28** (2000), no. 12, 5915–5929.

545. Roy Smith and Robert Varley, *Multiplicity g points on theta divisors*, Duke Math. J. **82** (1996), no. 2, 319–326.

546. Ernst Snapper, *Polynomials associated with divisors*, J. Math. Mech. **9** (1960), 123–139.

547. Martín Sombra, *A sparse effective Nullstellensatz*, Adv. in Appl. Math. **22** (1999), no. 2, 271–295.

548. Andrew Sommese, *On manifolds that cannot be ample divisors*, Math. Ann. **221** (1976), no. 1, 55–72.

549. ______, *Theorems of Barth–Lefschetz type for complex subspaces of homogeneous complex manifolds*, Proc. Nat. Acad. Sci. U.S.A. **74** (1977), no. 4, 1332–1333.

550. ______, *Concavity theorems*, Math. Ann. **235** (1978), no. 1, 37–53.

551. ______, *Submanifolds of Abelian varieties*, Math. Ann. **233** (1978), no. 3, 229–256.

552. ______, *Complex subspaces of homogeneous complex manifolds. I. Transplanting theorems*, Duke Math. J. **46** (1979), no. 3, 527–548.

553. ______, *Complex subspaces of homogeneous complex manifolds. II. Homotopy results*, Nagoya Math. J. **86** (1982), 101–129.

554. ______, *A convexity theorem*, Singularities (Arcata, 1981), Proc. Symp. Pure Math., vol. 40, Amer. Math. Soc., Providence, RI, 1983, pp. 497–505.

555. ______, *On the density of ratios of Chern numbers of algebraic surfaces*, Math. Ann. **268** (1984), no. 2, 207–221.

556. Andrew Sommese and A. Van de Ven, *Homotopy groups of pullbacks of varieties*, Nagoya Math. J. **102** (1986), 79–90.

557. Frauke Steffen, *A generalized principal ideal theorem with an application to Brill–Noether theory*, Invent. Math. **132** (1998), no. 1, 73–89.

558. ______, *Connectedness theorems for determinantal schemes*, J. Algebraic Geom. **8** (1999), no. 1, 169–179.

559. Andreas Steffens, *Remarks on Seshadri constants*, Math. Z. **227** (1998), no. 3, 505–510.

560. Irena Swanson, *Powers of ideals. Primary decompositions, Artin–Rees lemma and regularity*, Math. Ann. **307** (1997), no. 2, 299–313.

561. ______, *Linear equivalence of ideal topologies*, Math. Z. **234** (2000), no. 4, 755–775.

562. L. Szpiro, *Lectures on Equations Defining Space Curves*, Tata Institute of Fundamental Research, Bombay, 1979.

563. Shunsuke Takagi, *An interpretation of multiplier ideals via tight closure*, preprint.

564. Abdelhak Taraffa, *Conditions explicites d'amplitude liées au théorème de Faltings*, C. R. Acad. Sci. Paris Sér. I Math. **322** (1996), no. 12, 1209–1212.

565. Bernard Teissier, *Cycles évanescents, sections planes et conditions de Whitney*, Singularités à Cargèse, Soc. Math. France, Paris, 1973, pp. 285–362. Astérisque, Nos. 7 et 8.

566. ———, *Sur une inégalité à la Minkowski pour les multiplicités*, Ann. Math. (2) **106** (1977), no. 1, 38–44.

567. ———, *Bonnesen-type inequalities in algebraic geometry. I. Introduction to the problem*, Seminar on Differential Geometry, Annals of Math. Studies, vol. 102, Princeton Univ. Press, Princeton, N.J., 1982, pp. 85–105.

568. ———, *Variétés polaires. II. Multiplicités polaires, sections planes, et conditions de Whitney*, Algebraic Geometry (La Rábida, 1981), Lect. Notes in Math., vol. 961, Springer, Berlin, 1982, pp. 314–491.

569. ———, *Résultats récents d'algèbre commutative effective*, Séminaire Bourbaki, Année 1989/90, no. 189-190, 1990, pp. Exp. No. 718, 107–131.

570. Evgueni Tevelev, *Projectively dual varieties*, preprint, 2001.

571. Siu-Kei Tin, *Numerically positive polynomials for k-ample vector bundles*, Math. Ann. **294** (1992), no. 4, 579–590.

572. I-Hsun Tsai, *Dominating the varieties of general type*, J. Reine Angew. Math. **483** (1997), 197–219.

573. Yuri Tschinkel, *Fujita's program and rational points*, Higher dimensional varieties and rational points (Budapest, 2001), Bolyai Soc. Math. Stud., vol. 12, Springer, Berlin, 2003, pp. 283–310.

574. Hajime Tsuji, *On the structure of pluricanonical systems of projective varieties of general type*, preprint, 1999.

575. ———, *Numerically trivial fibration*, preprint, 2000.

576. Loring W. Tu, *The connectedness of symmetric and skew-symmetric degeneracy loci: even ranks*, Trans. Amer. Math. Soc. **313** (1989), no. 1, 381–392.

577. ———, *The connectedness of degeneracy loci*, Topics in Algebra, Part 2 (Warsaw, 1988), PWN, Warsaw, 1990, pp. 235–248.

578. Kenji Ueno, *Introduction to classification theory of algebraic varieties and compact complex spaces*, Classification of Algebraic Varieties and Compact Complex Manifolds, Lect. Notes in Math., vol. 412, Springer, Berlin, 1974, pp. 288–332.

579. Karen Uhlenbeck and Shing-Tung Yau, *On the existence of Hermitian–Yang–Mills connections in stable vector bundles*, Comm. Pure Appl. Math. **39** (1986), no. S, suppl., S257–S293.

580. Sampei Usui and Hiroshi Tango, *On numerical positivity of ample vector bundles with additional condition*, J. Math. Kyoto Univ. **17** (1977), no. 1, 151–164.

581. Bert van Geemen and Gerard van der Geer, *Kummer varieties and the moduli spaces of abelian varieties*, Amer. J. Math. **108** (1986), no. 3, 615–641.

582. Michel Vaquié, *Irrégularité des revêtements cycliques*, Singularities (Lille, 1991), London Math. Soc. Lecture Note Ser., vol. 201, Cambridge Univ. Press, Cambridge, 1994, pp. 383–419.

583. A. N. Varchenko, *Asymptotic Hodge structure on vanishing cohomology*, Izv. Akad. Nauk SSSR Ser. Mat. **45** (1981), no. 3, 540–591, 688.

584. ______, *Semicontinuity of the complex singularity exponent*, Funktsional. Anal. i Prilozhen. **17** (1983), no. 4, 77–78.

585. Wolmer Vasconcelos, *Arithmetic of Blowup Algebras*, London Mathematical Society Lecture Note Series, vol. 195, Cambridge University Press, Cambridge, 1994.

586. Jean-Louis Verdier, *Stratifications de Whitney et théorème de Bertini–Sard*, Invent. Math. **36** (1976), 295–312.

587. Peter Vermeire, *A counterexample to Fulton's conjecture on* $\overline{M}_{0,n}$, J. Algebra **248** (2002), no. 2, 780–784.

588. Eckart Viehweg, *Die Additivität der Kodaira Dimension für projektive Faserräume über Varietäten des allgemeinen Typs*, J. Reine Angew. Math. **330** (1982), 132–142.

589. ______, *Vanishing theorems*, J. Reine Angew. Math. **335** (1982), 1–8.

590. ______, *Weak positivity and the additivity of the Kodaira dimension for certain fibre spaces*, Algebraic Varieties and Analytic Varieties (Tokyo, 1981), North-Holland, Amsterdam, 1983, pp. 329–353.

591. ______, *Weak positivity and the stability of certain Hilbert points*, Invent. Math. **96** (1989), no. 3, 639–667.

592. ______, *Weak positivity and the stability of certain Hilbert points. II*, Invent. Math. **101** (1990), no. 1, 191–223.

593. ______, *Weak positivity and the stability of certain Hilbert points. III*, Invent. Math. **101** (1990), no. 3, 521–543.

594. ______, *Quasi-Projective Moduli for Polarized Manifolds*, Ergebnisse der Math. und ihrer Grenzgebiete (3), vol. 30, Springer-Verlag, Berlin, 1995.

595. Kari Vilonen, *Cohomology of constructible sheaves*, Master's thesis, Brown University, 1979.

596. Claire Voisin, *Courbes tétragonales et cohomologie de Koszul*, J. Reine Angew. Math. **387** (1988), 111–121.

597. ______, *On a conjecture of Clemens on rational curves on hypersurfaces*, J. Differential Geom. **44** (1996), no. 1, 200–213.

598. ______, *Green's generic syzygy conjecture for curves of even genus lying on a K3 surface*, J. Eur. Math. Soc. **4** (2002), no. 4, 363–404.

599. ______, *Hodge Theory and Complex Algebraic Geometry. I*, Cambridge Studies in Advanced Mathematics, vol. 76, Cambridge University Press, Cambridge, 2002, Translated from the French original by Leila Schneps.

600. ______, *Théorie de Hodge et Géométrie Algébrique Complexe*, Cours Spécialisés, vol. 10, Société Mathématique de France, Paris, 2002.

601. ______, *Green's canonical syzygy conjecture for generic curves of odd genus*, preprint, 2003.

602. Paul Vojta, *Mordell's conjecture over function fields*, Invent. Math. **98** (1989), no. 1, 115–138.

603. Jonathan Wahl, *A cohomological characterization of* $\mathbf{P}^n$, Invent. Math. **72** (1983), no. 2, 315–322.

604. R. O. Wells, Jr., *Differential Analysis on Complex Manifolds*, second ed., Graduate Texts in Mathematics, vol. 65, Springer-Verlag, New York, 1980.

605. Gerald E. Welters, *The surface* $C - C$ *on Jacobi varieties and 2nd order theta functions*, Acta Math. **157** (1986), no. 1–2, 1–22.

606. Burkhard Wilking, *Torus actions on manifolds of positive sectional curvature*, preprint.

607. P. M. H. Wilson, *On the canonical ring of algebraic varieties*, Compositio Math. **43** (1981), no. 3, 365–385.

608. ______, *Calabi–Yau manifolds with large Picard number*, Invent. Math. **98** (1989), no. 1, 139–155.

609. ______, *The Kähler cone on Calabi–Yau threefolds*, Invent. Math. **107** (1992), no. 3, 561–583.

610. Jarosław A. Wiśniewski, *On contractions of extremal rays of Fano manifolds*, J. Reine Angew. Math. **417** (1991), 141–157.

611. ______, *On deformation of nef values*, Duke Math. J. **64** (1991), no. 2, 325–332.

612. Bun Wong, *A class of compact complex manifolds with negative tangent bundles*, Complex Analysis of Several Variables (Madison, Wis., 1982), Proc. Sympos. Pure Math., vol. 41, Amer. Math. Soc., Providence, RI, 1984, pp. 217–223.

613. Geng Xu, *Curves in* $\mathbf{P}^2$ *and symplectic packings*, Math. Ann. **299** (1994), no. 4, 609–613.

614. ______, *Subvarieties of general hypersurfaces in projective space*, J. Differential Geom. **39** (1994), no. 1, 139–172.

615. Tamaki Yano, *b-functions and exponents of hypersurface isolated singularities*, Singularities (Arcata, 1981), Proc. Symp. Pure Math., vol. 40, Amer. Math. Soc., Providence, RI, 1983, pp. 641–652.

616. Sai-Kee Yeung, *Very ampleness of line bundles and canonical embedding of coverings of manifolds*, Compositio Math. **123** (2000), no. 2, 209–223.

617. ______, *Effective estimates on the very ampleness of the canonical line bundle of locally Hermitian symmetric spaces*, Trans. Amer. Math. Soc. **353** (2001), no. 4, 1387–1401.

618. M. G. Zaidenberg and V. Ya. Lin, *Finiteness theorems for holomorphic mappings*, Several Complex Variables III, Encyclopedia of Math. Sciences, vol. 9, Springer Verlag, Berlin, 1989, pp. 113–172.

619. F. L. Zak, *Projections of algebraic varieties*, Mat. Sb. (N.S.) **116(158)** (1981), no. 4, 593–602, 608.

620. ______, *Linear systems of hyperplane sections on varieties of small codimension*, Funktsional. Anal. i Prilozhen. **19** (1985), no. 3, 1–10, 96.

621. ______, *Severi varieties*, Mat. Sb. (N.S.) **126(168)** (1985), no. 1, 115–132, 144.

622. ______, *Tangents and Secants of Algebraic Varieties*, American Mathematical Society, Providence, RI, 1993.

623. Oscar Zariski, *The theorem of Riemann–Roch for high multiples of an effective divisor on an algebraic surface*, Ann. of Math. (2) **76** (1962), 560–615.

Glossary of Notation

Notation Introduced in Volume I

$f^{-1}\mathfrak{a}$	Pullback of ideal sheaf, I: 1
$S^k E$	Symmetric power of vector space or bundle, I: 1
$\mathrm{Sym} E$	Symmetric algebra of vector space or bundle, I: 1
E^*	Dual of vector space or bundle, I: 1
$\mathbf{P}(E)$	Projective bundle of one-dimensional quotients, I: 1
$\mathbf{P}_{\mathrm{sub}}(E)$	Projective bundle of one-dimensional subspaces, I: 1
pr_1, pr_2	The two projections of $X_1 \times X_2$ onto its factors, I: 2
$O(m^k)$	Rate of growth of function of m, I: 2
$\mathcal{M}_X$	Sheaf of rational functions on X, I: 8
$\mathrm{Div}(X)$	Additive group of Cartier divisors, I: 8
$\succcurlyeq$	Effectivity of difference of divisors, I: 8
$Z_k(X)$	Group of k-cycles on X, I: 9
$\mathrm{WDiv}(X)$	Group of Weil divisors on X, I: 9
$\mathrm{div}(f)$	Divisor of rational function, I: 9
$\mathrm{Princ}(X)$	Group of principal divisors on X, I: 9
$\equiv_{\mathrm{lin}}$	Linear equivalence of divisors, I: 9
$\mathrm{Pic}(X)$	Picard group of X, I: 10
ω_X	Canonical bundle of X, I: 11
K_X	Canonical divisor of X, I: 11
$\mathfrak{b}(\lvert V \rvert)$	Base ideal of linear series, I: 13
$\mathrm{Bs}(\lvert V \rvert)$	Base locus or scheme of linear series, I: 13
$\phi_{\lvert V \rvert}$	Morphism determined by linear series, I: 14

$(D_1 \cdot \ldots \cdot D_k \cdot V)$	Intersection number of divisors with subvariety, I: 15
$\int_V D_1 \cdot \ldots \cdot D_k$	Alternative notation for intersection number, I: 15
$(D_1 \cdot \ldots \cdot D_n)$	Intersection number of n divisors on n-fold, I: 16
$[V]$	Cycle of a scheme V, I: 16
$\equiv_{\text{num}}$	Numerical equivalence of divisors, I: 18
$\mathrm{Num}(X)$	Subroup of numerically trivial divisors, I: 18
$N^1(X)$	Néron–Severi group of X, I: 18
$\rho(X)$	Picard number of X, I: 18
$(\delta_1 \cdot \ldots \cdot \delta_k \cdot [V])$	Intersection of numerical equivalence classes, I: 19
$\mathrm{rank}(\mathcal{F})$	Rank of coherent sheaf, I: 20
$Z_n(\mathcal{F})$	Cycle of coherent sheaf, I: 20
ω_{FS}	Fubini–Study form on $\mathbf{P}^n$, I: 40
ω_{std}	Standard symplectic form on $\mathbf{C}^n$, I: 42
$\Theta(L, h)$	Curvature form of Hermitian line bundle, I: 42
$\mathrm{Div}_{\mathbf{Q}}(X)$	Group of $\mathbf{Q}$-divisors on X, I: 44
$N^1(X)_{\mathbf{Q}}$	Numerical equivalence classes of $\mathbf{Q}$-divisors, I: 45
$\mathrm{WDiv}_{\mathbf{Q}}(X)$	Group of Weil $\mathbf{Q}$-divisors, I: 47
$\mathrm{Div}_{\mathbf{R}}(X)$	Group of $\mathbf{R}$-divisors, I: 48
$N^1(X)_{\mathbf{R}}$	Numerical equivalence classes of $\mathbf{R}$-divisors, I: 48
$\mathrm{mult}_x C$	Multiplicity of curve at a point, I: 55
$\mathrm{Amp}(X)$	Ample cone of X, I: 59
$\mathrm{Nef}(X)$	Nef cone of X, I: 59
$N_1(X)_{\mathbf{R}}$	Numerical equivalence classes of real one-cycles, I: 61
$\mathrm{NE}(X)$	Cone of curves on X, I: 61
$\overline{\mathrm{NE}}(X)$	Closed cone of curves, I: 62
$D^{\perp}$, $D_{>0}$, $D_{\leq 0}$	Hyperplane, half-spaces determined by divisor, I: 62
$\mathcal{N}_V$	Null cone determined by subvariety, I: 82
$\mathcal{B}_X$	Nef boundary of X, I: 83
$\mathrm{Kahler}(X)$	Cone of Kähler classes on X, I: 84
$\overline{\mathrm{NE}}(X)_{D \geq 0}$	Subset of $\overline{\mathrm{NE}}(X)$ in D-non-negative halfspace, I: 86
$\mathrm{cont}_{\mathbf{r}}$	Contraction determined by extremal ray, I: 87
$e(\mathfrak{a}_1; \ldots; \mathfrak{a}_n)$	Mixed multiplicity of ideals, I: 91
$\mathrm{reg}(\mathcal{F})$	Regularity of coherent sheaf, I: 103
$d(\mathcal{I})$	Generating degree of ideal sheaf on $\mathbf{P}^r$, I: 111
$d(I)$	Generating degree of homogeneous ideal, I: 111
$\mathrm{reg}(I)$	Regularity of homogeneous ideal, I: 111

Cliff(A)	Clifford index of line bundle on a curve, I: 117
Cliff(X)	Clifford index of curve, I: 117
$\mathbf{N}(L)$, $\mathbf{N}(X,L)$	Semigroup of a line bundle, I: 122
$e(L)$	Exponent of a line bundle, I: 122
$\kappa(L)$, $\kappa(X,L)$	Iitaka dimension of a line bundle or divisor, I: 123
$\kappa(X)$	Kodaira dimension of a variety, I: 123
$R(L)$, $R(X,L)$	Section ring of line bundle or divisor, I: 126
$\mathbf{B}(D)$	Stable base locus of a divisor, I: 127
Big(X)	Big cone of X, I: 147
$\overline{\mathrm{Eff}}(X)$	Pseudoeffective cone of X, I: 147
$\mathrm{vol}(L)$, $\mathrm{vol}_X(L)$	Volume of line bundle or divisor, I: 148
vol, vol_X	Volume function on $N^1(X)_{\mathbf{R}}$, I: 153
$\mathrm{mult}_x \lvert V \rvert$	Multiplicity of linear series at a point, I: 165
$V_\bullet$	Graded linear series, I: 172
$R(V_\bullet)$	Section ring of graded linear series, I: 173
$V_\bullet \cdot W_\bullet$	Product of graded linear series, I: 174
$V_\bullet \cap W_\bullet$	Intersection of graded linear series, I: 174
$\mathrm{Span}(V_\bullet, W_\bullet)$	Span of graded linear series, I: 174
$V_\bullet^{(p)}$	Veronese of graded linear series, I: 174
$\mathbf{N}(V_\bullet)$	Semigroup of graded linear series, I: 174
$e(V_\bullet)$	Exponent of graded linear series, I: 174
$\kappa(V_\bullet)$	Iitaka dimension of graded linear series, I: 175
$\mathbf{B}(V_\bullet)$	Stable base locus of graded linear series, I: 175
$\mathrm{vol}(V_\bullet)$	Volume of graded linear series, I: 176
$\mathfrak{a}_\bullet$	Graded family of ideal sheaves, I: 176
$\mathrm{Rees}(\mathfrak{a}_\bullet)$	Rees algebra of graded system, I: 176
$\mathrm{Pow}(\mathfrak{a})$	Graded system of powers of ideal, I: 177
$\mathfrak{q}^{<k>}$	Symbolic power of radical ideal, I: 177
$\mathfrak{a}_\bullet \cap \mathfrak{b}_\bullet$	Intersection of graded families of ideals, I: 180
$\mathfrak{a}_\bullet \cdot \mathfrak{b}_\bullet$	Product of graded families of ideals, I: 180
$\mathfrak{a}_\bullet + \mathfrak{b}_\bullet$	Sum of graded families of ideals, I: 180
$\mathfrak{a}_\bullet^{(p)}$	Veronese of graded system of ideals, I: 181
$\mathrm{mult}(\mathfrak{a}_\bullet)$	Multiplicity of graded system of ideals, I: 182
L_ω	Cup product with Kähler form ω, I: 199
Sec(X)	Secant variety of projective variety, I: 215
Tan(X)	Tangent variety of projective variety, I: 215

$e_f(x)$	Local degree of branched covering f, I: 216
$\mathbf{T}_x$	Embedded tangent space to projective variety, I: 219
$\mathrm{Trisec}(X)$	Variety of tri-secant two-planes to X, I: 223
$\mathrm{except}(\mu)$	Exceptional locus of birational map μ, I: 241
$\Omega^1_X(\log D)$	Bundle of one-forms with log poles along D, I: 250
$\Omega^p_X(\log D)$	Bundle of p-forms with log poles along D, I: 250
$\mathcal{K}_X$	Grauert–Riemenschneider canonical sheaf, I: 258
$\mathrm{Pic}^0(X)$	Identity component of $\mathrm{Pic}(X)$, I: 261
$\mathrm{Alb}(X)$	Albanese variety of X, I: 262
alb_X	Albanese mapping of X, I: 262
$\varepsilon(L;x)$	Seshadri constant of L at x, I: 270
$s(B;x)$	Number of jets separated by B at x, I: 273
$B(\lambda)$	Open ball of radius λ in $\mathbf{C}^g$, I: 276
$w_G(M,\omega)$	Gromov width of symplectic manifold, I: 276
$\mathrm{mult}_x(F)$	Multiplicity of divisor F at a point, I: 282
$\mathrm{mult}_Z(F)$	Multiplicity of divisor F along subvariety Z, I: 282
$(d_1,\ldots,d_g)$	Type of polarization on abelian variety, I: 291
$m(A,L)$	Buser–Sarnak invariant of abelian variety, I: 291
(A,Θ)	Principally polarized abelian variety (PPAV), I: 292
$\mathcal{A}_g$	Moduli space of PPAVs, I: 292
(JC,Θ_C)	Polarized Jacobian of Riemann surface C, I: 293
$\varepsilon(A,L)$	Seshadri constant of polarized abelian variety, I: 293
$s_L(\mathcal{I})$	s-invariant of ideal sheaf, I: 303
$d_L(\mathcal{I})$	Generating degree of ideal sheaf w.r.t. line bundle, I: 304
$\deg_L(V)$	Degree of variety w.r.t. line bundle, I: 308
$\mathrm{reg}_L(\mathcal{I})$	Regularity of ideal sheaf w.r.t. line bundle, I: 310
$\mathcal{H}_i$	Homology sheaf of complex, I: 318
$K_\bullet(E,s)$	Koszul complex of section of vector bundle, I: 320

Notation Introduced in Volume II

$\mathbf{P}(E)$, $\mathbf{P}_X(E)$	Projective bundle over X, II: 7
ξ , ξ_E	Class of Serre line bundle on $\mathbf{P}(E)$, II: 8
V_X	Trivial vector bundle modeled on vector space V, II: 8
$\Gamma^\lambda E$	Bundle associated to E via representation λ, II: 15
$E{<}\delta{>}$, $E{<}D{>}$	Bundle twisted by $\mathbf{Q}$-divisor class, II: 21

$\delta(X, E, h)$	Barton invariant of a bundle, II: 25
$N_{X/M}$	Normal bundle to subvariety $X \subseteq M$, II: 27
$\widehat{M}_{/X}$	Formal completion of M along subvariety X, II: 31
E_f	Bundle associated to branched covering, II: 48
$\mu(E)$	Slope of vector bundle on curve, II: 57
$\mathrm{HN}_{\bullet}(E)$	Harder–Narasimhan filtration of bundle E, II: 59
E_{norm}	Normalized $\mathbf{Q}$-twist of vector bundle on curve, II: 60
$\mathrm{Zeroes}(s)$	Zeroes of section of vector bundle, II: 65
$D_k(u)$	Degeneracy locus of vector bundle map, II: 74
$W^r_d(C)$	Variety of special divisors on a curve, II: 82
$W^r(C; V)$	Special divisors associated to vector bundle, II: 84
$S^i(f)$	Singularity locus of a mapping, II: 86
$c_i(E{<}\delta{>})$	Chern class of $\mathbf{Q}$-twisted vector bundle, II: 102
$z(C, E)$	Cone class of cone in vector bundle, II: 106
$\mathbf{P}_{\mathrm{sub}}(C)$	Projectivization of cone, II: 107
$C_1 \oplus C_2$	Direct sum of cones, II: 107
$\overline{C} \subset \overline{E}$	Projective closure of cone in vector bundle, II: 107
$z(C{<}\delta{>}, E{<}\delta{>})$	$\mathbf{Q}$-twisted cone class, II: 110
$s_\lambda(c_1, \ldots, c_e)$	Schur polynomial, II: 118
$\ulcorner D \urcorner$	Round-up of $\mathbf{Q}$-divisor, II: 140
$\llcorner D \lrcorner$, $[D]$	round-down or integral part of $\mathbf{Q}$-divisor, II: 140
$\{D\}$	Fractional part of $\mathbf{Q}$-divisor, II: 141
$K_{X'/X}$	Relative canonical divisor, II: 146
$\mathcal{J}(D), \mathcal{J}(X, D)$	Multiplier ideal of a $\mathbf{Q}$-divisor, II: 152
$\mathcal{J}(\mathfrak{a}^c), \mathcal{J}(c \cdot \mathfrak{a})$	Multiplier ideal associated to ideal sheaf, II: 152
$\mathcal{J}(\mathfrak{a}_1^{c_1} \cdot \ldots \cdot \mathfrak{a}_t^{c_t})$	Mixed multiplier ideal, II: 153
$\mathcal{J}(c \cdot \lvert V \rvert)$	Multiplier ideal of linear series, II: 154
$\mathcal{J}(f, c \cdot \lvert L \rvert)$	Multiplier ideal of relative linear series, II: 158
$\mathrm{mult}_x(D)$	Multiplicity of $\mathbf{Q}$-divisor, II: 162
$\mathrm{ord}_Z(\mathfrak{a})$	Order of vanishing of ideal along subvariety, II: 164
$\Sigma_k(D)$	Multiplicity locus of $\mathbf{Q}$-divisor, II: 165
$\mathrm{lct}(D; x)$, $\mathrm{lct}(D)$	Log-canonical threshold of $\mathbf{Q}$-divisor, II: 166
$\mathrm{lct}(\mathfrak{a}; x)$, $\mathrm{lct}(\mathfrak{a})$	Log-canonical threshold of an ideal sheaf, II: 166
$b_f(s)$	Bernstein–Sato polynomial, II: 169
$P(\mathfrak{a})$	Newton polyhedron of monomial ideal, II: 171
$\mathbf{1} = (1, \ldots, 1)$	Vector of 1's, II: 171

$\mathfrak{J}(\varphi)$	Multiplier ideal associated to PSH function, II: 176
$\operatorname{adj}(D)$	Adjoint ideal of reduced divisor, II: 179
$\operatorname{Jacobian}(D)$	Jacobian ideal of divisor, II: 181
$\operatorname{Jacobian}_m(\mathfrak{a})$	Jacobian ideals of ideal sheaf, II: 181
(X, Δ)	A pair, II: 182
$\mathcal{J}((X,\Delta); D)$	Multiplier ideal of $\mathbf{Q}$-divisor on pair, II: 183
$\mathfrak{J}(h)$	Multiplier ideal of singular Hermitian metric, II: 192
D_H	Restriction of $\mathbf{Q}$-divisor to hypersurface, II: 195
$\mathfrak{a}_H$	Restriction of ideal sheaf to hypersurface, II: 197
$\mathfrak{a}_1 \stackrel{\circ}{+} \mathfrak{a}_2$	Exterior sum of two ideals, II: 206
$\overline{\mathfrak{a}}$	Integral closure of ideal, II: 216
$\mathbf{B}_+(L)$	Augmented stable base locus, II: 247
$\operatorname{Null}(L)$	Null locus of a nef divisor, II: 248
$\operatorname{LC}(D; x)$	Locus of log-canonical singularities, II: 255
$\mathcal{J}(c \cdot \Vert L \Vert)$	Asymptotic multiplier ideal of linear series, II: 271
$\mathcal{J}(c \cdot \mathfrak{a}_\bullet)$, $\mathcal{J}(\mathfrak{a}_\bullet^c)$	Asymptotic multiplier ideal of graded system, II: 276
$\operatorname{lct}(\mathfrak{a}_\bullet; x)$	Log-canonical threshold of graded family, II: 280
$\mathcal{J}(\mathfrak{a}_\bullet^c \cdot \mathfrak{b}_\bullet^d)$	Mixed asymptotic multiplier ideal, II: 281
$\mathcal{J}(c \cdot \Vert V_m \Vert)$	Asymptotic multiplier ideal of graded linear series, II: 281
$\mathcal{J}(c \cdot V_\bullet)$	Alternate notation for $\mathcal{J}(c \cdot \Vert V_1 \Vert)$, II: 281
$P_m(X)$	m^{th} plurigenus of X, II: 292
$\overline{\operatorname{Mov}}(X)$	Cone of movable curves, II: 307

Index

The small Roman numbers I and II refer to Volume I and II, respectively.

Printing: Strauss GmbH, Mörlenbach
Binding: Schäffer, Grünstadt